Coal Country

The Mean
Deindustrializa   nd

New Historical Perspectives is a book series for early career scholars within the UK and the Republic of Ireland. Books in the series are overseen by an expert editorial board to ensure the highest standards of peer-reviewed scholarship. Commissioning and editing is undertaken by the Royal Historical Society, and the series is published under the imprint of the Institute of Historical Research by the University of London Press.

The series is supported by the Economic History Society and the Past and Present Society.

Series co-editors: Heather Shore (Manchester Metropolitan University) and Jane Winters (School of Advanced Study, University of London)

Founding co-editors: Simon Newman (University of Glasgow) and Penny Summerfield (University of Manchester)

New Historical Perspectives Editorial Board
Charlotte Alston, Northumbria University
David Andress, University of Portsmouth
Philip Carter, Institute of Historical Research, University of London
Ian Forrest, University of Oxford
Leigh Gardner, London School of Economics
Tim Harper, University of Cambridge
Guy Rowlands, University of St Andrews
Alec Ryrie, Durham University
Richard Toye, University of Exeter
Natalie Zacek, University of Manchester

Coal Country

The Meaning and Memory of Deindustrialization in Postwar Scotland

Ewan Gibbs

LONDON
ROYAL HISTORICAL SOCIETY
INSTITUTE OF HISTORICAL RESEARCH
UNIVERSITY OF LONDON PRESS

Published by
UNIVERSITY OF LONDON PRESS
SCHOOL OF ADVANCED STUDY
INSTITUTE OF HISTORICAL RESEARCH
Senate House, Malet Street, London WC1E 7HU

© Ewan Gibbs 2021

The author has asserted his right under the Copyright, Designs and Patents Act 1988 to be identified as the author of this work.

This book is published under a Creative Commons Attribution-NonCommercial-NoDerivatives 4.0 International (CC BY-NC-ND 4.0) license. More information regarding CC licenses is available at https://creativecommons.org/licenses/.

Copyright and permissions for the reuse of many of the images included in this publication differ from the above. Copyright and permissions information is provided alongside each image.

Available to download free or to purchase the hard copy edition at https://www.sas.ac.uk/publications.

ISBNs
978-1-912702-54-1 (hardback edition)
978-1-912702-55-8 (paperback edition)
978-1-912702-58-9 (PDF edition)
978-1-912702-56-5 (ePub edition)
978-1-912702-57-2 (.mobi edition)

DOI 10.14296/321.9781912702589

New Historical
PERSPECTIVES

Cover image: Auchengeich Mining Disaster Memorial, Moodiesburn. © John O'Hara (2019).

Contents

	Abbreviations	vii
	Acknowledgements	ix
	Introduction: those who walked in the darkest valleys	1
1.	'Buried treasure': industrial development in the Scottish coalfields, c.1940s–80s	21
2.	Moral economy: custom and social obligation during colliery closures	55
3.	Communities: 'it was pretty good' in restructured locales	91
4.	Gendered experiences	119
5.	Generational perspectives	155
6.	Coalfield politics and nationhood	187
7.	Synthesis. 'The full burden of national conscience': class, nation and deindustrialization	225
	Conclusion: the meaning and memory of deindustrialization	251
	Appendix: biographies of oral history participants	259
	Bibliography	267
	Index	293

List of figures

Tables

1.1	Male employment in Lanarkshire.	24
1.2	Female employment in Lanarkshire.	24
1.3	Unemployment in Lanarkshire.	25
1.4	Scottish Coal employment.	25
5.1	Generation, temporality and employment structure in the Scottish coalfields.	157

Figures

0.1	RCAHMS, Map of the Scottish coalfields (2005).	xi
3.1	Auchengeich Mining Disaster Memorial, Moodiesburn.	108
6.1	Bob Starrett, 'Phase 3', *Scottish Marxist*, vi (1974), 24.	207
6.2	Miners' Gala Day 1969.	215
6.3	Gala Day 1969, featuring Joan Lester (Labour MP), Lawrence Daly, and Vietnamese Federation of Union representatives.	217
6.4	1988 Gala poster.	220

Abbreviations

ANC	African National Congress
BSR	Birmingham Sound Reproducers
CB	Coal Board
CCC	Colliery Consultative Committee
CEGB	Central Electricity Generating Board
COSA	Colliery Officials and Staff Association
CPGB	Communist Party of Great Britain
NATO	North Atlantic Treaty Organization
NCB	National Coal Board
NMMS	National Mining Museum Scotland archives, Newtongrange, Midlothian
NRS	National Records of Scotland
NUMSA	National Union of Mineworkers Scottish Area
ODNB	*Oxford Dictionary of National Biography*
SCDI	Scottish Council (Development and Industry)
SCEBTA	Scottish Colliery, Enginemen, Boilermen and Tradesmen's Association
SNP	Scottish National Party
SSEB	South of Scotland Electricity Board
SSHA	Scottish Special Housing Association
STUC	Scottish Trades Union Congress
TNA	The National Archives of the UK
UCS	Upper Clyde Shipbuilders
UMS	United Mineworkers of Scotland
WTFU	World Federation of Trade Unions

Acknowledgements

This book is the outcome of research that almost spanned a decade. Although I am solely responsible for its contents, others assisted me. First, I owe a great deal of thanks to the men and women who kindly gave up their time to discuss their life stories and family histories. In most cases, they did so with someone who was previously entirely unknown to them. I am also indebted to those who put me in touch with others. Thanks especially to Brendan Moohan and Willie Doolan who have become good friends since I interviewed them. I also owe thanks to patient archivists who made documents available to me at the National Records of Scotland, The National Archives and the National Mining Museum Scotland. Staff at Historic Environment Scotland helped me to obtain licences for the images reproduced in this volume. Carole McCallum of Glasgow Caledonian University archives assisted by providing a licence for the Bob Starrett cartoon used in chapter 6. John O'Hara very generously donated the picture of the Auchengeich Mining Disaster Memorial which is included on the cover. Jackie Kay kindly sent me her poem and Jim Monaghan allowed me to reproduce part of his. I was given financial support from a Royal Society of Edinburgh small grant, which enabled me to visit The National Archives. An Institute of Historical Research Scouloudi publication award supported the cost of image licensing.

I began the work behind this book as a master's student and then as a PhD candidate within the Economic and Social History subject area at the University of Glasgow. Jim Phillips and Duncan Ross were an excellent and good-humoured PhD supervision team. They supported my development as an independent-minded researcher while challenging me to explain the importance and relevance of my approach and findings. Other colleagues have provided guidance on my research as it developed: thanks to Jim Tomlinson, Ray Stokes and Jeff Fear for commenting on papers that have informed this work. Many thanks to Keith Gildart, Bill Knox and Catriona MacDonald, who examined full versions of the book as it progressed from thesis to monograph. Their insights have improved the completed manuscript.

The findings presented in this book were honed through several conferences and seminar papers and benefited from questions from audience members. Thanks especially to those who attended panels at the Business History Conference in 2015, the Economic History Society in 2016 and the Political Studies Association in 2017. Additionally, thank you to the Scottish Oral History Centre for offering me the privilege of delivering seminars on my research project in 2016 and 2019, and to the University of the West of Scotland Society, Politics, Governance and Justice Hub, the Yunus Centre and the University of Edinburgh's Scottish History seminar series for doing the same in 2017, 2019 and 2020. The Institute of Historical Research and the Royal Historical Society have also been of great assistance to a first-time author. Thank you to Simon Newman, Jane Winters, David Andress and Philip Carter for their guidance.

Between my time at Glasgow, UWS and meeting academics from other institutions, I have felt part of a supportive research community. Thank you to Andy Clark, Rory Scothorne, Felicity Cawley, Andrew Perchard, Dominic Reed, Ewan Kerr, Tom Montgomery, James Bowness, Diarmaid Kelliher, Henry Bell, Claire Harkins, Stephen Mullen, Duncan Hotchkiss, Christopher Miller, Scott Hames, Valerie Wright and Julie Clark for your friendship and collegiality. Thanks also to Nathan Blondel for helping me devise the title of this book.

Thanks to my mother and father, Maria and Neil Gibbs, for decades of support which made this possible.

Last but certainly not least. Thank you to Laura Dover, whose uncles Iain and Jimmy worked at Cardowan colliery, and whose love and encouragement made writing this book much happier.

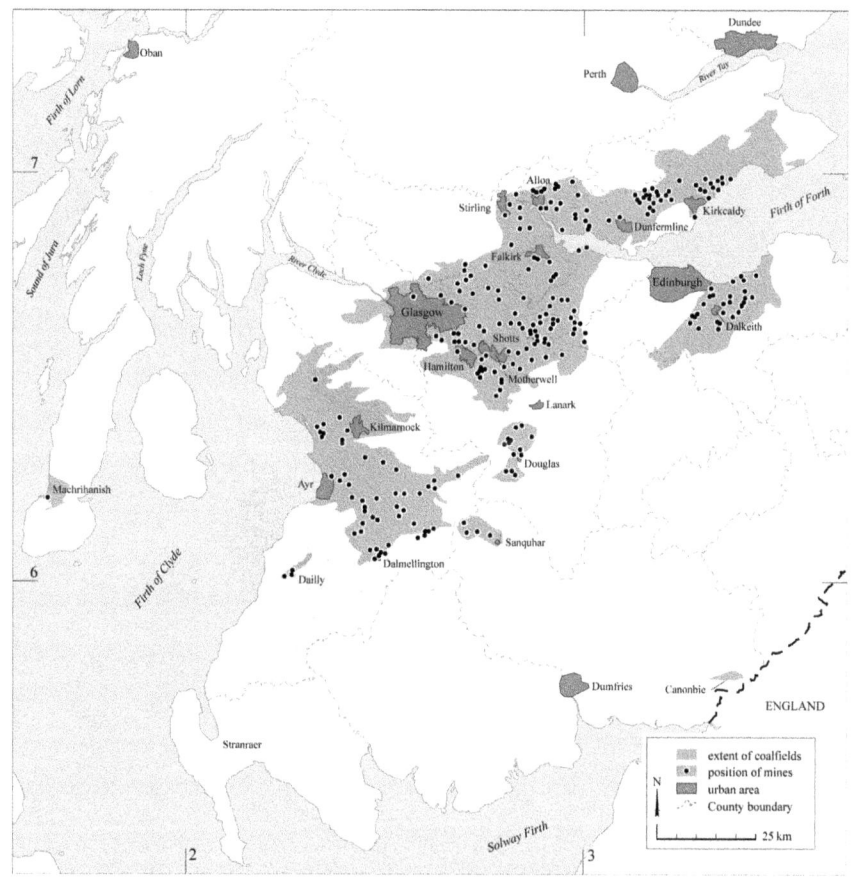

Figure 0.1. RCAHMS, Map of the Scottish coalfields (2005). ©Historic Environment Scotland.

Introduction:
those who walked in the darkest valleys

On Sunday 16 September 2018, several hundred people gathered by the Auchengeich Mining Disaster Memorial in the village of Moodiesburn, North Lanarkshire. They included retired miners, trade union representatives and local councillors alongside members of local football teams and a choir made up of schoolchildren. The annual memorial service is timed to coincide with the anniversary of the Auchengeich pit disaster in which forty-seven men lost their lives due to an underground fire on 18 September 1959.[1] Most of the people present had little direct experience of coal mining, and many participants in the service had come of age after Lanarkshire's last colliery, Cardowan, closed in 1983. During the service, Reverend Mark Malcolm, from the Church of Scotland's Chryston Parish, paid tribute to the fallen men by reading from Psalm 23, commemorating those that 'walk through the darkest valley'.[2] The annual service powerfully evokes collective memories of industry which continue to shape consciousness and identities in the towns and villages of the Scottish coalfields after deep mining itself has ceased. A permanent physical reminder is embedded in the landscape: the memorial statue of a 1950s-era British coal miner complete with cap lamp, battery and pick-axe. In a technical sense, deindustrialization can be understood as the decreasing contribution of industrial activities to gross domestic product and employment. But deindustrialization's impact was as keenly felt in cultural and political terms as it was economically. This volume traces the socioeconomic transformation brought to the coalfields by these developments. The closure of mines, steel mills and factories fundamentally altered livelihoods and associations by challenging a strongly held social order in towns and villages which had developed around coal mining. Colliery closures and the experience of labour market alterations have significantly contributed to the questioning of Scotland's position within the Union and the realignment of the politics of class and nationhood since the mid twentieth century.

[1] E. Gibbs and J. Phillips, 'Remembering Auchengeich: the largest fatal accident in Scottish coal in the nationalised era', *Scottish Labour History*, liv (2019), 47–57.

[2] Observation notes from Auchengeich colliery memorial service, 16 Sept. 2018; Psalm XXIII:4 (New International Version).

'Introduction: those who walked in the darkest valleys', in E. Gibbs, *Coal Country: The Meaning and Memory of Deindustrialization in Postwar Scotland* (London, 2021), pp. 1–19. License: CC-BY-NC-ND 4.0.

These changes were not sudden. They unravelled over several decades. Between 1965 and 1967, the workforce at Auchengeich and the adjacent Western Auchengeich collieries faced a different threat as the National Coal Board's (NCB) closure programme intensified. Union representatives, with the support of junior managerial staff, attempted to convince the Board that the coal reserves in the area were a profitable proposition.[3] From mid century onwards, coal faced intensifying competition from petroleum and nuclear fuels. Contraction was incremental and phased, as well as regionally varied. The Scottish coalfields were permanently reconfigured. Its settlements had principally grown during the previous 150 years under the impetus of industrialization.[4] From the onset of coal's nationalization in 1947, mining developed to match the shifting priorities of the NCB. The pace of investment and divestment was dictated by coal's place within UK energy policies, especially as they related to electricity generation. Although Moodiesburn's local pits had closed by the late 1960s, miners from the surrounding area continued to work in the dark valleys of Scotland's collieries for decades after. They commuted increasing distances across the central belt before deep mining ceased entirely following flooding at the Longannet complex during 2002. Longannet was comprised of highly modern drift mines in Fife and Clackmannanshire. They were initially developed during the late 1960s to feed the large power station which shared its name. By the time of its closure, Longannet's workforce was drawn from across the Scottish coalfields. Connections between small-scale territorial communities and workplaces were weakened, but a more intimate Scottish national coalfield community was also established in the process.[5] This is symbolized by the presence of former miners from across the Scottish coalfield at the Auchengeich commemoration. At the sixtieth anniversary in 2019, banners from collieries in Fife and Midlothian were present on a procession to the Memorial along with the National Union of Mineworkers Scottish Area (NUMSA) banner.[6]

Willie Doolan, who leads the memorial ceremony on behalf of the Moodiesburn Miners Memorial Committee, was among the last of the men employed in Scottish coal mining when Longannet closed. Willie

---

[3] National Records of Scotland, Edinburgh (NRS), Coal Board (CB) 207/14/3, L. Johnston, National Union of Mineworkers (Scottish Area) (NUMSA), Report on Wester Auchengeich coal reserves (1967).

[4] R. Duncan, *The Mineworkers* (Edinburgh, 2005), p. 145.

[5] J. Phillips, 'The moral economy of deindustrialization in post-1945 Scotland', in *The Deindustrialized World: Ruination in Post-Industrial Places*, ed. S. High, L. MacKinnnon and A. Perchard (Vancouver, 2017), pp. 313–30, at p. 314.

[6] Observation notes from Auchengeich colliery memorial service, 15 Sept. 2019.

remembers the Auchengeich disaster, which occurred when he was four years old. In the aftermath, he began school with classmates who had lost fathers at Auchengeich. His own father attended the disaster as a mines rescue worker. During an interview in 2019, Willie recalled that, in contrast to the highly advanced computerized form of coal-getting pioneered at Longannet, his father had worked 'hand stripping' coal faces with a pick and shovel. Under the 'pillar and stoop' system, colliers left 'pillars' of coal to support roofs, while mining from the chambers between them. When Willie began work in 1971, Cardowan operated under a longwall system of advancing face mining using more developed technology, but conditions remained adverse. Men worked in two feet-high seams using a coal cutter – known as a 'plough' – to extract increments from sections that were 600 feet long.[7]

Pat Egan, a colleague of Willie's at the Longannet complex, also recalled difficult conditions in Lanarkshire. At Bedlay colliery during the late 1970s and early 1980s, Pat and his workmates wore protective kneepads while 'crawlin about on oor haunds and knees'. They cut coal by hand in damp seams of a similar size to those at Cardowan. Upon transfer to the Longannet complex's Solsgirth mine, Pat remembered being surprised by the noise and shocked at the sight of mechanized coal shearers 'the size of a single-decker bus' which operated in seams seven feet high.[8] The transition to an increasingly technologically developed production system, and complex division of labour, increased mutual workforce dependency. It also engendered a shared sense of precarity as coalfield reorganization led to the shrinkage of the workforce and pit closures. Nicky Wilson, like Willie, started work at Cardowan but later transferred to the Longannet complex. He recalled traditions of workgroup solidarity whereby younger and fitter men would 'muck in' so that older or injured miners could retain better-paid jobs underground.[9] The creation of a more cosmopolitan Scottish mining workforce tended to encourage workforce unity across traditional parochial divisions. In the context of a British nationalized industry, deindustrialization stimulated a Scottish national consciousness around coal which grew in importance as the industry's future became more strongly tied to electricity production.

[7] Willie Doolan, interview with author, The Pivot community centre, Moodiesburn, 14 June 2019.

[8] Pat Egan, interview with author, Fife College, Glenrothes, 5 Feb. 2014.

[9] Nicky Wilson, interview with author, John Macintyre Building, University of Glasgow, 10 Feb. 2014.

Developments in the Scottish coalfields reveal that 'deindustrialization does not just happen'.[10] The incremental contraction of Scottish coal mining was not the natural outcome of either market forces or geological exhaustion. Instead, it was the human power relations between miners, employers and the state that dictated pit closures and the provision (or otherwise) of economic security for the workers and communities affected. These factors are once again to the fore in discussions about the future of energy production and increasingly urgent calls to abandon fossil fuels. The organization of a 'just transition' depends on achieving environmental and social justice through arranging meaningful alternative employment for the workforces affected.[11]

When deindustrialization first became the subject of academic study during the immediate aftermath of final closures and strikes and lockouts in the 1980s and 1990s, scholarship often resembled a 'body count' of lost factories and jobs. The analysis in this study uses the vantage provided by distance from these events to look 'beyond the ruins' of industrial Scotland.[12] Emphasizing the protracted process of coalfield contraction challenges the dominance of the closures of the 1980s within prevailing historical accounts. This is apparent in both the 'instant post-industrialization' reading of Scottish economic development, and the predominant focus upon the pit closure programmes of the 1980s and early 1990s within British coalfield literature.[13] Both centre on the ramifications of the 1984–5 miners' strike and the economic policies of the Thatcher government, losing sight of long-term dynamics. These narratives overlook UK energy policy, which tended to favour alternatives to coal from the 1950s onwards, and the nationalized industry's concentration on investment in 'super pits' within the most

---

[10] B. Bluestone and B. Harrison, *The Deindustrialization of America: Plant Closings, Community Abandonment, and the Dismantling of Basic Industry* (New York, 1982), p. 14.

[11] R. Popp, 'A just transition of European coal regions: assessing the stakeholder positions towards transitions away from coal', *E3G Briefing Paper* (2019) <https://www.e3g.org/showcase/just-transition/> [accessed 20 Nov. 2019].

[12] J. Cowie and J. Heathcott, 'Introduction: the meanings of deindustrialization', in *Beyond the Ruins: The Meanings of Deindustrialization*, ed. J. Cowie and J. Heathcott (Ithaca, N.Y., 2003), pp. 1–15, at p. 6.

[13] C. Harvie, *No Gods and Precious Few Heroes: Twentieth Century Scotland* (Edinburgh, 1998), p. 164; T. Strangleman, 'Networks, place and identities in post-industrial communities', *International Journal of Urban and Regional Research*, xxv (2001), 253–66; M. Foden, S. Fothergill and T. Gore, *The State of the Coalfields: Economic and Social Conditions in the Former Mining Communities of England, Scotland and Wales* (Sheffield, 2014), pp. 13–14; P. Ackers, 'Review essay: life after death: mining history without a coal industry', *Historical Studies in Industrial Relations*, i (1996), 159–170, at pp. 160–1.

productive coalfields.[14] The great strike for jobs was not the harbinger of deindustrialization. It began due to the breakdown of the comparatively consensual approach to closures which the NCB had developed in partnership with the NUM between the 1940s and the 1970s.[15]

As the memorial service at Auchengeich indicates, memories and associations formed through industry continue to reverberate in the Scottish coalfields. This annual event is an example of deindustrialization's 'half-life': the lingering influence of industrial society's 'historically mediated structures, action and experiences' upon the culture and politics of deindustrializing localities.[16] In place of a neat series of economic eras, deindustrialization instead suggests a longer, contested and more painful transition. By examining the Scottish coalfields, rather than a single location or industry, this study follows the suggestion that scholars of deindustrialization ought to 'shift our focus from industrial ruins to the wider processes of ruination'.[17] An extended time-period – spanning the second half of the twentieth century and into the present – allows for the development of a *longue durée* analysis. It reveals the 'submerged history' of major alterations in basic economic functions and social routines which deindustrialization entails. In the case of coal mining, these relate to pivotal questions of how societies choose to obtain their heat and light.[18] The rundown of colliery employment led to the wholesale reconstruction of labour markets in the areas affected by incremental closures over several decades.

Towns and villages within the Scottish coalfields developed in tandem with the coal mining industry, which often pivotally determined their social structure and associational culture. Yet this was never a monoculture: other sectors such as metals and heavy engineering also evolved alongside Scottish coal mining.[19] As coal employment contracted, replacement activities in

---

[14] W. Ashworth, *The History of the British Coal Industry*, v: *1946–1982: the Nationalized Industry* (Oxford, 1986), pp. 38–9, 163; M. K. Oglethorpe, *Scottish Collieries: an Inventory of the Scottish Coal Industry in the Nationalised Era* (Edinburgh, 2006), p. 20.

[15] J. Phillips, *Collieries, Communities and the Miners' Strike in Scotland, 1984–5* (Manchester, 2012) pp. 20–81.

[16] T. Strangleman, 'Deindustrialisation and the historical sociological imagination: making sense of work and industrial change', *Sociology*, li (2017), 466–82, at p. 467.

[17] S. High, L. MacKinnnon and A. Perchard, 'Introduction', in High et al, *The Deindustrialized World*, pp. 3–22, at p. 18.

[18] F. Braudel, *The Mediterranean and the Mediterranean World in the Age of Philip II* (2 vols., London, 1972), i. 16.

[19] P. Payne, *Growth and Contraction: Scottish Industry, c.1860–1990* (Glasgow, 1992), pp. 20–1.

mass production industries were directed towards the coalfields through UK regional policy. These sectors developed through inward investment, especially by American multinationals. Assembly goods plants were popularly understood in terms of an exchange in forms of employment. Policymakers presented coalfield reconstruction and inward investment as substantial improvements which could sustain affluent lifestyles and safer working conditions.[20] A comprehensive study of deindustrialization in the Scottish coalfields therefore requires analysis of the transition towards a diversified industrial structure between the 1940s and the 1960s.

During the late 1960s, the transition from the 'solidaristic' environment of traditional workplaces embedded in community structures to the 'instrumental' motivations behind labour on assembly lines was theorized in the *Affluent Worker* study.[21] Recent studies of industrial relations in engineering sectors have contested 'privatist' understandings of class in mid and late twentieth-century Britain, pointing to retained occupational identities, recurring cultures of solidarity, and expressions of collective trade union 'voice'.[22] This book adds to these conclusions by exploring how exchanging employment in traditional industries for assembly factories and more advanced collieries shaped workers' perspectives. In the Scottish coalfields, links between work and community were retained through extensive economic and social reconstruction, including slum clearances and significant public house building. New jobs were understood as having been paid for at the cost of older workplaces closing. As a result, employment in factories and new or redeveloped collieries were viewed as the property of communities. These factors shaped workplace relations and encouraged opposition to closures as deindustrialization intensified during the 1980s.

Since 1945, the UK's landscape has been transformed by the protracted transition from an economy centrally configured around industrial production towards one dominated by services. Deindustrialization's long-term effects have provided explanations for deepening political polarization during the 2010s, especially the 2016 Brexit referendum.[23]

[20] E. Gibbs and J. Phillips, 'Who owns a factory? Caterpillar tractors in Uddingston, 1956–1987', *Historical Studies in Industrial Relations*, xxxviii (2018), pp. 111–37, at pp. 114–15.

[21] J. Goldthorpe et al, *The Affluent Worker: Industrial Attitudes and Behaviour* (London, 1968), pp. 39–41.

[22] J. Boyle, B. Knox and A. McKinlay, '"A sort of fear and run place": unionising BSR, East Kilbride, 1969', *Scottish Labour History*, liv (2019), 103–25, at p. 117; J. Saunders, *Assembling Cultures: Workplace Activism, Labour Militancy and Cultural Change in Britain's Car Factories, 1945–82* (Manchester, 2019), pp. 4–5.

[23] J. Tomlinson, *Managing the Economy, Managing the People: Narratives of Economic Life in Britain from Beveridge to Brexit* (Oxford, 2017), p. 107.

However, the distinctions between patterns of voting at the referendum, principally the fact that deindustrialized areas of Scotland voted to 'remain', should caution against simplistic conclusions about the relationship between deindustrialization and right-wing 'populism'.[24] Scotland's renegotiated relationship with the rest of the UK can also be explained by deindustrialization. This volume is not primarily concerned with explaining developments in terms of party politics, including the electoral success of the Scottish National Party since the establishment of the Scottish parliament in 1999. Instead, it underlines the importance of industrial developments to understanding the growing political significance of Scottish national identity during the closing decades of the twentieth century. The experience of both falling industrial employment and the growing importance of London-based policymaking to the Scottish economy were significant ingredients in popularizing demands for political autonomy. Support for devolution – then more commonly known as 'home rule' – coalesced around the assumption that Scotland was an 'industrial nation,' which was shared across class divides.[25] The failure of both Labour and Conservative governments to sustain this reality encouraged dislocations between Scottish and British national identities. Deindustrialization also removed direct connections between Scottish communities and the state, especially in the context of nationalized industries. Industrial closures further undermined links between different parts of the UK that had been furnished through occupational solidarities.

Approaches to deindustrialization

In recent years, a historical account of deindustrialization in Scotland has developed which underlines the rundown of coal mining employment from the late 1950s onwards.[26] Andrew Perchard and Jim Phillips pioneered moral economy perspectives on Scottish deindustrialization, emphasizing how manual workers' understandings of economic justice shaped

---

[24] S. L. Linkon, *The Half-Life of Deindustrialization: Working-Class Writing about Economic Restructuring* (Ann Arbor, Mich., 2018), pp. xiii–xvi; 'EU referendum results by region: Scotland', *Electoral Commission*, 29 Sept. 2019 <https://www.electoralcommission.org.uk/who-we-are-and-what-we-do/elections-and-referendums/past-elections-and-referendums/eu-referendum/results-and-turnout-eu-referendum/eu-referendum-results-region-scotland> [accessed 19 Nov. 2019].

[25] J. Tomlinson and E. Gibbs, 'Planning the new industrial nation: Scotland, 1931 to 1979', *Contemporary British History*, xxx (2016), 584–606.

[26] J. Phillips, 'Deindustrialization and the moral economy of the Scottish coalfields, 1947 to 1991', *International Labor and Working-Class History*, lxxxiv (2013), 99–115.

practices and assessments of colliery closures.[27] Moral economy analyses of deindustrialization are centrally concerned with 'non-economic norms affecting commercial interactions', principally community and workforce attitudes to the ownership of jobs and workplaces.[28] They are influenced by critiques of industrial capitalism that developed between the 1920s and 1970s, based on its moral or spiritual effects rather than material outcomes. Karl Polanyi and E. P. Thompson were two key theorists who rejected a naturalized view of liberalized market orders, insisting on their social contingency.[29] Polanyi's theory of the double movement illuminates the conflict between market forces and protective social agents which shaped deindustrialization in the Scottish coalfields. Thompson's moral economy assists in conceptualizing the customs and expectations that determined workforce and community perceptions of legitimacy or illegitimacy during the closure of industrial workplaces.

Polanyi conceived of industrialization and capitalist production as history's 'great transformation ... Instead of the economy being embedded in social relations, social relations are embedded in the economic system'. This transformation swept away traditional 'safeguards' provided by customs and legislation which 'protect the prevailing economic organization of society from interference on the part of market practices'.[30] However, Polanyi contended that industrial capitalism was characterized by a 'double movement' that created renewed safeguards: episodes of economic liberalization are met by protective 'countermovements' which act to preserve social bonds from the erosive commodifying effects of market forces.[31] Industrial capitalist economies are never fully 'disembedded' from societal mores but are subject to a continual renegotiation between market and social pressures. Liberalization strains national societies to the point of a protective restriction on 'the market mechanism'. It is these tendencies towards the 'integration' of political and economic spheres – overcoming their socially disastrous formal separation and disintegration – which lie

---

[27] A. Perchard and J. Phillips, 'Transgressing the moral economy: Wheelerism and management of the nationalised coal industry in Scotland', *Contemporary British History*, xxv (2011), 388–9.

[28] T. Rogan, *The Moral Economists: R. H. Tawney, Karl Polanyi, E. P. Thompson and the Critique of Capitalism* (Princeton, N.J., 2017), p. 7.

[29] Rogan, *Moral Economists*, pp. 1–15.

[30] K. Polanyi, *The Great Transformation: the Political and Economic Origins of Our Time* (Boston, Mass., 2001), pp. 60–5.

[31] Polanyi, *Great Transformation*, p. 70.

at the heart of the double movement.[32] The living, breathing status of workers and the inherently political status of labour means that it can only be commoditized fictitiously, and destructively. State regulation and the activities of workers organized in trade unions and labour movement political parties serve 'a wider interest than their own' in struggles to protect labour from commodification. These are necessitated because 'the organization of capitalist production itself has to be sheltered from the devastating effects of a self-regulating market'.[33]

Coal mining provides an archetypal example of the double movement. The industry's development included early and significant contributions from state regulators as well as organized labour in the recurring 'clash of the organizing principles of economic liberalism and social protection'.[34] Dangers present in the industry combined with paternalistic Victorian social norms to encourage initial legislation such as the 1842 Mines Act, which regulated the labour market by prohibiting women and boys under the age of ten from working underground.[35] By the early twentieth century, this had passed into other areas including health and safety and the length of the working day. The state legislated for minimum conditions which were highly contested by the coal owners and trade unions.[36] Coal mining's subsequent nationalization in 1947 can be understood as a key example of economic and political integration which restored social objectives over economic processes.[37] In Karl Polanyi's terms, nationalization can be viewed as a partial attempt 'to take labour out of the market' by enhancing the role of the state and trade unions in determining production, wages and conditions.[38] The privately owned industry's legacy of conflict and failure to provide employment stability moulded expectations of how contraction would be managed under state stewardship. Rather than a single event, the double movement should be viewed as a recurring conflict between

[32] K. Polanyi, *The Present Age of Transformation: Five Lectures by Karl Polanyi, Bennington College, 1940* (Northbrook, 2017), pp. 15–22.

[33] Polanyi, *The Present Age of Transformation*, pp. 15–22.

[34] Polanyi, *The Present Age of Transformation*, p. 140.

[35] C. Mills, *Regulating Health and Safety in the British Mining Industries, 1800–1914* (Farnham 2010), pp. 56–66.

[36] R. Church, 'Employers, trade unions and the state, 1889–1987: the origins and decline of tripartism in the British coal industry', in *Workers, Owners and Politics in Coal Mining: an International Comparison of Industrial Relations*, ed. G. Feldman and K. Tenfelde (New York, 1990), pp. 12–73, at p. 28.

[37] J. Tomlinson, 'A "failed experiment"? Public ownership and the narratives of post-war Britain', *Labour History Review*, lxxiii (2008), 228–43, at p. 235.

[38] Polanyi, *The Great Transformation*, p. 258.

capital accumulation and societal order. The operation of liberalized markets encouraged protective responses phrased in a moral language of social responsibility. This continued under nationalization. Financial and productionist pressures competed with commitments to workforce security and dialogue with trade unions in the industry's governance.

Moral economy analysis of human agency in the coalfields has its origins in E. P. Thompson's account of eighteenth-century riots against the liberalization of foodstuffs marketing in England. The 'crowd' of plebian consumers defended and enforced customary practices of open selling through disciplined direct action, including price setting. Their activities were shaped by a moral economy consciousness which 'taught the immorality of any unfair method of forcing up the price of provision by profiteering upon the necessities of the people'.[39] The customs which informed Thompson's food rioters are an example of a countermovement invoking safeguards to protect social structures from economic disruption.[40] Thompson later summarized his conception of the moral economy at a theoretical level. It centred on claims of 'non-monetary rights' to resources predicated on traditions of 'community membership [which] supersedes price as a basis of entitlement'.[41] The coalfield moral economy also had a basis in popular custom and expectation regarding the provision of economic resources but related to the provision of industrial employment rather than food consumption. Its contentions centred on an interpretation of the social responsibilities of elites, NCB officials and government policymakers, to provide security against economic instability. These agents were obligated to fulfil the nationalized industry's promise of worker 'voice' through consultation with trade unions. Perchard and Phillips defined the coalfield moral economy's contours as:

> Joint regulation, through agreement between managers and union representatives, of workplace affairs, including pit closures, job transfers, substantial alterations to production and the labour process; and guaranteed economic security, so that miners displaced by colliery closures could find equally well-remunerated alternative employment, at other pits or elsewhere in industry.[42]

---

[39] E. P. Thompson, *The Making of the English Working Class* (Middlesex, 1968), pp. 67–72.

[40] E. P. Thompson, 'The moral economy of the English crowd', *Past and Present*, i (1971), 76–136, at pp. 77–9, 95–6.

[41] E. P. Thompson, *Customs in Common* (London, 1991), pp. 338–9.

[42] Perchard and Phillips, 'Transgressing the moral economy', p. 389.

Introduction: those who walked in the darkest valleys

Like Thompson's moral economy of plebeian consumers, the coalfield moral economy was instigated by communitarian claims to economic resources: collieries and the employment they sustained. It was discernible in the application of customary measures: the observation of procedure regarding the management of closure through discussion with trade union representatives; ensuring transfers for the workforce within travel distance; the provision of transfers to appropriate jobs, especially for high-earning faceworkers; the provision of either retirement or suitable positions for elderly and disabled workers. Those practices were rarely followed in full, and their application evolved over the nationalized period. Nevertheless, these expectations structured workforce attitudes during colliery closures. While it was largely practised through workforce and management dialogue between the 1940s and 1970s – before it was defended during the year-long strike that began in March 1984 – the moral economy was animated by the collective memory of class struggles, especially the crisis-ridden 1920s and 1930s. Reference points included the 1921 miners' lockout and the 1926 general strike and lockout, as well as the blacklisting of workers and forced relocation of mining families which often followed. These moments of mobilization and recrimination were conjoined with other memories of social violence associated with the private industry, such as its poor health and safety record, and mass unemployment. The interwar years were remembered in families, passed on in communities and workplaces and institutionalized by the NUMSA.

Coal's centrality to Britain's industrial economy, and the government's regulation of mining, contributed to the state's central role in the moral economy. Polanyi understood that 'modern nationalism is a protective reaction against the dangers inherent in an interdependent world'.[43] Deindustrialization's territorial politics developed within these parameters. Localized objections to closure were given the standing of national crises where they reached a critical mass. Objections to closures in Scotland were voiced within the framework of the Union, but they also stimulated demands for increased autonomy. The countermovement's national grounding built on earlier responses to economic crises, including when national and class prerogatives merged during the agitation against unemployment in the 1930s.[44] Consciousness that developed from community and workplace experiences served as 'building blocks' upon which broader occupational

---

[43] Polanyi, *The Present Age of Transformation*, p. 7.

[44] H. McShane, 'The march: the story of the historic Scottish hunger march', *Variant*, xv (2002), 30–4, at p. 30.

and class-based solidarities were sustained.[45] These dimensions were apparent in the fusion of occupation, class and nationhood which shaped the political perspectives of Scottish miners and their trade union. The countermovement to the dislocation of communitarian norms experienced in colliery closures and industry-wide contraction was substantially framed in terms of national identity. Moral economy sentiments came to rest on expectations of action from devolved, and therefore supposedly more sympathetic, Scottish authorities.

The practices which shaped class consciousness in the Scottish coalfields can be encompassed by the concept of 'labourism'. As theorized by historians of Welsh coalfield politics, labourism is an actively constructed expression of working-class consciousness and practical politics, rather than an aberration from the development of an overtly socialist political culture.[46] This is contrary to the dominant accounts of labourism which have their origins in the 'new left' analysis of the 1960s, and its successors, which distanced itself from both orthodox Marxism and social democracy. Within the new left view, labourism has broadly been understood as the product of a deliberately moderated and reformist politics. Labourism is primarily a parliamentary effort to mobilize the existing apparatus of the British state to improve incrementally the well-being of an organized working-class constituency.[47]

In the Scottish coalfields, a heterogeneous coalition of political forces maintained a Labourist culture, beyond these boundaries of moderation. Neighbourhood organization, workplace conflict and cooperation, trade unionism, and local government shaped a variegated manual working-class consciousness. The NUMSA was central to these experiences, and Communist Party of Great Britain (CPGB) activists played a prominent role. Communist orthodoxies predominated within Scottish mining trade unionism at least from the election of Abe Moffat as president of the National Union of Scottish Mineworkers in 1942. CPGB influence continued to grow after the union became the Scottish Area of the NUM in 1945. The office of NUMSA president was filled successively by CPGB members for over five decades until the post was abolished following

---

[45] H. Barron, *The 1926 Miners' Lockout: Meanings of Community in the Durham Coalfield* (Oxford, 2009), p. 54.

[46] K. Gildart, *North Wales Miners: a Fragile Unity, 1945–1996* (Cardiff, 2001), pp. 4–5; C. Williams, *Capitalism, Community and Conflict: the South Wales Coalfield, 1898–1947* (Cardiff, 1998), p. 5.

[47] G. Elliot, *Labourism and the English Genius* (London, 1993), p. vii; R. Miliband, *Parliamentary Socialism: a Study in the Politics of Labour* (London, 1973); J. Saville, 'Labourism and the Labour government', *Socialist Register*, iv (1967), 43–71.

George Bolton's retirement in 1996, by which point the union had become a marginal presence as the industry shrank.[48] NUMSA policy accorded with the CPGB's perspective of supporting Scottish cultural and political autonomy while developing class unity across the UK. Under CPGB stewardship, the NUMSA pioneered a deepening commitment to political devolution through a Scottish parliament that gained influence across the wider Scottish labour movement.

Figures such as Bolton and Moffat embodied Raphael Samuel's perception of Scottish miners' 'own distinctive version of militancy – more "educated", more Communist, more statesman-like' than its English counterpart.[49] Their form of radical labourism, or on occasion respectable militancy, was the product of the Area's autonomy and the Scottish coalfields' institutional political environment. This enabled the NUMSA to exercise influence through the Scottish Trades Union Congress (STUC) and to consciously develop cadres of successor generations from the ranks of young activists. The Scottish miners also had strong connections with the Labour party to which it was affiliated. To this extent, the NUMSA – and the CPGB activists who shaped its political positioning – broadly adhered to 'shared labour movement ethics'.[50] Yet the NUMSA's persistent and sometimes successful agitation on international as well as constitutional questions cannot be explained in this framework alone. The prolonged contraction of coal mining employment was central to shaping arguments for enhanced Scottish political autonomy. These sustained developments were influenced by the complex status of the NCB and the NUM as having unitary structures with significant devolved elements. Devolutionary sentiment was further magnified by the centrality of coal mining employment to the economic welfare and cultural identity of coalfield areas.

However, the political reach of the CPGB-aligned activists in the NUMSA was always limited in the coalfields. They faced opposing tendencies within the labour movement, including organized Catholic and Protestant factions which asserted sectarian identities as well as social conservatism. Complex

---

[48] P. Ackers, 'Gramsci at the miners' strike: remembering the 1984–1985 Eurocommunist alternative industrial strategy', *Labor History*, lv (2014), 151–72, at pp. 160–4; R. Robertson, 'Rob Robertson meets the union leader on the coal face to save Monktonhall', *Herald*, 16 May 1997 <http://www.heraldscotland.com/news/12324728.Rob_Robertson_meets_the_union_leader_working_at_the_coal_face_in_the_fight_to_save_Monktonhall_Digging_deep_for_survival/ > [accessed 20 Nov. 2019].

[49] R. Samuel, 'Introduction', in *The Enemy Within: Pit Villages and the Miners' Strike of 1984–5*, ed. R. Samuel, B. Bloomfield and G. Boanas (London, 1986), pp. 1–38, at p. 19.

[50] G. Andrews, *Endgames and New Times: the Final Years of British Communism, 1964–1991* (London, 2004), p. 86.

and conflicting political forces shaped identifications and social relations in the Scottish coalfields across the twentieth century. But both more radical and moderate currents were broadly unified in support for a politically and industrially united labour movement. They shared a working-class consciousness embedded in manual occupational identities and public housing tenancies. Labourism was therefore politically fractured, acting as a container and manager of divisions, including religious sectarian ones, and differences over political alignment. But it also acted as a unifying force, providing a common cultural and organizational repertoire and reference points grounded in the coalfield moral economy. Samuel's observations affirm the gendered status of the moral economy, indicating its origins in communities structured around male employment and industrial and political leadership. The moral economy was renegotiated as coalfield labour markets changed. Working-class affluence was often secured through married women's wages, and the assembly factories which partially replaced colliery employment also provided work for women. These factors contributed to an extension of the coalfield moral economy in sectoral and gender terms.

Sources and structure

The analysis in this volume combines oral testimonies from coalfield respondents with archival research relating to energy policy, the management of colliery closures and regional policy. Depth is added to existing accounts of deindustrialization in Scotland through shifts in theoretical lenses, temporal and geographical focus and detailed attention to dimensions of policymaking and application at Scottish and UK levels. The source base incorporates extensive archival material from the nationalized coal industry, government and trade unions, spanning from between the 1940s and 1980s. These include the minutes of Colliery Consultative Committee (CCC) meetings which were held during closure procedures and featured input from trade union representatives, as well as correspondence and reports from Scottish and UK NCB headquarters related to forward planning. Scottish Office and the Board of Trade records reveal dimensions of regional policy application and inward investment. Records from the Ministry of Fuel and Power and its successors shed light on the priorities and conflicts that guided energy policy and power station investment. These reveal a politicized anti-coal bias within the Ministry. Officials and government ministers from the mid 1950s to the mid 1970s frequently pursued the liberalizing impulse of the double movement. They supported uneconomic experimental nuclear power projects and championed replacing coal with oil as the UK advanced from a single- to a multi-fuel energy economy. These developments anticipated the assertion of financial prerogatives under the Thatcher governments. Minutes from the NUMSA's executive committee and annual conferences give voice to both trade union officials and elected colliery delegates who contested these decisions. Further access to union perspectives is provided through STUC annual reports, including conference contributions made by NUMSA representatives which shed light on the countermovement's national framing.

The archival findings are supplemented by testimonies from an oral history project. The testimonies provide distinctive perspectives which assist in recovering experiences of social change, but they also crucially reveal deliberations on the cumulative impact of deindustrialization. Interviews were conducted with over thirty men and women in a life-story format, in addition to two focus groups. The interviewees were recruited through a variety of methods. Most participants were located through 'snowballing' from existing contacts, but some were also found through press advertisements.[51]

Recruitment was conducted with the aim of collecting testimonies from a range of voices in gender and generational terms. Former trade union activists were over-represented. This was a product of the contact networks that snowballing was conducted through. The predominance of former trade unionists reflects the tendency of activists to have the strongest narratives and retain social and emotional connections with movements they took part in.[52] However, narratives were also gathered from former miners who were not supporters of the NUMSA leadership's communist-influenced politics. These include individuals with a history of being active within loyalist or Unionist organizations such as the Orange Order, as well as former NCB managers. Despite these differences, a fusion of class and national framings was present within the testimonies. This resonated with the development of the double movement and the protective countermovement's basis in Scottish national consciousness. These tendencies were encouraged by contemporary politics through the consolidation of a centre-left understanding of 'social justice' associated with Scotland's contemporary political self-image. Perceptions of Scotland's comparatively egalitarian

---

[51] P. T. Knight, *Small-Scale Research: Pragmatic Inquiry in Social Science and the Caring Professions* (London, 2008), p. 65.

[52] N. Fielding and H. Thomas, 'Qualitative interviewing', in *Researching Social Life*, ed. N. Gilbert (London, 2008), pp. 245–65, at p. 260; L. Stizia, 'Telling Arthur's story: oral history relationships and shared authority', *Oral History*, xxvii (1999), 58–67, at pp. 63–4; J. Kirk, *Class, Culture and Social Change: on the Trail of the Working Class* (Basingstoke, 2007), pp. 163–5.

ethos have been reinforced since the establishment of a devolved parliament in 1999.[53]

Dialogue developed around the interviewee's life-story, with questions emphasizing meanings and understandings of community, connections to the mining industry, and experiences of closure. Women's narratives highlighted changes in gender roles and family life connected with increasing female participation in the workforce and rising male unemployment. Generational distinctions were another crucial source of differentiation in experiences of labour market changes. The testimonies are deployed with an emphasis on the role familial influences and communal experiences have played in shaping perspectives on job losses and workplace closures. Critical nostalgia – a reflective attitude towards negative aspects of industrial experiences that nevertheless mourns the loss of collective bonds and economic resources – has framed recollections of class, occupation and community within former coalfield localities. The 'radical imagination' offers a valuable critique of the past through a lens that considers what has been lost and gained in complex socioeconomic changes.[54] Respondents identified injustices and conservatism in traditional coalfield communities, but also questioned deindustrialization's historical inevitability and critiqued contemporary labour markets and societal inequalities

One focus group was conducted with a retired miners' group in Moodiesburn, centring on experiences of industrial relations and pit closures. Another comprised of a mixed gender group in Shotts in eastern Lanarkshire, which focused on community and social changes. Both followed a semi-structured format, with dialogue largely evolving through participant interaction. The retired miners' group, which continues to meet regularly, was recruited through a contact. The Shotts focus group was recruited through contacting a local history group which principally involved men and women from mining backgrounds. In Moodiesburn, the discussion revolved around workplace experiences, with an emphasis on patterns of colliery closures and memories of industrial relations and the history of dangers and disasters. The group in Shotts focused on community, emphasizing shared social activities. There was minimal crossover between the focus groups and testimonies, but one Shotts participant, Willie Hamilton, also took part in a separate interview along with his wife, Marian.

---

[53] G. Hassan, 'Back to the future: exploring twenty years of Scotland's journey, stories and politics', in *The Story of the Scottish Parliament: the First Two Decades Explained*, ed. G. Hassan (Edinburgh, 2019), pp. 1–27, at pp. 5–13.

[54] A. Bonnett, *Left in the Past: Radicalism and the Politics of Nostalgia* (New York, 2010), pp. 1–3.

Both the life-story interviews and focus groups displayed elements of collective memory and evidenced the tendency for communal experiences to be retold through myths and legends consolidated by family, community and labour movement connections.[55] These forces shape the operation of a coalfield 'cultural circuit' which sets the terms through which life-experience is mediated. It was common for analogies to be drawn with the strife and dislocation of the private coal industry during the first half of the twentieth century.[56] In the focus groups, the collective construction of memories was evident. The dialogue was often fragmentary and revolved around shared elements of everyday social life in a community and workplace setting, or communal experiences of events such as gala days, pit closures or mining disasters. Participant interaction drove dialogue and encouraged the 'generative' connections through which the cultural circuit operates.[57]

The geographical focus of the oral history project and the archival research into individual colliery closures and regional policy investment primarily related to the Lanarkshire coalfield. However, individuals who lived and worked in Scotland's other major coalfield areas – Ayrshire, the Lothians, Clackmannanshire and Fife – were also interviewed. Lanarkshire affords an important vantage given its centrality to the processes that reshaped society and the economy in the Scottish coalfields. It was Scotland's largest coalfield upon nationalization in 1947 but underwent extensive contraction during the late 1940s and 1950s as capital and manpower were shifted eastwards. Towns and villages which had either been born or developed around mining experienced profound alteration due to the closure of local collieries before comparable settlements did elsewhere in Scotland. The moral economy practice of offering transfers to miners within travel distance, and objections to migration upon colliery closures which came to prominence during the 1950s, were formed through this experience. These norms had origins in local experiences of deindustrialization, which took place while Scottish coal mining employment was still expanding. The focus on Lanarkshire also affords a study of inward investment and the diversification of industrial employment through analysis of documentation from the Board of Trade and Scottish Office departments responsible for applying regional policy and liaising with plant management. Additionally, several of the

[55] S. A. Crane, 'Writing the individual back into collective memory', *American Historical Review*, cii (1997), 1372–85, at p. 1373; A. Portelli, *The Death of Luigi Trastulli and Other Stories: Form and Meaning in Oral History* (Albany, N.Y., 1991), pp. 1–2.

[56] L. Abrams, *Oral History Theory* (London, 2010), p. 6.

[57] H. Finch and J. Lewis, 'Focus groups', in *Qualitative Research Practice: a Guide for Social Science Students and Researchers*, ed. J. Ritchie and J. Lewis (London, 2003), pp. 170–98, at p. 171.

interviewees had worked in assembly goods factories or had relatives who had. Lanarkshire reveals the long-term consequences of deindustrialization, as well as the transmission of culture, occupational identities and senses of place invested in the industrial past across generations. The annual commemoration at Auchengeich exemplifies these processes.

The first two chapters of this volume comprise a detailed analysis of the processes that shaped deindustrialization in the Scottish coalfields. They are primarily, but not exclusively, based on archival records. Chapter 1 is themed around industrial developments, providing a top-down perspective on changes in labour market structures. It traces coal's changing position in the pecking order which determined power station investment and analyses the application of regional policy through a case study of the Lanarkshire coalfields. In the second chapter, the perspective is altered to a bottom-up analysis. It assesses how the practices of the coalfield moral economy developed between the 1940s and 1980s. Chapter 2 includes three geographical case studies of colliery closures in Lanarkshire. It emphasizes how the early experiences of community abandonment in the Shotts area of Eastern Lanarkshire during the late 1940s and early 1950s shaped more careful management practices in future closures. During the 1980s, NCB management abandoned moral economy obligations.

The following four chapters provide a reflective account of deindustrialization's long-term social, cultural and political consequences. They heavily profile oral testimonies, which are supplemented by archival sources. Chapter 3 analyses the processes of community reconstruction associated with deindustrialization. It distinguishes between the public house-building and slum clearances which accompanied industrial diversification between the 1940s and 1970s with the private house-building that accompanied intensified deindustrialization in later years. During the earlier period, new industrial communities formed through shared manual working-class identities and public housing tenancies while older ones were sustained by them. These communities have become the subject of critical nostalgia as suburbanization and accelerated workplace closures disembedded locational connections to industrial production. The fourth chapter considers the dimensions which shaped perceptions of deindustrialization by analysing how gender conditioned men's and women's experiences of labour market restructuring. Male feelings of social redundancy through job loss are contrasted with the expansion of women's labour market opportunities in assembly plants and public services. Women's direct experience of industrial job loss are highlighted with reference to the moral economy status that work in the factories brought to Scotland between the 1940s and 1960s had attained by the 1980s.

Chapter 5 switches focus by examining how formative episodes moulded the outlooks of three discrete cohorts between nationalization during the 1940s and displacement during the closures of the 1980s and 1990s. The sixth chapter develops how national identity was reshaped by deindustrialization through investigating the construction of a Scottish coalfield 'imagined community', drawing heavily on the NUMSA's records.[58] It reveals how the predominantly communist-leaning leadership of the union developed a distinctive political culture which merged aspirations for home rule on the one hand with support for a UK-wide nationalized industry on the other. Chapter 7 provides a synthesis, linking the fusion of class and Scottish national consciousness through the complex process of coalfield reconstruction and colliery closures. The conclusion reflects on how this study's findings relate to the scholarship of deindustrialization, underlining the challenges and fruitfulness behind reconciling intensely personal, human experiences with deep structural transformation.

---

[58] B. Anderson, *Imagined Communities: Reflections on the Origin and Spread of Nationalism* (London, 2009).

1. 'Buried treasure': industrial development in the Scottish coalfields, c.1940s–80s

Scotland's post-Second World War economic development was characterized by an evolving relationship between energy generation and industry. This chapter presents a chronological assessment from the inauguration of major structural changes during the mid 1940s, to the demise of mass industrial employment in the 1980s. The analysis centres on how UK fuel and regional economic policies vastly altered the labour market in the coalfields. Over four decades, nationalized coal and power generation sectors were central to reconfiguring the basic economic structure which had evolved during the previous century. A much-altered geographical distribution of coal mining employment emerged through the concentration of production within technologically advanced collieries in eastern Scotland. Their markets were increasingly dominated by electricity production. These developments were augmented by the application of regional policy, which sought to diversify the nation's industrial base through inward investment. Initially, modernizing civil servants sought to supplement employment in heavy industry by developing new mass production sectors. By the mid 1960s, these more modest aims had been displaced by the objective of pursuing economic growth through diverting labour from traditional sectors to light engineering.

The closure of collieries was largely managed consensually between the 1940s and the 1970s. Policymakers met moral economy obligations by engaging in dialogue with trade union representatives and ensured labour market security through linking divestment with the creation of jobs in new plants. This was not a straightforward process. There was a dynamic relationship between fuel policy and power generation on the one hand, and the application of regional policy on the other. As coalfield employment shrank, and the industry was disembedded from the towns and villages which had grown around it, a protective countermovement developed and articulated its objections in Scottish national terms. This became increasingly forthright when the nationalized industry's rationalized structure was threatened by the closure of redeveloped pits during the 1960s. Those concerns were generalized as the logic of Scotland's new political economy revealed itself. Coal employment had become more and more dependent

---

'"Buried treasure": industrial development in the Scottish coalfields, c.1940s–80s', in E. Gibbs, *Coal Country: The Meaning and Memory of Deindustrialization in Postwar Scotland* (London, 2021), pp. 21–54. License: CC-BY-NC-ND 4.0.

upon power station burning, while multinationals became increasingly important in manufacturing. Demands for Scottish political autonomy grew within the labour movement, becoming increasingly enmeshed with criticisms of the centralizing tendencies of the National Coal Board (NCB) and the failure of regional policy to provide adequate replacement for lost employment in traditional industries. Coalfield experiences were formative, both to the major alterations in Scotland's economic structure, but also in shaping the political response to these changes.

In 1945, Scotland's economy would have looked familiar to a Victorian observer. Its industrial heartland, Clydeside, was characterized by family firms engaged in the 'staple' sectors of textiles, metals, heavy engineering and coal. In more backward collieries, workers still used pick axes and shovels to 'hand strip' faces, leaving 'pillars' of coal in place to maintain structural integrity.[1] Steel production remained concentrated in plate, much of which was used by a shipbuilding sector where bespoke products and craft skills predominated.[2] After the Second World War, a fundamental shift in political economy took place under state direction. It substantially altered both ownership structures and employment and production patterns. This transition included the growth of assembly goods plants, incorporating significant investment by American multinationals. Scottish industrial employment peaked in the mid 1950s, with considerable contraction following in the staple sectors. However, deindustrialization was not a straightforward process. Manufacturing employment peaked over a decade later, spurred by the expansion of sectors that grew through inward investment, before these sectors also shrank during the 1970s.[3]

The coalfields were at the heart of these changes. Scotland's largest coalfield, Lanarkshire, experienced extensive colliery closures as the industry was rebalanced towards more productive seams to the east. Under nationalization, the workforce was concentrated in fewer, larger, highly mechanized pits.[4] Over the same period, coal became increasingly dependent on a single market: power generation. Between 1950 and financial year

---

[1] J. Phillips, *Scottish Coal Miners in the Twentieth Century* (Edinburgh, 2019), p. 64; Willie Doolan, interview with author, the Pivot Community Centre, Moodiesburn, 14 June 2019.

[2] C. Harvie, *No Gods and Precious Few Heroes: Scotland's Twentieth Century* (Edinburgh, 1998), p. 55.

[3] J. Phillips, 'The moral economy of deindustrialization in post-1945 Scotland', in *The Deindustrialized World: Ruination in Post-Industrial Places*, ed. S. High, L. MacKinnnon and A. Perchard (Vancouver, 2017), pp. 313–30, at p. 314.

[4] R. Smith, 'New towns for Scottish miners: the rise and fall of a social ideal (1945–1948)', *Scottish Economic and Social History*, ix (1989), 71–9, at pp. 73–6.

1972–3, power station coal consumption rose from thirty-three to sixty-nine million tonnes, from 16 per cent to 53 per cent of the UK's annual coal use.[5] As the steel industry contracted over the 1970s and early 1980s, dependency was further consolidated. By 1981, the NCB was selling 80 per cent of its product to the electricity market.[6] Concurrently, government supported the development of alternative fuels for power generation. This judgment related to the requirements of maximizing productivity in Britain's full employment economy, concerns about ability of the NCB to provide for expanding energy use, and to political anxieties over the power of organized labour within coal production and distribution.[7] These developments created tension between miners and the NCB on the one hand and the Ministry of Fuel and Power and its successors, and the nationalized power generators, on the other. Although policymaking took place in a unitary, UK-wide, framework, the NCB's Scottish division, and the South of Scotland Electricity Board (SSEB) secured significant devolved dimensions to administration.

The release of labour from coal was welcomed by an influential policymaking community who were empowered by the Scottish Office's autonomy in the application of UK regional policy. A modernizing agenda developed through the Scottish Council (Development and Industry) (SCDI), a body dominated by business representatives, but which was also subscribed to by local authorities and trade unions.[8] Civil servants and their allies within business and academia developed a strong commitment to applied economics. This stemmed from an analysis of Scotland's industrial depression during the 1920s and 1930s. Modernizers emphasized the negative effects of Scotland's overdependence on export-oriented heavy industries and capital goods production. They underlined the absence of mass production consumer durables sectors that proved more resilient in the English midlands and south-east.[9]

---

[5] The National Archives (TNA), Coal 31/168, Coal Demand (draft section of CIE Interim Report) (1974).

[6] TNA, Coal 31/135 Part 2, Plan for Coal Review, 1981.

[7] J. Walker 'The road to Sizewell: the origins of the UK nuclear power programme', *Contemporary Record*, i (1987), 33–50, at p. 44.

[8] E. A. Cameron, 'The stateless nation and the British state since 1918', in *The Oxford Handbook of Modern Scottish History*, ed. T. M. Devine and J. Wormald (Oxford, 2013), pp. 620–34, at p. 629.

[9] J. Tomlinson and E. Gibbs, 'Planning the new industrial nation: Scotland, 1931 to 1979', *Contemporary British History*, xxx (2016), 584–606.

Table 1.1. Male employment in Lanarkshire

Year	Mining	(%)	Metal Manufacturing	(%)	Engineering*	(%)	Combined (%)
1951	20,225	15.5	23,403	18.0	12,821	9.9	43.4
1961	13,440	9.3	23,340	16.2	28,010	19.4	44.9
1966	6,610	4.6	24,030	16.8	33,380	23.7	45.1
1971**	3,720	2.8	21,810	16.2	34,500	25.6	44.6
1981	1,060	0.9	10,680	9.4	24,449	21.6	31.3

Sources: Census 1951 Scotland vol. iv: Occupation and Industries (Edinburgh, 1956), pp. 433–69; *Census 1961 Scotland: Occupation and Industry County Tables: Glasgow and Lanark, Leaflet no. 15* (Edinburgh, 1966), pp. 16–21; *Sample Census 1966 Scotland: Economic Activity: County Tables: Leaflet no. 3. Glasgow and Lanark* (Edinburgh, 1968), pp. 33–44; *Census 1971 Scotland: Economic Activity, County Tables part 2* (Edinburgh, 1976), table 3; *Census 1981 Scotland: Economic Activity: 10% Strathclyde Region* Microfiche (Edinburgh, 1983), Table 3.

*Engineering in 1951 includes 3,915 constructional engineers who were omitted from later figures. The 1961, 1966, 1971 and 1981 engineering figures include engineering, vehicle manufacturing and other metal goods manufacturing.

**The 1971 and 1981 figures are based on a 10 per cent sample. As there are no county figures available for 1981, the figures are an amalgamation of Clydesdale, Cumbernauld and Kilsyth East Kilbride, Hamilton, Motherwell, and Monklands District Council figures.

Table 1.2. Female employment in Lanarkshire

Year	Textiles	(%)	Engineering	(%)	Clothing	(%)
1951	3,820	8.1	4,226	8.9	2,760	5.8
1961	4,070	7.1	8,548	15	3,320	5.8
1966	2,990	4.3	13,090	18.9	4,540	6.5
1971	2,200	2.8	15,290	19.5	3,560	4.6
1981	1,000	1.2	6,920	8.6	4,040	5

Source: As table 1.1.

Table 1.3. Unemployment in Lanarkshire

Year	Male	(%)	Female	(%)
1951	7,301	4.3	2,342	3.4
1961	9,450	5.4	2,990	4.0
1966	6,690	3.9	4,530	4.9
1971	11,350	6.3	5,580	5.4
1981*	28,940	17.7	14,131	13.5

Sources: Census 1951 Scotland vol. iv: Occupation and Industries (Edinburgh, 1956), p. 190; *Census 1961 Scotland vol. vi, part III* (Edinburgh, 1966), p. 10; *Sample Census 1966 Scotland, Economic Activity: County Tables: Leaflet no. 3, Glasgow and Lanark* (Edinburgh, 1968) p. 5; *Census 1971 Scotland: Economic Activity, County Tables part 2* (Edinburgh, 1976) Table 3; *Census 1981 Scotland: Economic Activity: 10% Strathclyde Region* Microfiche (Edinburgh, 1983), Table 1.

*The 1971 and 1981 figures are based on a 10 per cent sample. As there are no county figures available for 1981, the figures are an amalgamation of Clydesdale, Cumbernauld and Kilsyth, East Kilbride, Hamilton, Motherwell, and Monklands District Council figures. Temporary sick in receipt of benefits are included within 1961 and 1981 figures.

Table 1.4. Scottish Coal employment

Year	Employment (thousands)
1947	77
1957	82
1967	32
1977	21
1987	6

Source: M. K. Oglethorpe, *Scottish Collieries: an Inventory of the Scottish Coal Industry in the Nationalised Era* (Edinburgh, 2006), p. 20.

State policy was central to Scotland's protracted deindustrialization. Coalfield contraction, and the later rundown of steel employment, were managed through nationalization, which cemented a relatively consensual approach to closures and their integration with regional policy.[10] The closure of collieries and steelworks made available the space and labour necessary for renewal in light engineering. Coordinated investment and divestment was secured by regional policy efforts which utilized incentives and coercion through Industrial Development Certificates to 'steer' investment from the 'congested' South East and Midlands to 'peripheral' areas of the UK.[11] Tables 1.1 and 1.2 demonstrate that industrial employment rates remained high in Lanarkshire between the 1940s and early 1970s. Assembly engineering replaced lost male jobs in coal and steel and women's work in textiles. Overall, male industrial employment rates remained approximately proportionally static in the context of a growing workforce, while the female industrial workforce grew absolutely and proportionally. Unemployment rates were relatively low until employment in assembly engineering fell from the late 1970s onwards, as is demonstrated in table 1.3.

Coal employment expanded during the early years of public ownership, as demonstrated in table 1.4, but in an uneven manner. Table 1.1 shows a sharp fall in male coal mining employment between 1951 and 1961, indicating the impact rationalization had in the closure of smaller and less productive collieries in Lanarkshire. The 1945 Labour government's case for taking coal into public ownership rested on the benefits of planning: 'the nation's most precious raw material' had, for a quarter of a century, been 'floundering chaotically under the ownership of many hundreds of independent companies'. Public ownership promised 'great economies' through modernizing investment and the redeployment of labour.[12] By 'humanising' industrial relations, the nationalized industry could secure partnerships between workers and management and heal the scars of interwar conflict.[13] Efficiency through industrial renewal was clearly enunciated by the NCB Scottish division in *Scotland's Coal Plan*. The 1955 prospectus outlined a £100-million plan for future development, including fifteen new colliery sinkings and thirty-nine major reconstructions. It was announced as 'the greatest single industrial venture ever undertaken for the benefit of the Scottish people'.[14]

---

[10] J. Tomlinson, 'A 'failed experiment'? Public ownership and the narratives of post-war Britain', *Labour History Review*, lxxiii (2008), 228–43, at p. 238.

[11] G. C. Cameron, *Industrial Movement and the Regional Problem* (Edinburgh, 1966), p. 1.

[12] *British General Election Manifestos, 1900–1974*, ed. F. W. S. Craig (London, 1975), p. 127.

[13] Tomlinson, 'Failed', p. 239.

[14] NCB Scottish Division, *Scotland's Coal Plan* (Edinburgh, 1955), pp. 6–7.

In the eight years prior to 1955, the Board had already closed 67 of the 275 pits that the division inherited from the private sector when ownership was transferred on vesting day, 1 January 1947. Despite these closures, the NCB had succeeded in maintaining production levels.[15] The plan entailed a significant further redistribution of employment. Lanarkshire was to continue to contract, while investment was to be focused on the 'buried treasure' within the Limestone coal reserves, 800 to 3,000 feet beneath ground. New developments were primarily concentrated in Fife and Midlothian in Eastern Scotland, and Clackmannanshire in the centre of the country. Showpiece projects included the new sinking at Rothes in Fife, where a new town, Glenrothes, received migrants from the Lanarkshire coalfield, as well as the construction of a major drift mine complex at Glenochil in Clackmannanshire. The Scottish division's strategy was based on consolidating the industry into larger modern, mechanized, collieries that would generally have an output of 4,000 to 6,000 tonnes per day. This was over five times a typical pre-nationalization unit. Mechanization entailed moves towards a 'continuous mining' approach, providing a consistent flow of coal through simultaneous cutting and loading. This was achieved through new machinery for coal-getting including scraper box ploughs, as well as roof controls, face conveyors, power loading machinery and electronic winding engines.[16]

Coalfield-wide reconstruction initially enjoyed the support of the NUMSA, which was committed to ensuring that nationalization succeeded. Union leaders across the UK concurred, including those on the left. Arthur Horner, who was the union's general secretary and a Welsh Communist, declared in 1948 that 'old pits have to die'.[17] Horner's fellow party member and NUMSA president, Abe Moffat, also supported rationalization. At the NUM's Scottish conference in 1947 Moffat told delegates that:

> The question of reorganization within the Scottish pits would be one of the most difficult tasks within the next five years and delegates would have to face up to their responsibilities. There might come a time when they would have to explain unpopular policy ... We had to reorganize the most inefficient industry this country had, and we must ensure that this reorganization was carried out as speedily as possible.[18]

---

[15] NCB Scottish Division, *Scotland's Coal Plan*, p. 6.

[16] NCB Scottish Division, *Scotland's Coal Plan*, pp. 6–16.

[17] A. Taylor, *The NUM and British Politics*, i: *1944–1968* (Aldershot, 2003), p. 92.

[18] National Mining Museum Scotland archives, Newtongrange, Midlothian (NMMS), National Union of Mineworkers Scottish Area (NUMSA), Minutes of Executive Committee and Special Conferences, 8 July 1946 to 11 June 1947, pp. 6, 37–8.

Moffatt's perspective concurred with the modernizers' view of the need for a major reorganization of the Scottish economy. The *Clyde Valley Regional Plan* of 1946, commissioned by the Scottish Office, stated that mining in Lanarkshire was 'fast dying' and prescribed a solution that combined miners migrating to other coalfields with inward investment. Lanarkshire's steelmaking capacity would also be lost over this period as the 'heavy industry centre of gravity' moved south-west of Glasgow. The plan argued that a coastal-based steel strip mill would be better suited for an industry which relied on imported iron ore.[19] In place of collieries and steelworks, 'several substantial new industries employing thousands of men' were required. Large-scale external investment, 'rather than the setting up of concerns merely to supplement the basic industrial structure' would be brought to the region to achieve progress.[20] A modernizing perspective for comprehensive industrial rejuvenation can be seen in the plan's designation of state policy as the nodal point for economic development: improved transport infrastructure would connect new publicly owned industrial estates that were to house light industries characterized by 'footlooseness', such as the manufacture of ball bearings, vacuums and electric registers. The achievement of such a 'reasonable measure of industrial balance' depended on attracting inward investment to develop new industries which would provide improved employment opportunities, especially for women workers.[21]

These economic aims were strongly tied to social goals. Both industrial and population congestion would be abated through establishing new towns, including two in Lanarkshire: Cumbernauld to the northeast and East Kilbride to the south. They were to be in part populated by the 'overspill' population from Glasgow's slum clearances, offering work in a large range of sectors to avoid the plight of the 'one-industry town' which had suffered so greatly in the interwar period.[22] The *Clyde Valley Regional Plan's* aim of redevelopment, first through two new towns in Lanarkshire and second by replacing traditional industries with external investment in state provisioned industrial estates, was principally fulfilled. However, a significant divergence from the plan's proposals was the relocation of the steel industry. Ultimately, the Macmillan government, which faced political

[19] NMMS, NUMSA, Minutes of Executive Committee and Special Conferences, 8 July 1946 to 11 June 1947, pp. 79–84.

[20] P. Abercrombie and R. H. Matthew, *The Clyde Valley Regional Plan, 1946* (Edinburgh, 1949), p. 96.

[21] Abercrombie and Matthew, *Clyde Valley Regional Plan*, pp. 94–100.

[22] Abercrombie and Matthew, *Clyde Valley Regional Plan*, p. 7.

constraints in a region characterized by high rates of male unemployment, maintained production in Lanarkshire and pressurized Colvilles to invest in a strip mill at Ravenscraig. This was part of an effort to provide the materials required for assembly engineering activities.[23] The continuity in steel employment between the early 1950s and early 1970s (visible in table 1.1) was tied to modernization through inward investment. Ravenscraig reoriented Scottish steel production towards the requirements of consumer durables manufacturing.

Modernizing industrial investment developed concurrently with the incremental replacement of coal by oil and nuclear fuels, which contributed to concerns about employment levels in contracting coalfields. As is demonstrated in table 1.4, coal employment contracted swiftly from the mid 1950s. This was the result of deliberation by Conservative government ministers and civil servants who held pronounced opposition to coal on economic and political grounds. Officials in the Ministry of Fuel and Power encouraged coal's displacement, reflecting both their commitments to market liberalization and their hostility towards organized labour. A briefing prepared for the minister of power, George Lloyd, before he met oil company representatives during the summer of 1955, prescribed 'public encouragement to a rapid conversion from coal to oil in the field of general industry and commerce'. It went on to emphasize that a railway workers' strike, and unofficial action by Yorkshire miners, had produced a 'psychological moment' to argue for greater oil use.[24] Ministry civil servants and Conservative ministers regularly identified coal and its labour relations as a justification for transition to alternative fuel sources. During discussions with the Central Energy Authority, Ministry of Fuel and Power officials posited that 'the Government must loosen miners' stranglehold' on energy policy.[25] The same phrase had been used the previous year in discussions over supporting the nuclear power programme.[26] These remarks indicate the importance of political perspectives to guiding the fuel policy decisions which led to the closure of collieries during the second half of the 1950s and early 1960s.

Imposing market competition was another key objective. This was openly asserted by Lloyd, who had a background in oil administration, when he addressed parliament on the 1955 coal price increase. Lloyd described the price rise as 'the most powerful stroke for fuel efficiency since the war'.

---

[23] P. Payne, *Growth and Contraction: Scottish industry, c.1860–1990* (Glasgow, 1992), p. 38.

[24] TNA, POWE 33/2156/54, brief for the minister's meeting with the oil companies, 7 June 1955.

[25] TNA, POWE 33/2156/9, responses to ministry paper in dialogue with CEA, 1955.

[26] Walker, 'Sizewell', p. 48.

He stated that the NCB 'lack true cost consciousness'. Its managers were trained in technical mining competency but not in entrepreneurship, which was crucial for the sector's future.[27] Lloyd's speech mirrored the sentiments of his civil servants, but they used more direct language. C. L. Wilkinson, a Ministry official, saw nuclear and oil as means to 'enforce the measure of discipline that is needed' on miners and the NCB. These policies challenged the Scottish division's plans for expansion and its assumption that electricity markets would continue to be dominated by coal.[28] Politicized hostility towards coal miners was twinned with suspicion towards the nationalized industry's monopoly position in the production of the UK's basic energy source. These both encouraged the adoption of fuel diversification policies, including the construction of financially uncompetitive nuclear power stations, which were formative to deindustrialization in the coalfields from the late 1950s.

The liberal market perspective of civil servants jarred with the NCB's commitment to planning the sector's future, and provoked concerns within coalfield areas over future employment. Dimensions of the double movement are apparent in these differences where financial priorities competed with communitarian perspectives on coal mining employment. Within its own operations, the NCB pursued productionist imperatives and adherence to cost prerogatives. This outlook competed with moral economy obligations to the workforce in industry governance, especially colliery divestment and investment. However, when facing government and the generating boards, the NCB often shared interests with mining trade unions in making the case for maximizing coal output and asserting obligations to utilize national resource endowments for economic gain. During 1957, a senior Scottish officer at the Ministry of Power wrote to the Coal Division in London expressing concern that the conversion of power stations on the Thames to oil will 'result in surplus of electricity smalls and loss of traditional seaborne UK market'. This was considered a significant threat. The areas most affected by this change were adjacent to the Firth of Forth. It would affect developments in East Fife and the Lothians which were at the forefront of the NCB Scottish division's investment programme.[29]

The agency of policymakers in fuel transitions affirms the importance of long-term investment decisions to determining episodes of

[27] TNA, POWE 33/2156, Ministry of Fuel and Power, Debate on the National Coal Board report for 1954: note for minister's speech, 20 July 1955.

[28] TNA, POWE 33/2156/38, C. L. Wilkinson, Note to minister: coal policy, 13 July 1955.

[29] TNA, POWE 14/857, teleprinter message from senior Scottish officer, Edinburgh, to A. B. Powell, Coal Division, 1957. Electricity smalls refers to the variety of coal most commonly burned in power stations.

deindustrialization.[30] Britain's move towards oil in industry and multi-fuel power generation was not the straightforward result of technological development. Rather, it was a product of political economy, incorporating resistance and conflict.[31] Plentiful cheap oil imports, which seemed dependable for the long term were formative to this process, as was policymaker faith in the capacity of nuclear power to deliver a low cost energy future, favourable in terms of productivity and the balance of payments. Another key element was suspicion of coal due to its reliance on organized labour and a desire to impose market forces onto the nationalized industry, which was impossible from a monopoly fuel position. The widespread adoption of oil underpinned a comparatively egalitarian 'democratic capitalist' economic management in developed economies after 1945. However, investment in oil infrastructure was motivated by miners' capacity to disrupt or 'sabotage' capital accumulation.[32]

High modernization

From the early 1960s, Scottish modernizers actively promoted the contraction of traditional sectors.[33] This would free up the labour and space required for the mass production industries necessary to match UK growth rates, and overcome persistently higher rates of unemployment.[34] The construction of Ravenscraig coincided with the peak of Scottish heavy industrial employment in the late 1950s, after which coal, steel and shipbuilding shed tens of thousands of jobs.[35] During 1961, the SCDI published an influential study actively welcoming this, and arguing that economic growth could be achieved through releasing labour from heavy industry to provide manpower for light engineering. The report was produced by an inquiry chaired by Sir John Toothill. Toothill was also the chairman of Ferranti, an English electronics firm which relocated to Edinburgh during the Second World War.[36]

---

[30] B. Bluestone and B. Harrison, *The Deindustrialization of America: Plant Closings, Community Abandonment, and the Dismantling of Basic Industry* (New York, 1982), p. 15.

[31] R. Stokes, *Opting for Oil: the Political Economy of Technological Change in the West German Chemical Industry, 1945–1961* (New York, 1994), p. 96.

[32] T. Mitchell, *Carbon Democracy: Political Power in the Age of Oil* (London 2013), p. 15.

[33] P. Smith and J. Brown, 'Economic crisis, foreign capital and working-class response, 1945–1979', in *Scottish Capitalism: Class, State and Nation from before the Union to the present*, ed. T. Dickson (London, 1980), pp. 287–320, at p. 292.

[34] J. N. Randall, 'New towns and new industries', in *The Economic Development of Modern Scotland, 1950–1980*, ed. Richard Saville (Edinburgh, 1985), pp. 245–69, at p. 245.

[35] R. J. Finlay, *Modern Scotland, 1914–2000* (London, 2004), pp. 259–65.

[36] J. Foster, 'The twentieth century', in *The New Penguin History of Scotland: from the Earliest Times to Present Day*, ed. R. A. Housing and W. W. Knox (London, 2001), pp. 417–96, at p. 469.

The Toothill Report warned against 'supporting the inefficient and seeking to postpone for a little by means of subsidy or control the decay of industries and districts that had no prospect of achieving independent prosperity or growth', which would waste manpower. Growth would be attained through releasing labour to 'the newer industries' via the 'build-up of industrial complexes and centres which offer prospects of becoming zones of growth'.[37] The focus of these complexes was to be 'chiefly in large-quantity production,' which signalled a departure from Scottish industry's traditional orientation towards bespoke production for niche markets. Ravenscraig was welcomed as the 'prerequisite, almost, of the development of modern mass-production engineering-based consumers' durables industry'.[38] These arguments were relayed by the Conservative secretary of state for Scotland, Michael Noble, at the opening of a new factory in East Kilbride in 1964. Noble noted the important role Ravenscraig was to play in supplying materials for producing motor vehicles, which put it at the heart of Scotland's economic rejuvenation.[39]

Toothill had considerable influence over Scottish policymaking. In 1964, the report's emphasis on industrial complexes was referred to by a Ministry of Labour representative as justifying investment in infrastructure to support the 'fast-growing and specialised' electronics industry. Toothill's perspective warranted investment in workforce reskilling for contracting coalfield areas, with Motherwell and East Kilbride listed as potential locations for electronic engineering training centres alongside areas in West Lothian and Fife, which had previously been designated an expanding coalfield.[40] The report's impact was also apparent when the Scottish Development Department was established in 1962, fulfilling a key recommendation.[41] Toothill's objectives closely aligned with Labour governments' objectives. In 1967, Prime Minister Harold Wilson summarized his understanding that 'the fundamental problem facing this country is to secure and maintain a

[37] J. N. Toothill, *Inquiry into the Scottish Economy, 1960–1961: Report of a Committee Appointed by the Scottish Council (Development and Industry) under the Chairmanship of J. N. Toothill* (Edinburgh, 1961), pp. 38, 155.

[38] Toothill, *Inquiry into the Scottish Economy*, pp. 36–8.

[39] National Records of Scotland, Edinburgh (NRS), SEP 4/567/42, Board of Trade, Scottish Office, 21 Apr. 1964, some notes relating to industrial development in Scotland as a contribution to the secretary of state's speech at the opening of BNJ's new factory at East Kilbride.

[40] NRS, SEP 17/70/15, training arrangement in the growth areas: note of a meeting held by the minister of state at 11am on Tues. 4 Feb. 1964.

[41] I. Levitt, 'The origins of the Scottish Development Department, 1943–1962', *Scottish Affairs*, xiv (1996), 42–63, at p. 59.

high level of economic growth'. Wilson further emphasized Britain's failure to perform in export markets relative to other 'advanced countries', giving a distinctly 'declinist' tone to the modernization effort.[42] These motifs were especially marked in the Scottish context where indigenous industrial failures were seen to require rectification through inward investment. Accordingly, the movement of labour from contracting heavy industries to export assembly goods sectors became an economic priority.

The modernizers achieved hegemony in Scottish economic policy between the 1940s and 1960s. Initially, their objectives centred on the diversified industrial structure envisioned by the *Clyde Valley Regional Plan*. These were later refined and extended to focus on releasing labour from heavy industry to establish the complexes of engineering plants that Toothill called for. Trade union representatives were initially agreeable to the modernizing perspective. They tolerated the decline of employment in heavy industry in return for manufacturing inward investment. As colliery closures intensified from the late 1950s, the NUMSA's leadership were relatively accepting of the approach outlined by Toothill. In late 1961, the NUMSA president Alex Moffat – who had succeeded his brother Abe earlier in the year – made it clear that his priority was employment stability and *not* the maintenance of coal jobs. At an Area conference, Moffat bluntly stated, 'The government could shut down all the pits if every miner had a reasonable job to go to'.[43] The 1964–70 Wilson governments encouraged this approach. At a speech in Glasgow in 1970, the minister of state for the Ministry of Technology, Eric Varley – who was a former coal-mining craftsman and NUM-sponsored MP – advocated modernization in now familiar terms. Varley was critical of those who lamented the decline of shipbuilding and coal, which he referred to as 'my own industry'. The minister praised the 'remarkable progress' made by Honeywell, the American office machinery manufacturers who were delivering employment and product expansion at the Newhouse, Bellshill and Uddingston industrial estates. With employment at the American firm due to expand to 7,000, North Lanarkshire had a 'bright future'.[44]

Varley's comments came in the wake of the landmark 1967 *Fuel Policy* white paper, which welcomed the development of a four-fuel (coal, oil, nuclear

---

[42] TNA, PREM 13/1610, Ministry of Power, 27 Sept. 1967, notes for the prime minister's meeting with the NUM on 29 Sept.; J. Tomlinson, 'Inventing "decline": the falling behind of the British economy in the postwar years', *Economic History Review*, xlix (1996), 731–57.

[43] NMMS, NUMSA, Minutes of Executive and Special Conferences, 12 June 1961 to 6/8 June 1962, p. 283.

[44] NRS, SEP 4/1629, 3 Feb. 1970, Glasgow: report of a visit by Minister of State; G. Kaufman, 'Varley, Eric Graham, Baron Varley', *ODNB* <https://doi.org/10.1093/ref:odnb/100192> [accessed 20 Oct. 2020].

and natural gas) energy economy in Britain. It prominently announced 'the coming of age of nuclear power as a potential major source of energy', despite a decade of disappointments with the sector's development.[45] The NCB were 'the only influential opponents' of the decision to invest in an extensive advanced gas cooler reactor programme during the 1960s. Nuclear power was judged by politicians and civil servants as having near limitless potential, but they also considered oil to be dependable, cheap and secure in supply.[46] The predominance of these outlooks in government contributed to conflict with the NCB, whose senior officials contested the assumptions behind nuclear projections and advised caution over anticipations of a future of ever-affordable oil. Scottish autonomy within the NCB was considerably reduced as employment shrank, especially in the sustained rundown of the 1960s. During 1967, the Divisional structure was replaced by less autonomous 'Areas', which confirmed the centralization of power to the Board's headquarters at Hobart House in London.[47] Several major Scottish pit development failures between the late 1950s and early 1960s anticipated these later administrative changes, which only confirmed existing alterations to internal power relations. Centralization was encouraged by employment contraction and the policy consensus that the industry's production and workforce would continue to fall. It also allowed the NCB's core management to oversee a major and coordinated overhaul of the UK coalfields, which privileged the 'central' coalfields of the English Midlands, Nottinghamshire and Yorkshire in the distribution of manpower and investment.[48]

A Scottish coal lobby emerged from the shock of modernized collieries being closed between 1959 and 1962. It persisted into the imposition of liberalized market logic on the industry by Conservative governments during the 1980s. The lobby combined political interest from coalfield areas. Protests were voiced from within the coal industry in explicitly Scottish national terms, anticipating both the Nationalist electoral breakthrough and high-profile arguments over Scotland's constitutional status later in the decade. While closures during the 1950s had been largely accepted and even embraced by the NUMSA, there was a distinct change of attitude as some

[45] *Fuel Policy* (Parl. Papers 1967 [Cmnd. 3438]), p. 1.

[46] S. Taylor, *The Fall and Rise of Nuclear Power in Britain: a History* (Cambridge, 2016), p. 19.

[47] R. S. Halliday, *The Disappearing Scottish Colliery: a Personal View of some Aspects of Scotland's Coal Industry since Nationalisation* (Edinburgh, 1990), p. 7.

[48] W. Ashworth, *The History of the British Coal Industry, v: 1946–1982: the Nationalized Industry* (Oxford, 1986), p. 250.

of the projects upon which Scottish coal's future had been staked failed. Major examples included the closure of Devon in Clackmannanshire in 1959, with nearby Glenochil following in 1962 and the prestige project at Rothes in Fife also closing the same year. This was followed by the cessation of production at the redeveloped Barony colliery in Ayrshire following a shaft collapse on 8 November 1962 that killed four men.[49]

The announcement of Devon colliery's closure in June 1959 was met by resistance from the workforce, who undertook an unofficial 'stay down' strike for three days, gaining considerable support in a walkout across the Scottish coalfield. Coal Board officials explained that miners' anger was stimulated by an extensive rundown of manpower in the area, with nearby Policy and Meta having closed the same year, and because Devon was a modern colliery.[50] Devon's closure also followed the abandonment of the nearby Airth colliery, which had previously been projected to be a major coking coal producer. Furthermore, the announcement at Devon came alongside the closure of nineteen other pits in Scotland. These were part of a 'special' economic closure programme across the UK, but which was highly concentrated in Scotland.[51] Ministry officials noted that the workforce's feeling was inflamed by the rate of closures, which 'means that the openings for miners' sons who had hoped to go into mining would be much reduced'.[52]

These sentiments indicate that contraction across the Scottish coalfield presented threats to both communal identities and economic security, which contributed to a developing sense of crisis. Deindustrialization was experienced as a threat to social order and stability as well as livelihoods. The effects of closures, which had already been felt in dislocation at local levels through coalfield reconstruction, became recalibrated as a national problem as the industry's modernized collieries became vulnerable. These fears were consolidated by the announcement of closure at Rothes and Glenochil in 1962, following severe faltering and failure to meet performance expectations. Objections were encouraged by the experience of restructuring: modernized pits had been paid for at the price of the closure of older collieries. The Board had assured miners of secure employment in modern mines. Board officials

---

[49] E. Gibbs and J. Phillips, 'Remembering Auchengeich: the largest fatal accident in Scottish coal in the nationalised era', *Scottish Labour History*, liv (2019), 47–57, at pp. 47–9.

[50] TNA, POWE 37/481, 24 June 1959, Ministry of Power Coal Division, note on the strike at Devon Pit.

[51] TNA, POWE 37/481, R. Corley, Draft brief for the parliamentary secretary's meeting with the delegation from the Scottish NUM on Thurs. 26 Feb. 1959.

[52] TNA, POWE 37/481, R. Corley, Draft brief.

noted that as late as the closure of Devon, the NUMSA had been informed that its take could be worked from Glenochil, and furthermore that 850 miners had transferred into Scottish Special Housing Association (SSHA) houses in the area built for the Glenochil development. Most incomers had migrated from Lanarkshire. In response to these failures within the Scottish coalfields, the NCB proposed transfers to England. Offers were made to men in short-life collieries as well as closing ones.[53] The NCB operated a system which differentiated collieries between categories A, B and C with category A pits regarded as 'long-life' viable prospects, B as intermediate and C 'short-life' or likely to close. These categories were meant to provide security by integrating individual collieries within the Board's plans. But the sudden closure of units that were previously regarded as viable had the opposite effect by encouraging perceptions that the nationalized industry was governed by distant authorities following short-term economic logic.

The Board's admission of long-term contraction across Scotland was accompanied by significant centralization within the NCB. Rothes's closure was disputed by the Scottish Division, who felt that the pit had the potential to economically work its best reserves with reduced manpower.[54] The Division were successful in winning approval from the Board to continue developing drivages during 1961. However, their activities were closely monitored. E. H. Browne, a Board member, was charged with investigating high expenditure levels. Browne noted that after a visit to Rothes in June 1961, 'HQ representatives left the Divisional Board in no doubts that they could not see any prospect of accepting the Divisional Board's proposal'. He emphasized that the Scottish division were nevertheless determined to continue operations and began to incur expenditure over and above awarded allocations on two occasions later in 1961. Browne concluded that there had been an overspend of just over £40,000 and put this down to the speeding up of drivage operations. He argued, in contrast to the Division, that 'so pronounced a change should not have occurred'.[55]

These findings accord with Halliday's memories of employment in the NCB Scottish Division's production department, which he published in 1990. Halliday recalled the closures of Rothes and Glenochil as 'a watershed in relations between the Divisional reconstruction team and headquarters

[53] TNA, Coal 31/96, 25 Apr. 1962, Glenochil drift mine (aide memorandum for the board's side).

[54] TNA, Coal 31/96, 29 March 1962, Memorandum: Glenochil and Rothes Closures.

[55] TNA, Coal 31/96, 16 March 1962, E. H. Browne, Rothes colliery: investigation into expenditure on development drivages; Coal 31/9612, March 1962, A. J. O Davis, Further report into the finance on expenditure against authorisation on five feet foot dook drives, Rothes colliery, East Fife Area, Scottish Division.

in Edinburgh and London', referring to a 'battle' between the Scottish Division and the NCB's UK headquarters at Hobart House.[56] Rothes's closure had ramifications beyond the mining industry itself and was understood as a threat to Scotland's industrial prestige. The Scottish Trades Union Congress (STUC) general secretary, George Middleton, wrote to Alf Robens, the chairman of the NCB, objecting to Rothes's closure, which he described as 'a serious blow to Scotland'. Middleton expressed doubt about the NCB's 'rather facile inclination to talk about alternative employment … for those becoming redundant'. Prefiguring the keener mobilization of Scottish national identity by trade union representatives over the following decade, Middleton highlighted Roben's geographical and social distance from Fife by adding that 'this does not impress us very greatly in Scotland'.[57] The theme of discontent with UK economic management was articulated more vociferously by the Scottish National Party's leader, William Wolfe. Following the announcement of Barony's closure, Wolfe wrote a letter to Robens which criticized the Board's commitment to Scotland:

> To close it permanently in light of evidence available to the public would be an act of national sabotage. And the fact that, in this context, national means Scotland does not detract on what from the immensity of the crime should it be committed, regardless of Hobart House's insistence on looking at things from an overall United Kingdom point of view.[58]

While he was the object of these protests, Robens made use of fears over unemployment and southwards migration when agitating for Scottish power station investment. In August 1962, before the disaster at Barony, Robens addressed a meeting of the SCDI. He made the case for a coal-fired power station in central Scotland based on providing work: 'if they had another nuclear station in Scotland not only would it have to be subsidized, but it would cost several thousand miners their jobs'. J. L. Warrander reported to the Ministry of Power that 'this suggestion was not unfavourably received'.[59] During March 1963, the NCB publicly increased pressure on government and the SSEB as part of a successful campaign for coal-fired power station investment. Following the Barony disaster, it released a public statement that underlined that only investment in a new coal-fired station could justify

---

[56] Halliday, *Disappearing*, p. 3.

[57] TNA, Coal 31/96, NCB, London, 14 March 1962, G. Middleton, STUC, Glasgow to Lord Robens.

[58] TNA, Coal 31/96, NCB, London, 1963, William Wolfe, SNP, to Lord Roben.

[59] TNA, POWE 14/1495/6, Ministry of Power, London 16 Aug. 1962, J. L. Warrander, Ministry of Power, Edinburgh to R. E. Dearing.

spending £2.5 million to sink a new shaft and raise 500,000 tonnes per year. Reopening the pit would restore employment to 1,600 men, including the 800 miners who were still redundant following the disaster. Furthermore, the Board claimed that 'unless the longer-term demand for coal in Scotland is established many men now employed in the mining industry in Ayrshire will find permanent work in the mining industry only by transferring to English collieries. The Board recognize that this would have a serious effect on the level of employment in Ayrshire and particular in Auchinleck and Cumnock'.[60] Officials in the Ministry of Power were furious with the Board's communiqué, noting that they had not even had the 'courtesy' to inform the minister first: 'As things are, however, the announcement has been made in terms which are liable to cause embarrassment to many people: and perhaps this was the intended effect. If the decision goes against the coal-fired station in Scotland it will be alleged that those who so decided are responsible for continuing unemployment in the Barony area'.[61]

Widespread support in Scottish industry and political circles for the NCB's campaign to ensure investment in a major power station facility delivered material results when the SSEB commissioned Longannet power station in 1964. It was supplied by a drift mine complex, which spanned across Fife and Clackmannanshire. The NCB's case built on the findings of the Mackenzie Committee on the Generation and Distribution of Electricity in Scotland, which reported that Scotland required a coal-fired power station with over two megawatts of capacity.[62] W. S. McKinnell of the Ministry of Power reported 'a strong coal lobby in the committee', who were eager to see further investment in Scotland beyond the commissioning of Kincardine in Fife and Cockenzie in the Lothians. Some committee members favoured Ayrshire as the site for a third investment.[63] In giving evidence before the committee, senior NCB officials, including Robens, his future successor, Derek Ezra, and the chair of the Scottish Divisional Board, Ronald Parker, argued that the Scottish coalfields had the capacity to support another power station. Robens alluded to the politically unpalatable alternative: 'I do not think that public opinion would stand long for the idea of closing pits in Scotland and sending Scots miners down to Yorkshire to mine coal to send back for power stations in Scotland'.

---

[60] TNA, POWE 14/1495/5/3, 11 March 1963, NCB Ayrshire Press Office, 'Barony colliery'.

[61] TNA, POWE 14/1495, 13 March 1963, M. Stevenson to M. Cairns.

[62] TNA, POWE 14/1495, 6 Feb. 1963, Ministry of Power General Division, TUC and Fuel and Power policy brief for minister's meeting on 12 Feb. 1963.

[63] TNA, Coal 74/1287, 25 Jan. 1962, W. J. S. McKinnell to secretary, Memorandum: departmental committee on electricity in Scotland.

While emphasizing these social dimensions, the Board also pressed home their productionist agenda for consolidation. Parker argued that rather than Ayrshire, it was Clackmannan and Fife which would best provide coal for power station use.[64] This was consistent with the Board's long-term aims of rebalancing the workforce towards the most productive coalfields in the east of Scotland.

However, Longannet's designation as a coal-fired power station also amounted to a major change in mining strategy. Instead of focusing on deep Limestone coals, Scottish coal mining was reoriented towards readily accessible low-quality Hirst coal, which was suited to power station requirements.[65] Longannet's drift mines became the core of a much-reduced coal sector that was primarily made up of long-life pits associated with electricity production. New and modernized Scottish collieries received over £75 million of investment during the 1950s and 1960s. Their workforce continued to grow through miners transferring from the contracting Lanarkshire coalfield and other older collieries. At Longannet, the extent of modernization included pioneering computerized technology, the Remotely Operated Longwall Face ('ROLF').[66] ROLF was an ultimately abortive attempt to apply self-steering and censor technology to coal cutting that would allow miners to more safely operate machinery at distance. It was deployed alongside electronic indicative and signalling equipment which eventually inspired the Mines Operating System (MINOS) that the NCB developed during the 1970s. These developments involved considerable growth in surveillance, threatening faceworkers' skills and autonomy. Traditionally, hewers had maintained their social status and economic power through their physical distance from management overseers as well as their physical strength and knowledge of coal cutting techniques.[67] After coal was cut in Longannet's drift mines, it was mechanically conveyed directly to the power station above ground.[68] This zenith of late twentieth-century Scottish coal mining was a dramatic transformation from the small

---

[64] TNA, Coal 74/1287, 1 May 1962, Departmental Committee on electricity in Scotland: oral evidence by the NCB.

[65] TNA, POWE 52/17/38A, Digest of report by Merz and McLellan on proposed new station in Scotland, 1963.

[66] TNA, Coal 31/120, Lauriston House, Edinburgh, 22 June 1967, NCB note: the pattern of output of coal for electricity generation in the 1970s from the Scottish coalfields; R. Saville, 'The coal business', *Scottish Economic and Social History*, viii (1988), 93–6, at p. 93.

[67] J. Winterton and R. Winterton, *Coal, Crisis and Conflict: the 1984–85 Miners' Strike in Yorkshire* (Manchester, 1989), pp. 11–17.

[68] TNA, POWE 52/278, *Scotsman*, 25 July 1968, 'Productivity in coal industry continues to rise'.

collieries operating under the pillar and stoop methods that the NCB had inherited, and largely closed, since 1947. Extensive sectoral reorganization and technological transformation was achieved over the course of a single working lifetime under public ownership, while maintaining broadly consensual industrial relations.

Even the Board's modern investments were threatened by alternative sources of electricity generation. After 1965, British coal production targets were incrementally reduced. An initial decrease from 175 to 155 million tonnes gave way to a target of 135 million tonnes of deep-mined coal by 1970.[69] The NCB objected to this being declared publicly on the grounds that the new target would have 'a serious impact on the morale of men and management', because it would lead to the closure of collieries that had been previously judged economically viable.[70] In planning for up to 35,000 job losses per annum, the Board noted that Scotland would be worst affected part of the UK. At some collieries, transfers would only be offered to workers prepared to move to expanding English central coalfields.[71] These outcomes were especially concerning to Robens and the NCB because they threatened the modernized industry assembled during the previous decade of rationalization. Robens unsuccessfully argued for further coal-fired power stations investment, in preference to the government's decision to invest at a nuclear facility at Hunterston in Ayrshire. He underlined the Board's success in concentrating production within high productivity units. Failure to invest in coal-fired generation would threaten new 'pits on which large capital sums have been spent in recent years specifically to provide for the needs of the electricity industry in Scotland'.[72] Pessimism over the industry's future was confirmed by a Ministry official, K. M. Tait, in 1971, on her return from a tour of Scottish collieries:

> When I was visiting pits in Scotland recently it was the mine managers' opinion that the men were normally hardly aware of what was happening in the next pit, let alone in the country as a whole and were not sensitive to national trends. And as you have said, no amount of window-dressing will deceive the miners if the outlook is really bad. On the employment side we could accept

---

[69] TNA, Coal 31/123, *Times*, 'Sign of a new battle over coal', in NCB Press Office, What the papers say, 8 May 1967; Coal 31/96, 25 Apr. 1962, Note: Glenochil drift mine (aide memorandum for the board's side).

[70] TNA, Coal 30/629, Manpower and social implications (1967).

[71] TNA, Coal 30/629, preliminary report of the working party on fuel policy (1967).

[72] TNA, Coal 31/123, A. Robens, NCB, London to R. Marsh, Ministry of Power, London, 12 June 1967.

I suppose that there is some maximum rate at which the industry can be run down without the risk of disintegration. But I am not sure that our ideas on this are soundly based.[73]

Tait's image of endemic parochialism among Scottish miners overlooked what they had in common: a shared experience of incremental contraction over the previous decade, which had diminished their confidence in the future of coal mining. Parochial divisions were eroded by the experience of commuting growing distances to work in 'cosmopolitan' collieries and the perceived threat of further closures. This encouraged the development of a more pronounced collective *national* identity across Scotland's coalfield territories. National objections became more pronounced as unemployment grew and the modernizers' ambitions appeared to falter during the late 1960s and early 1970s. Simultaneously, the relationship between coal and electricity generation policy became increasingly politicized.

Stabilization

Moral economy responsibilities were incurred by policymakers overseeing the application of regional policy concurrent with colliery closures. Obligations towards communities with high unemployment rates competed with development goals in the allocation of inward investment, despite support for the aims of the Toothill report. Scottish Office officials who were charged with directing inward investment operated regional policy to ameliorate the worst excesses of unemployment created by coalfield contraction. In September 1969, H. J. Henson of the Board of Trade Office for Scotland wrote to the department of industry expressing concern at the 'seriously aggravated' male unemployment that he expected to develop in Kilsyth, North Lanarkshire, over the following months. Henson detailed the expected closure of Cardowan colliery, with the immediate redundancy of 1,200 men, due to the pit being put onto 'jeopardy' status regarding its financial losses. The adjacent Bedlay colliery, which employed Kilsyth men too, was also expected to close due to a gas problem. There were only limited employment opportunities in the area. Henson noted that just less than half, 80 of 180, available local jobs were classified as 'male' and most of the local advance factories were oriented towards 'women's work'. These deliberations confirm the moral economy's grounding in the highly gendered understandings of the social status ascribed to colliery employment, as well as commitments to localized coalfield communities. The letter concluded

---

[73] TNA, POWE 14/2501, Miss K. M. Tait to Mr Emmett, draft cabinet paper on energy policy, 7 Oct. 1971.

that 'when drawing up any future programme of advance factories, the needs of Kilsyth will be borne in mind', given the 'effect that the continued rundown in the coal-mining industry is likely to have on employment'.[74]

These concerns typified the development of regional policy as objectives shifted during the late 1960s and early 1970s from economic growth to maintaining employment.[75] Unemployment in Kilsyth remained a concern. A 1972 department of industry report about the Burroughs office machinery plant in nearby Cumbernauld noted that Kilsyth was suffering from redundancies at the plant as well as worries over the future of Bedlay and Cardowan. The male unemployment rate of 19.7 per cent was over double the Scottish rate of 8.5, itself twice the post-1945 average. Despite the relatively small numbers involved (444 men), Scottish Office officials argued that male joblessness constituted a major problem in the locality.[76] A broader concern was emerging over the modernizers' strategy, centred on its failure to embed viable new sectors in the Scottish economy. It was in this context that coal's employment contraction slowed, as demonstrated in tables 1.1 and 1.4. This materialized in North Lanarkshire through Bedlay and Cardowan's survival into the 1980s.

From the 1940s to early 1970s, inward investment was focused towards the Glasgow Planning Area. It principally came from American multinationals and was concentrated in the new towns and industrial estates built in the contracting Lanarkshire coalfields. This area accounted for 56,700 of the 82,600 US subsidiary employees in Scotland during 1972, predominantly in mechanical, electrical and instrument engineering. By 1972, US multinationals accounted for 26.5, 39.5 and 45.6 per cent of employment in these sectors within Scotland.[77] The SCDI concluded that American inward investors were making 'a most important contribution to Scotland's economic revival' through the introduction of new products and processes. Export from American plants grew in value from £75 million to £367.1 million between 1964 and 1972. Even after accounting for inflation,

---

[74] NRS, SEP/4/4251/5, H. J. Henson, Board of Trade Office for Scotland, Glasgow, to J. H. Brown, Department of Industry Division of the Board of Trade, London, 22 Sept. 1969.

[75] P. Scott, 'Regional development and policy', in *The Cambridge Economic History of Modern Britain*, iii: *Structural Change and Growth, 1939–2000*, ed. R. Floud and P. Johnson (Cambridge, 2004), pp. 332–67, at pp. 353–4.

[76] NRS, SEP/4/3791/26/3, background note for the meeting between the minister for industry and R. W. Macdonald, president of the Burroughs Corporation, to take place on 24 July 1972.

[77] Scottish Council Research Institute, *US Investment in Scotland* (Edinburgh, 1974), pp. 3, 8.

this was a significant increase.[78] Less optimistically, a Scottish Office official noted with concern that between 1962 and 1967, engineering redundancy rates more than doubled, from below to above average for the Scottish workforce. The biggest concentration was in the 'other machinery' category, which largely consisted of electronics and electrical engineering. Between 1962 and 1966, Scottish employment in electronics increased by over 50 per cent but only by 14 per cent in Britain as a whole. However, Scotland's share of industrial R&D fell relatively and absolutely. This led to the pessimistic conclusion that 'at least to some extent the trends result from structural factors such as the large number of branch factories in Scotland in these growth sectors of industry controlled from headquarters elsewhere'. These trends were particularly 'disturbing because of the great emphasis we have attached to the long-term change in the Scottish industrial structure, resulting from the increase in modern science-based engineering and electrical industries'.[79]

This outlook was rejected by the modernizers. Willie Ross was the member of parliament for Kilmarnock, within the Ayrshire coalfields, and secretary of state for Scotland from 1964 to 1970 and 1974 to 1976. He keenly endorsed the commitment to achieving higher economic growth rates by encouraging the development of assembly goods industries held by the Wilson governments elected in 1964 and re-elected in 1966.[80] In January 1970, Ross made a speech at the unveiling of an extension to the Hoover plant in Cambuslang in South Lanarkshire, the seventh since its opening in 1946, which profiled the plant as a major success. Rising employment and production made it 'a model operation' based around growing exports. This was facilitated by government investment in transport infrastructure across land, sea and air, including a new inland clearance depot at Coatbridge, to the north-east of Cambuslang, which connected Lanarkshire to upgraded port facilities on both the east and west coasts.[81] The modernizers emphasized the positive aspects of job creation and industrial development associated with inward investment, while the negative aspects of dependency on multinational enterprises were ignored by politicians and policymakers.

This stance increasingly lost credibility with labour movement representatives who had previously been supportive of industrial

---

[78] Scottish Council Research Institute, *US Investment in Scotland*, pp. 2, 8.

[79] NRS, SEP 4/2337, A. W. Teel to Mr Grant, SEPB, Edinburgh, subject: redundancies and unemployment change, 6 Sept. 1967.

[80] S. Holland, *The Regional Problem* (London, 1976), pp. 29–36.

[81] NRS, SEP 4/13/32, E. Reoch, SEPD, Edinburgh to J. A. Scott, SEPD, Edinburgh, subject: secretary of state's engagement at Messrs Hoover, 23 Jan. 1970.

diversification policies, even at the cost of losing jobs in traditional industry. After mounting speculation, the announcement of 5,800 steel redundancies in Lanarkshire during June 1972 was described as 'a greater crisis' than the threatened closure of Upper Clyde Shipbuilders (UCS) by an STUC spokesperson.[82] The UCS yards were occupied by the workforce between June 1971 and October 1972 in opposition to Conservative prime minister Ted Heath's policy to close the yards and other 'lame duck' industrial enterprises. The work-in's eventual success in forcing the Heath government to 'U-turn' by providing investment that preserved shipbuilding employment in three of the four yards signalled a change in government policy. Inward investment's inability to maintain employment levels in areas experiencing further contraction in traditional industries destabilized the modernization agenda. It had enjoyed legitimacy as long as it appeared to deliver relative labour market stability. Government commitment to preserving industrial employment was affirmed in the 1972 Industry Act, which included a more favourable view of heavy industry.[83]

Changes in energy's international political economy associated with tensions in the Middle East profoundly altered prevailing policy assumptions during the early 1970s by creating severe fluctuations in oil prices. This exposed the UK's reliance on imported oil which had developed during the 1950s and 1960s. A mixed-fuel energy policy developed during the early 1970s, which incorporated a substantial guarantee of coal's place in power generation. Changes in policy were triggered by a shift in the trade-offs behind energy policymaking. These most readily related to the oil spike of 1973–4, and oil-price instability, as well as the failure of nuclear to meet politicians and civil servants' expectations. Rather than the economic costs of price increases, it was the perception of international vulnerability and insecurity which led British policymakers to lessen reliance on oil.[84] Coal was bolstered by the assertion of power by miners during UK-wide national strikes in 1972 and 1974, which – although formally fought over wages – were underpinned by the experience of contraction and demands for sectoral security. The stabilization of coal's place in energy markets was highly interrelated with protective labour market action and a renewed social embedding of coal industry. The incoming Wilson–Callaghan Labour government adopted the *Plan for Coal* in its resolution of the 1974 dispute. It enshrined a commitment to securing coal's long-term place in British

---

[82] NRS, SEP/4/3550, *Scotsman*, 21 June 1972.

[83] J. Phillips, *The Industrial Politics of Devolution: Scotland in the 1960s and 1970s* (Manchester, 2008), pp. 109–10.

[84] M. Chick, *Electricity and Energy Policy in Britain, France and the United States since 1945* (Cheltenham, 2007), p. 25.

power generation and renewing the industry. Annual capacity was to be maintained at 120 million tonnes per annum through to 1985, by replacing 42 million tonnes of capacity, and then working towards up to 170 million deep-mined tonnes per annum under *Plan 2000*. Contemporaneously, North and Spooner emphasized that the experience of the oil spike 'really threw into disarray existing conceptions about the energy base of the British economy'.[85] They noted that 'the militant miners' and the NCB now had a common interest in overturning the norms which prevailed during 'the dreadful decade' of the 1960s by asserting the value of coal and obtaining investment in the industry.[86] The plan re-established the hope of the early nationalized period, holding out the promise of an orderly transition to productive units and securing life-time employment.

Policymakers significantly altered their assessments of coal, including their perceptions of industrial relations conflicts. Their reflections confirm that a conflict between liberalizing economic imperatives and a protective countermovement was formative to the alterations in energy policy in the 1970s. During the previous decade, the NCB had viewed pit closures as an effective mechanism to discipline the workforce by ensuring they felt market pressures. In May 1962, J. H. Plumptre, the general manager for the Board's South East Division, had written to Robens congratulating him on pressing for the closure of Glenochil and Rothes. Plumptre claimed these measures had helped improve performance at Betteshanger and other Kent collieries that traditionally had a combative workforce, 'because NUM spokesmen have referred publicly to the danger that the sort of things that have been happening in Scotland might also occur in Kent unless results improved'.[87]

By the early 1970s, officials within the Department for Trade and Industry were critical of nuclear power stations' failure to match the economically viable performance which had been forecast in the 1950s and 1960s. One civil servant, writing in response to cabinet papers on energy policy, used Prime Minister Heath's language of 'lame ducks' to describe nuclear power as an industry which could not perform without uneconomic state subsidies. K. M. Tait was also keen to argue that coal had a similar status, asserting – after the withdrawal of government funding for UCS and Rolls Royce

---

[85] J. North and D. Spooner, 'The great UK coal rush: a progress report to the end of 1976', *Area*, ix (1977), 15–27, at pp. 15–19.

[86] North and Spooner, 'The great UK coal rush', pp. 15–19.

[87] TNA, Coal 31/96, J. H. Plumptre, NCB South East Area, Dover, to A. Robens, NCB, London, 11 May 1962; J. McIlroy and A. Campbell, 'Beyond Betteshanger: Order 1305 in the Scottish coalfields during the Second World War, part 1: politics prosecutions and protest', *Historical Studies in Industrial Relations*, xv (2003), 27–72.

– that 'the maintenance of employment at public expense has manifestly not been a government aim in some other areas'.[88] Tait also questioned the insistence that coal was more secure than oil, following the beginning of significant production in the North Sea. This stance was shared by others commenting on the papers. One official argued that 'more North Sea gas and oil enable us to phase out coal without loss of security'.[89] In light of these findings, a 1973 STUC annual conference contribution from the NUMSA's general secretary and STUC general council member, Bill McLean, seems particularly perceptive. McLean moved a NUMSA motion in support of a coordinated energy policy operating within a long-term conception of national requirements rather than market fluctuations. He argued this was a pressing concern given expanding North Sea production. Oil could have been a valuable resource in a diversified energy sector. Instead, the fuel was as a threat to coalfield communities: 'it was unfortunate indeed that every time a new oil strike was announced a shudder ran through the mining industry. This resulted from the lack of a definite government fuel strategy'.[90]

However, despite the discovery of deposits in the North Sea shelf, by late 1971 policymakers were actively pursuing 'alternatives to oil dependency', emphasizing the build-up of political pressures in the Middle East and OPEC's willingness to use oil prices for political and economic gain. These developments changed perspectives on coal. Industrial relations conflicts in the coalfields were viewed as a threat to economic security, just as they had been before the development of significant alternative fuel options during the late 1950s.[91] The experience of the 1972 and 1974 strikes, alongside the 1973–4 oil spike, consolidated shifts towards a policy that strengthened coal's position in the UK's energy supply. As early as February 1971, the prime minister's advisors had already noted the diplomatic prescience of reducing oil dependency to provide 'a demonstration to the oil producing countries that they are in danger of killing, or at least weakening, the goose that lays the golden eggs'.[92] During January 1972, the month in which the first of the two major coal disputes of the Heath premiership began, the Central Policy Review Staff argued that the government should take steps to preserve up to 120 million tonnes of annual coal production. This was justified to maintain

---

[88] TNA, POWE 14/2501, Miss K. M. Tait to Mr Emmett, subject: draft cabinet paper on energy policy, 7 Oct. 1971.

[89] TNA, POWE 14/2501, B. D. Emmet to Mr Thomas, subject: energy policy, 5 Oct. 1971.

[90] Scottish Trades Union Congress (STUC), *Annual Report 1972–1973*, lxxvi (1973), 286.

[91] TNA, POWE 14/2501/38, B. Jones, Draft report D, 30 Sept. 1971.

[92] TNA, PREM 15/1144, note for Prime Minister: energy policy, 25 Feb. 1971.

employment and the balance of payments. Officials also noted that coal production could not be speedily restarted, which necessitated supporting a level of capacity judged to be of a suitable economic quality.[93] Notably, the figure of 120 million tonnes of deep-mined coal was the one that appeared in the *Plan for Coal* two years later following a change of government and remained central to coal production planning throughout the decade.[94]

The NUM president, Joe Gormley, carried these arguments into tripartite meetings on energy policy. These were established under the Wilson–Callaghan governments as forums between government departments, trade unions representing coal and energy workers, and the NCB and the generating boards. In 1976, Gormley argued that coal mining productivity had only increased during the 1960s through 'chopping limbs off,' and that energy policy 'should be as self-sufficient as possible'.[95] Indicating the relatively favourable environment for coal during the mid 1970s, a Central Energy Generating Board (CEGB) representative accused the NCB of using its power to 'jack up' prices in a captive market. Frank Tombs, chair of the SSEB, similarly complained that Scottish electricity coal prices had doubled in recent years, but that the Electricity Board was compelled to purchase it while other users were not.[96] At the same meeting, the NUMSA president, Michael McGahey, who had succeeded Alex Moffat in 1967, argued in favour of 'a Scottish energy plan'. McGahey cited the distinct electricity markets and generating boards as justifying this approach. Scottish and Welsh Office representatives also attended the meeting, indicating the intersection of energy policy with constitutional concerns in the context of advanced discussions on devolution proposals.[97]

A distinct Scottish coal interest was strongly communicated under the 1970s Labour governments, principally through secretary of state for energy Tony Benn's receptiveness to the NUMSA, but also by the voices of sympathetic MPs and Scottish Office and Coal Board officials. Coal

---

[93] TNA, PREM 15/1144/10A, Central Policy Review Staff, A coal strategy for the United Kingdom, Jan. 1972.

[94] TNA, Coal 31/166, NCB memorandum: Coal industry tripartite discussion: review of *Plan for Coal*, 6 Oct. 1976.

[95] TNA, Coal 31/166/13, Department of Energy paper 14: Tripartite energy consultations, London, 20 Feb. 1976.

[96] TNA, Coal 31/166/13, Department of Energy paper 14: Tripartite energy consultations, London, 20 Feb. 1976.

[97] TNA, Coal 31/166/13, Department of Energy paper 14: Tripartite energy consultations, London, 20 Feb. 1976; A. Ebke, 'The decline of the mining industry and the debate about Britishness of the 1990s and early 2000s', *Contemporary British History*, xxxii (2018), 121–41, at pp. 129–32.

concerns were strongly related to power station investment and fuel choices. A tripartite meeting on Scottish power generation during 1976 discussed the threat to 10,000 coal jobs from power stations under construction which had been ordered before alterations in energy policy: the nuclear station at Hunterston, and oil-fired stations at Inverkip and Peterhead. Benn concurred with both the NCB and NUM that it was 'intolerable and unacceptable' to allow large-scale contraction in the Scottish coalfields. Furthermore, he argued that Scottish coal was in fact 'relatively cheap compared to world prices'.[98]

Benn was supported by Alex Eadie, who was minister of state for energy throughout the 1974–9 Labour governments. Eadie was the member for Midlothian, and a 'miners' MP' from an NUM background. He shared Benn's sentiments towards the promotion of coal. Under Benn's and Eadie's stewardship, some tangible benefits were provided in a situation where the Scottish coal lobby's capacity was strengthened. At the 1978 NUMSA conference, Eadie was commended by McGahey for having 'done so much for the coal mining industry in pushing forward our interests'. The Area president singled out the government's provision of a support grant to the SSEB to preserve 8 million tonnes of annual coal burn as a major contribution to stabilizing the sector. This development laid the basis for the development of further modernized coal production in Scotland. A prominent example was extensive NCB investment in a project linking the Midlothian coalfield with the Longannet complex via Musselburgh, which facilitated access to coal reserves under the Firth of Forth. Another development connected Kinneil colliery in West Lothian to the Longannet drift mine at Castlebridge. However, noting the continuation of a long-term concern, McGahey also underlined that the Scottish nuclear sector was continuing to develop with considerable public investment in a new facility at Torness in East Lothian.[99] Although there were some significant reversals of policies which had undermined coal's position during the 1960s, the *longue durée* trends towards power station diversification and the contraction of coal use in other sectors of the economy continued in the 1970s. North Sea oil represented a heightened threat, given it held out the prospect of secure domestic supplies in a major competitor fuel. Miners' attitudes had shifted to a more assertive approach towards colliery closures. The commitments in the *Plan for Coal* had embedded the moral economy within the operation of energy policy.

[98] TNA, Coal 31/166, Tripartite meeting on electricity coal burn in Scotland, 5 Feb. 1976.

[99] NMMS, NUMSA, Executive Committee Minutes, 27 June 1977 to 14–16 June 1978, p. 601.

Accelerated deindustrialization

The election of the Thatcher government in 1979 brought a new set of priorities to bear on the coal industry. This agenda involved open antagonism towards trade unions and social democratic economic infrastructure, of which the nationalized mining industry was a prime example. Its initial objectives included ending government subsidies for coal, eroding trade union influence through redistributing workplace authority towards management in the medium term, and finally the long-term goal of privatizing the industry. A phased approach towards coal sat within the Thatcherite 'stepping stones' strategy to eliminating trade union power by isolating and defeating groups of workers one by one.[100] Meeting the first objective began with the implementation of the 1980 Coal Industry Act which moved towards the complete removal of grant funding (other than social grants) by 1983–4.[101] The second aim was achieved following the government's and NCB's decisive victory in the 1984–5 miners' strike. Privatization followed in 1994.[102] The build up to the strike was decisive in shaping the trajectory upon which policy later followed. It included a decisive reduction in NCB autonomy vis-à-vis government, in part through the strict application of External Funding Limits. This placed severe limitations on the Board's expenditure, restricting its ability to sustain loss-making collieries or invest in new developments.

As with other restrictions on public spending and subsidy that are characteristically associated with neoliberalism, these measures had origins in Jim Callaghan's premiership and the ramifications of the reforms implemented after the IMF bailout of 1976. Deflationary measures included public spending cuts and tax and interest rate rises. Unemployment resultantly rose to unprecedented levels in the post-Second World War era, but the economy was in fact recovering by the time of the 1979 general election.[103] The effects of neoliberal prescriptions became more pronounced as an instrument of energy policy transformation after 1979.[104] At workplace level upwards, the imposition of market logic was accompanied by the assertion of managerial prerogative and the elimination of customary trade union rights. These developments contributed to incrementally worsening

---

[100] J. Hoskyns and N. Strauss, 'Stepping Stones' (1977) *Centre for Policy Studies* <https://www.cps.org.uk/files/reports/original/111026104730-5B6518B5823043FE9D7C54846CC7FE31.pdf> [accessed 21 Nov. 2019].

[101] TNA, Coal 31/138, Coal Industry Act (1980).

[102] M. Parker, *Thatcherism and the Fall of Coal* (Oxford, 2000), p. 210.

[103] J. Meadhurst, *That Option No Longer Exists: Britain, 1974–76* (Arelsford, 2014), p. 2.

[104] Chick, *Electricity*, p. 103.

relations between the union and the Board and government. Relations between the NCB and government also soured before Ian MacGregor was appointed as the Board's chair in 1983, with a remit for aggressive anti-trade unionism.[105]

The first Thatcher government came to power in the context of a second oil spike following the Iranian Revolution, which renewed concerns over oil supplies. These developments appeared to provide a context for continuity in energy policy. During 1979, the NCB produced 83 million tonnes of coal for power stations, which was its highest ever sale. This required a 'special effort' to achieve deep-mined output.[106] The Board were working to a medium-term development plan which assumed that power station demand would roughly maintain itself at this level, if not expand, up to the year 2000. Although approximately stagnant, maintaining this level of production required extensive replacement of older capacity. Under the *Plan for Coal*, the NCB aimed to develop 38 million tonnes of annual production by 2000, or approximately 2 million tonnes per annum. F. B. Harrison, a Board official, was keen to stress that, rather than a policy of 'coal at any price' in the context of high demand, the NCB was in fact concentrating production within economically viable collieries.[107] Indicating continuity with the Benn period, during June 1979, Derek Ezra wrote to the new secretary of state for energy, David Howell, to confirm attendance at a tripartite meeting which would discuss 'the role of coal in helping to diminish the consumption of oil'.[108] However, the meeting's focus on the coal industry's finances was an indicator of the priorities which Howell and his successors would pursue over the decade and a half to follow.[109]

Before the strike began in 1984, energy policy was shaped by deliberations over the tolerable extent of colliery closures and the imposition of more stringent financial expectations onto the NCB within an increasingly liberalized market setting. Until the NCB's anti-union turn was consolidated under MacGregor, there was a similar joint position between unions and the Board to when they had articulated its opposition to the direction of energy

---

[105] J. Phillips, 'Containing, isolating and defeating the miners: the UK cabinet ministerial group on coal and the three phases of the 1984–85 strike', *Historical Studies in Industrial Relations*, xxxv (2014), 117–41, at p. 125.

[106] TNA, Coal 31/138, F. B. Harrison, *Plan for Coal* review: summary of main points for discussion with M. J. Parker, 13 Feb. 1979.

[107] TNA, Coal 31/138, Short-term proposals for the industry, 4 July 1979.

[108] TNA, Coal 31/138, D. Ezra, NCB, London, to D. Howell, Department of Energy, London, 20 June 1979.

[109] TNA, Coal 31/138, D. Ezra, NCB, London, to D. Howell, Department of Energy, London, 20 June 1979.

policy during the 1960s. These contentions related to views of finances, but also policies on coal importation: with restrictions lifted, the British Steel Corporation had begun to replace British-mined coal with foreign coke. Michael McGahey expressed outrage at economic vandalism which threatened viable collieries and undermined commitments to national self-sufficiency, stating that 'it would be like Alice in Wonderland to replace dependency on oil with dependency on Polish coal'.[110] Howell insisted that the Coal Industry Act represented a continuation of the *Plan for Coal* when he met the Board and unions in April 1980. The first Thatcher government's scaling back of regional policy, including the end of regional grants for industrial enterprises, was another concern to both the other parties. Ezra emphasized that the NCB made significant contributions to Development Areas, emphasizing its Scottish presence, and underlining that its operations were threatened by the removal of support from government. He was echoed by Gormley.[111]

On 10 February 1981, a large unofficial walkout across the British coalfields was triggered by the NCB's announcement of closures which aimed to achieve significant financial savings. The list was withdrawn in the aftermath of the strike. NUM representatives did not object to pit closures in principle, but rather to the disruption of the nationalized industry's consultative machinery and threats to coal's place in a mixed energy policy. Gormley asserted fifteen days after that, 'Our aim is to have the *Plan for Coal* fully implemented and for the industry to be able to sell all it produces, and we are having talks with the Government to this end'.[112] By this point, the commitment to coal upon which tripartite arrangements had been founded during the 1970s was being severely stretched. During the summer of 1981, the future of the Ayrshire coalfield was called into question when the Northern Irish Electricity Service was granted the option to use imported coal from outside the UK.[113]

By 1983, these developments had led the Coal Board under MacGregor to assume that both major remaining collieries in Ayrshire, Barony and Killoch, would close in the immediate future. The Board worked to an agenda of closing seventy-one collieries and eliminating 25 million

[110] TNA, Coal 31/138, D. Ezra, NCB, London, to D. Howell, Department of Energy, London, 20 June 1979.

[111] TNA, Coal 31/138, note of a meeting held at 3pm on Wednesday 23 Apr. 1980 in House of Commons, 30 Apr. 1980.

[112] TNA, Coal 31/138, meeting held at 11am on Wednesday 25 Feb. 1981, Thames House South, London.

[113] TNA, Coal 31/38, meeting held at 11am on Tuesday 16 July 1981, Thames House South.

tonnes of production per annum by 1987–8. As part of this forecast, the NCB also foresaw the closure of a swathe of modernized electricity pits in the Lothians and Fife, as well as Lanarkshire's only remaining colliery, Cardowan.[114] These steps provided the tinder for the 1984–5 miners' strike. In 1982, the NUM's newly elected left-wing president, Arthur Scargill, referred to the closure plans as a 'hit list'.[115] These projections originated in the government's rejection of the *Plan for Coal*'s emphasis upon the use of indigenous resources and the effect which the deep recession of 1979–81 had upon the energy demand projected during the mid 1970s. The Thatcher government's disregard for concerns with the balance of payments and preserving industrial production within the UK were pivotal to shaping the emergent energy policy regime.[116] A symptomatic example was the closure of one of Scotland's principal energy users, the aluminium smelter at Invergordon, in December 1981. NCB analysts noted the closure had a long-term downwards impact on anticipated electricity demand.[117]

The effects of these policies are underlined by the unemployment rates shown in table 1.3, and the steep reduction in coal employment between 1977 and 1987 in table 1.4, as well as the reduction in steel and engineering employment (tables 1.1 and 1.2). Given major changes to the census industrial occupation categories, it is not possible to generate comparative figures for 1991, but it is clear the trends apparent in the 1981 figures continued. Large-scale steel production in Scotland ceased in 1992 with the closure of Ravenscraig strip mill in Motherwell, North Lanarkshire, which was anticipated by a series of major layoffs in the preceding years.[118] The engineering sector was affected by the 'retreat' of multinationals from Scotland. Major 1980s factory closures in North Lanarkshire alone included Burroughs' office machinery in Cumbernauld in 1986 and Caterpillar's tractor factory at Tannochside in 1987.[119]

---

[114] TNA, Coal 31/138, R. J. Price, S. Medley and K. Hunts, Memorandum: NCB general purposes committee area: five year strategies, 21 July 1983.

[115] TNA, Coal 31/433/5, Second report from the Energy Committee: pit closures, 21 Dec. 1982.

[116] J. Tomlinson, 'De-industrialisation not decline: a new meta-narrative for post-war British history', *Twentieth Century British History*, xxvii (2016), 76–99, at p. 87.

[117] TNA, Coal 101/580, The effects of introducing Scottish Coal into the CEGB System in 1982–3.

[118] P. Payne, 'The end of steelmaking in Scotland c.1967–1993', *Scottish Economic and Social History*, xv (1995), 66–84, at pp. 76–7.

[119] C. Woolfson and J. Foster, *Track Record: the Story of the Caterpillar Occupation* (London, 1988), p. 33; N. Hood and S. Young, *Multinationals in Retreat: the Scottish Experience* (Edinburgh, 1982).

Debates over the development of Scotland's economic structure had become enmeshed with the growth of support for autonomy via 'home rule' or devolution through the experience of deindustrialization. An initial stimulus was provided by the onset of Scotland-wide coalfield contraction and the closure of modernized collieries during the 1950s. From the late 1960s, the inability of regional policy to successfully generate employment at a rate surpassing the loss of jobs in heavy industry generalized concerns. The effect of intensified deindustrialization accentuated demands for Scottish autonomy. McGahey explicitly linked the two at the 1983 NUMSA conference and made use of the term deindustrialization, indicating a view that Scotland was now enduring the loss of its industrial base rather than just sectoral contraction. He stated that Conservative 'policies had led to the deindustrialization of the country, to the decimation of the coal, steel and railway industries, to the threat to the shipbuilding industry and to the attacks on social services'. In light of the 1983 election results, where a majority Conservative government was elected across the UK while Labour topped the poll in Scotland, McGahey stated that, '[t]he Conservative government does not represent Scotland'. He called on the Scottish Labour party leadership 'to reconsider their position of refusing to take part in a broad-based campaign for a Scottish Assembly' which could mobilize 'all progressive and democratic opinion in Scottish society'. A motion in favour of a Scottish Assembly was passed unanimously. Articulating a sentiment which strengthened over the 1980s, David Hamilton, the NUM delegate for Monktonhall colliery in Midlothian, argued that '[t]he Tories did not have a mandate in Scotland'.[120] Hamilton's words are significant in anticipating the more prominent articulation of the 'mandate' argument following the 1987 general election, when Conservative representation fell from twenty-one to ten Scottish seats.[121]

By 1983, as the great strike for jobs approached, the NUMSA articulated its opposition to Thatcherism in terms that fused national and industrial or class-centred perspectives which were conjoined through the coalfield moral economy. This outlook was the outcome of the long experience of economic restructuring in the Scottish coalfields, and the product of industrial relations within the nationalized industry, especially its consultative structure and consensual culture. The application of a national framing to Scottish coalfield development had origins at least as far back as the NCB itself, in the form of its relatively autonomous Scottish Division. Opposition to closures from the late 1950s onwards took on an increasingly

---

[120] NMMS, NUMSA, Executive Committee Minutes, July 1982 to June 1983, pp. 702–3.

[121] D. McCrone, *The New Sociology of Scotland* (London, 2017), p. 121.

pronounced national, rather than merely localized, mantle as the future of the coal mining industry itself was questioned. These developments also converged with the growing domination of coal consumption by power station usage, which firmly connected the sector's fate to key government policy decisions. Both the Scottish national and class facets of the NUMSA's opposition to Thatcherism were grounded in an understanding that colliery closures were illegitimate, which bolstered its opposition to the broader accelerating process of deindustrialization. Closures were imposed without popular assent, ignoring establishing customs, and withdrawing the safeguarding of communal economic security that both the governance of the nationalized industry and the regional policy regime had broadly provided between the 1940s and 1970s. It was the transgression of these norms which differentiated the 1980s from the four previous decades. This culminated not only in increasing colliery job losses, but also the closure of assembly plants brought to Scotland through UK regional policy. These experiences disembedded industrial employment from the communitarian norms through which coalfield deindustrialization had been managed under public ownership hitherto.

2. Moral economy: custom and social obligation during colliery closures

The nationalized coal industry was subject to its workforce's expectation that industrial governance would involve dialogue with organized labour and the provision of employment security. Those assumptions were powerfully shaped by the collective memory of social dislocation in the coalfields during the first half of the twentieth century, especially interwar class conflict and mass unemployment. Nationalization was understood by miners and their families as an imposition of social order which would displace liberal market logic. Colliery closures were a moment that tested the nationalized industry's distinctions from its privately owned predecessor. During the early years of the nationalized industry, customary measures developed during colliery closure processes to protect economic security: the observation of procedure regarding the management of closure through discussion with trade union representatives; ensuring transfers for the workforce within travel distance; the provision of transfers to appropriate jobs, especially for high-earning faceworkers; the provision of either retirement or suitable positions for elderly and disabled workers. Moral economy conditions were rarely met in full, but these customs formed the basis of how pit closures were managed between the 1940s and 1970s. The National Coal Board's (NCB's) repudiation of its moral economy responsibilities provoked the 1984–5 miners' strike, while the conflict's decisive outcome confirmed the demise of joint regulation and the exercise of a significant trade union 'voice' within the industry. This chapter begins by outlining memories of the nineteenth and early twentieth century, before providing an overview of the coalfield moral economy's development over the nationalized period. It emphasizes the continually contested nature of moral economy practices and their evolution over time as the geographical focus of closures shifted.

Three area case studies from the Lanarkshire coalfields provide an overview of the moral economy's evolution under nationalization. They draw on the records of consultative committee meetings held between management and trade union representatives during closure processes as well as oral testimonies from former miners who were affected by closures. The first case study analyses Lanarkshire's 'eastern periphery', the area around Shotts that experienced major closures during the early nationalized period. These

---

'Moral economy: custom and social obligation in colliery closures', in E. Gibbs, *Coal Country: The Meaning and Memory of Deindustrialization in Postwar Scotland* (London, 2021), pp. 55–89. License: CC-BY-NC-ND 4.0.

experiences took place while overall Scottish coal mining employment was still rising (see table 1.4) and were formative to the moral economy as it was later practised across the Scottish coalfields. Later closures were organized around the principle of offering work within travelling distance following the reluctance of Shotts miners to uproot their families by migrating eastwards to Fife or the Lothians. Lanarkshire's 'southern area' included dedicated mining settlements which were somewhat distant from the more diversified industrial areas to the north. Closures between the mid 1950s and early 1970s were managed by offering transfers to increasingly distant workplaces. These practices stretched local community connections to workplaces but broadly upheld moral economy expectations.

In Lanarkshire's heavily industrialized 'northern core', the closure of redeveloped collieries on economic grounds during the 1960s and early 1970s provoked objections from miners who had often experienced earlier rounds of rationalization and been ensured of employment security at their now threatened workplaces. These episodes contributed to the renegotiation of the moral economy through miners' industrial action and energy market fluctuations in the early 1970s. Struggles for wages in 1972 and 1974 were framed as highly moral arguments for the recognition of miners' centrality to Britain's industrial economy. The northern core case study culminates in an analysis of the closure of Lanarkshire's last collieries during the early 1980s. It contrasts the relatively consensual closure at Bedlay during 1982, with the highly contested closure of Cardowan, a year later. These closures exemplify the centrality of customary expectations to the moral economy, especially joint regulation and the provision of collective employment security. At Cardowan, the Board's commitment to dialogue was undermined when management incentivized workers to accept redundancy packages without the NUM's agreement. Events at Cardowan during 1983 were formative to the miners' strike which began a year later. The NCB's aggressive assertion of managerial prerogatives and financial objectives at the expense of industrially viable units during the 1980s closely resembled the liberalizing pressures of the double movement. Trade unionists' insistence on moral economy customs formed a protective countermovement against the disruption of social order by corrosive market forces. But as the case studies demonstrate cumulatively, these factors were continually present in the nationalized industry, especially during the governance of pit closures.

The moral economy's development was shaped by a sense of both the recent and the more distant past. Shifting practices under nationalization were epitomized by dialogue between the NCB's chairman, Derek Ezra, and the NUM's president, Joe Gormley during the summer of 1979. Ezra tried to reassure Gormley, who cited earlier periods of large-scale colliery closures.

In the Board's view the industry was in 'a completely different situation now from the 1960s when it was contracting. Now, closures take place only when a pit has reached the end of its useful life'.[1] The oral testimonies recorded for this study include appeals to a longer sense of coalfield history. Historical experience is retold through a non-linear relationship between personal experience and public narratives. Recollection involves an 'interactive construction'; testimonies entail an act of cultural production shaped by hegemonic and highly politicized versions of historical memory.[2]

A cultural circuit of coalfield memory counterpoised the obligations of the nationalized industry with the deprivations and injustices of private ownership which moulded the moral economy expectations placed on the NCB during colliery closures. The circuit's 'conceptual and definitional effects' shaped memories. When recalling events which have remained politically contentious, interviewees' 'memories were entangled with the myth[s]' which political and media representations have generated.[3] Counter-hegemonic narratives can be sustained by appeals to the memory of 'particular publics'.[4] Coalfield communities and left-wing political activists have sustained distinct understandings of miners' experiences. Particular generations of workers have obtained a war veteran-like association with key episodes of industrial conflict that embed narratives of community resolve against employer and state hostility.[5] An influential example is the 'folk memory' of the 1926 miners' lockout. This enabled striking miners to see 'themselves as part of the collective story of both their own family and the community in which they lived' during the 1984–5 strike and was a source of historical framing of their struggle and ultimate defeat afterwards.[6] Before this, these experiences were formative to the workforce's understanding of the social responsibilities held by NCB managers.

---

[1] The National Archives (TNA), Coal 31/138, Minutes of the meeting held on 12 July 1979 at Thames House South.

[2] Popular Memory Group, 'Popular memory: theory, politics, methodology', in *The Oral History Reader*, ed. R. Perks and A. Thomson (London, 2006), p. 44; L. Abrams, *Oral History Theory* (London, 2010), p. 18.

[3] P. Summerfield, *Reconstructing Women's Wartime Lives: Discourse and Subjectivity in Oral Histories of the Second World War* (Manchester, 1998), p. 14.

[4] A. Thomson, 'Anzac memories: putting popular memory theory into practice in Australia', in Perks and Thomson, *The Oral History Reader*, pp. 300–10, at pp. 301–2.

[5] D. Nettleingham, 'Canonical generations and the British left: narrative construction of the British miners' strike, 1984–85', *Sociology*, li (2017), 850–64, at pp. 852–3.

[6] D. Leeworthy, 'The secret life of us: 1984, the miners' strike and the place of biography in writing history "from below"', *European Review of History: Revue europeenne d'histoire*, xix (2012), 825–46, at p. 828.

A cultural circuit of coalfield memory was visible in the repertoires of public events such as the annual Scottish miners' gala, the rhetoric of union leaders, and historical literature. It was also transmitted at a more localized level. The circuit had a powerful basis in connecting family histories with communal remembrances of major events such as industrial disputes and mining disasters. Collective memories were developed with institutional support. The NUMSA projected a Scottish mining community defined through a representation of the traditions and histories of localized mining communities within a national and class political framing. Union leaders were keen to place themselves within a longer history of miners' struggles against injustice despite the union having only been established in its current form during the 1940s. Robert Page Arnot, a communist activist and sometime NUM official, published *A History of the Scottish Miners* in 1955.[7] Arnot foregrounded Scottish miners' 'record of suffering and of heroic struggle against the soulless mine-owner', including their eighteenth-century campaigns to abolish serfdom, which eventually succeeded in 1799. He also detailed the conditions of the early industrial revolution, using quotations from the 1840 Children's Employment Commission to illustrate the horrific conditions in which children as young as seven years old worked up to thirteen hours per day.[8] Arnot's analysis profiled the achievement of trade union consultation within the NCB's social democratic infrastructure. He counterpoised nationalization with the employer hostility and divisions within mining trade unionism that had characterized the 1920s and 1930s. The wartime construction of a Scotland-wide, and then subsequently Britain-wide, miners' union was both a key achievement of 'men who were not prepared to be put off by the difficulties and obstacles that had baffled their predecessors', and an accompaniment to nationalization.[9]

Personal, familial and community experiences were reinforced by an awareness of a longer mining history transmitted through the cultural circuit. Peter Downie was raised in a mining family in Greengairs, North Lanarkshire, before starting work at the local colliery, Glentore, during the 1950s. His father was involved in an accident at Bedlay colliery in 1938 and not provided with work to suit his condition. As a result, Peter's family grew up in poverty and he was 'raised on the parish'. In Peter's view this was part of a broader history of economic insecurity suffered by miners:

> When you go and take your history from the 1840s, the 1840s onwards, they were living in deprivation. The miners were living in deprivation because the

[7] R. P. Arnot, *A History of the Scottish Miners: from the Earliest Times* (London, 1955).
[8] Arnot, *A History of the Scottish Miners*, pp. 12–28.
[9] Arnot, *A History of the Scottish Miners*, pp. 252–3.

situation was that they couldnae feed the weans that they haved. And they were living in wooden shacks, stane flares and, the weans had rickets, born weakness, and the people who helped them was very, very, little. The coal owners gave them nothing.

Peter summed up his perspective in another discussion by stating that 'up to 1947 the men were living in deprivation'.[10] His reflections are given added significance in that he was not sympathetic to the communist orthodoxies that predominated within the NUMSA. Peter has a background of involvement in the Orange Order, a loyalist or Unionist organization which is suspicious of socialist politics, and he rose to become an overman official within the NCB. Nevertheless, Peter shares an investment in the moral economy that was shaped by the transmission of the cultural circuit. Another appeal to the cultural circuit took place in the same focus group of retired mineworkers in Moodiesburn, North Lanarkshire. Billy Maxwell articulated an understanding of coal's centrality to industrial societies. He worked at Cardowan colliery between 1957 and 1979, having been brought up in a mining household in Muirhead, North Lanarkshire. Billy stated that, 'miners for centuries, as I said to you before, were fuelling the industry ae Great Britain. They were the most important ingredient in the making of wealth'. Earlier in the focus group, Billy claimed that 'deep mining in the industrial revolution financed the world we have today', and underlined this was 'at the cost ae a lot of miners'. An understanding of the true 'price of coal', juxtaposing arguments relating to financial costs to that of lost miners' lives, was compounded by the view that miners did not abuse their power: 'the miners could ae held the country to ransom at that time [during the 1950s] and didnae dae it cause they were too decent a people'.[11] Awareness of coal's strategic importance was combined with an understanding of the dangers endured by the workforce. This anchored a moral economy perspective based on the obligation of the state to reciprocate the 'decent' conduct of miners.

Memories of 1920s and 1930s class conflicts were pivotal to framing nationalization. Jessie Clark grew up in a mining family in the coalfield village of Douglas Water, an isolated mining village in the far south of Lanarkshire, during the 1920s and 1930s. She recalled the Coltness Iron Company's employment practices as punishing trade unionists and socialists such as her father and encouraging social divisions within the village:

---

[10] Moodiesburn focus group, retired miners' group, The Pivot Community Centre, Moodiesburn, 25 March 2014. 'Weans' is a Scots word meaning children.

[11] Moodiesburn focus group, retired miners' group, 25 March 2014.

My father was a union man. And during the thirties when I was growing up, my father was unemployed quite a lot. You know, it was a question of first out, you know, last in. And I have no doubt, I have no doubt there were people who. I'm afraid there's two things that, the enemies, that my father always talked about was the likes of the landowners and the Masonic Lodge.[12]

Both Jessie, her father, and her husband, Alex, who was also a miner, were Communist Party of Great Britain (CPGB) members. Jessie worked in the pit canteen at Douglas Castle during the 1940s when coal was taken into public ownership. She understood nationalization as a social advance and a victory in class struggle terms. It entailed 'getting rid of the coal owners and there was going to be a bit more democracy you know within the working area … At least there wasn't some eh lazy so-and-so drawin the money from your, the interest from your labour'. Jessie's articulation of nationalization entailing 'a bit more democracy', indicates the rootedness of the moral economy conception of collieries as communal resources.

The nationalized industry

The nationalization of British coal mining is an emblematic example of developed economies' societal embedding after the Second World War: 'social control was restored over the economy' as the objectives of full employment and economic security underpinned an enhanced role for state intervention.[13] Rather than a once and for all act, the double movement describes an ongoing process. Conflict between the priorities of narrowly financially defined performance objectives and trade union consultation and employment security was a continual feature within the contested management of closure. Nevertheless, nationalization was viewed as a major achievement in all the testimonies collected for this study. It granted miners independence from the coal owners and secured improvements in pay and conditions. These findings affirm Jones et al's conclusion that enthusiasm for nationalization was not confined to a minority of trade union activists.[14] Ina Zweingier-Bargielowska's work revised earlier findings on nationalization's significance by emphasizing the importance of regional and colliery-level distinctions to understanding the meaning of public ownership.[15] The three territorial case studies detailed below are structured

---

[12] Jessie Clark, interview with author, residence, Broddock, 22 March 2014.

[13] K. Polanyi Levitt, *From the Great Transformation to the Great Financialization: on Karl Polanyi and other essays* (London, 2013), p. 100.

[14] B. Jones, B. Roberts and C. Williams, '"Going from darkness into light": South Wales miners' attitudes towards nationalisation', *Llafur*, vii (1996), 96–110, at p. 102.

[15] I. Zweiniger-Bargielowska, 'South Wales miners' attitudes towards nationalization: an essay in oral history', *Llafur*, vi (1993), 70–84, at p. 78.

in this vein. They indicate that rather than doubt over the principles of nationalization, discontent with the operation of the publicly owned industry tended to follow the double movement's logic whereby the Board's pursuit of market ends was held to conflict with its responsibilities towards its workforce and coalfield communities. These conflicting perspectives shaped the development of closure practices within the nationalized industry. One important innovation was the commitment to ensuring the localized community's integrity by offering transfers within travelling distance. This was strengthened following the experience of divestment in the eastern periphery early in the nationalized period.

In the official history of the nationalized industry, Ashworth claimed that the priorities miners accorded public ownership were 'first and foremost a guarantee of a better working life'.[16] This was verified by testimonies from those with direct memories of the changes associated with nationalization. Comments from a focus group in Shotts demonstrated a commonly held association between public ownership and improved remuneration and health and safety standards by men who did not identify as having been trade union activists or socialists:

> Willie Hamilton: It improved the miners' conditions tremendously y'know. As I say it was practically slavish y'know wi the private owners then you had a bit of independence after that. The money wisnae great mind you it could have been better.
>
> Ella Muir: Did you have more security?
>
> Willie Hamilton: Very much more. And as I said they opened new pits and that the miners moved away.
>
> Bill Paris: It became safer as well.
>
> Willie Hamilton: The safety side aw the mines was greatly improved, especially the support in the roofs and so forth. With the private owner, he skimped on the material used to support the roof. But when the Coal Board come in, they upgraded everything.[17]

Within these recollections, improvements in conditions and living standards were accompanied by employment practices which emphasized social priorities. In addition to material improvements, the NCB provided employment suitable for men with learning difficulties and those who had been injured and made disabled during industrial accidents. Pat Egan, who

---

[16] W. Ashworth, *The History of the British Coal Industry*, v: *1946–1982: the Nationalized Industry* (Oxford, 1986), p. 129.

[17] Shotts focus group, Nithsdale Sheltered Housing Complex, Shotts, 4 March 2014.

followed his father into coal mining employment by starting at Bedlay during the late 1970s, recalled:

> There also at that time that you looked you employed quite an amount ae people wi learning disabilities and stuff that wouldnae of got work elsewhere. They would be employed in the surface or the baths. They were mentored or coached but I mean they got a job and it gied them a bit ae a feeling ae self-worth, a wee bit of boost to their confidence and they were looked after basically by the community. Very looked after. Quite a few people who had Down syndrome and stuff, but they were always looked after.[18]

However, such accommodation was not always forthcoming from the NCB. The moral economy was negotiated between the liberalizing and countermovement forces of the double movement and was consistently challenged by the NCB's financial pressures. In 1959, during the closure of Douglas Castle colliery in South Lanarkshire, an NUM representative on the pit's Colliery Consultative Committee (CCC) commented that the 139 redundancies at the colliery left him 'astounded at the callous manner in which elderly and disabled workmen had been discarded after serving a life time in the industry and it would appear that human relationship had been ignored completely by the Board'. Notably, the meeting secured some moral economy concessions with up to nineteen 'extreme hardship cases' to be reinstated.[19]

Coordinated practices evolved to defuse tensions through the organization of transfer to specific positions. In 1966, the closure of the Garscube pit in Maryhill, Glasgow, which placed transferees in facework and power-loading jobs, was agreed through meetings with the NUM branches at Garscube and Cardowan, and the Cardowan colliery manager, John Frame. Mr Callaghan, an NUM representative on the Garscube CCC, stated that having been initially 'suspicious' about closure, he was now 'quite happy' to accept the transfers.[20] As will be discussed below, seventeen years later Frame was commended by union activists for his opposition to the closure of Cardowan, and the victimization he suffered as a result. Frame's earlier actions cemented his reputation through the cultural circuit as a manager who operated within, and defended, the moral economy.

---

[18] Pat Egan, interview with author, Fife College, Glenrothes, 5 Feb. 2014.

[19] National Records of Scotland, Edinburgh (NRS), CB 280/30/1, Douglas Castle colliery: summary of manpower as at 6 Jan. 1959; CB 280/30/19, Notes of a meeting held in the manager's office, Douglas Castle colliery, on 4 Feb. 1959.

[20] NRS, CB 298/6/1, Minutes of meeting held at Robertson Street, Glasgow, Wednesday 3 Aug. 1966.

Confrontations between the NUM and NCB over Cardowan's future from 1968 to 1972 epitomized the dynamics of the moral economy and its fusion with national concerns over Scottish industrial and political autonomy. Over this period, the NUMSA succeeded in pursuing appeals processes that saved the pit from closure. Discussions within the NCB drew attention to the strategic importance of Cardowan as a coking coal provider and the political ramifications of leaving Scottish steelworks reliant on imports from England during a period of high local unemployment.[21] The NUMSA's case for maintaining Cardowan was an example of the countermovement prevailing over the logic of market forces. During 1971, J. B. Burton, an NCB official, noted in a memorandum that the colliery's 'high losses' could not justify its continuation, and therefore 'the decision now as to its possible removal from jeopardy has to be taken therefore on grounds other than financial [ones]'.[22]

Early closures and migration

The eastern periphery of the Lanarkshire coalfields, centred on Shotts, was designated for closure during the late 1940s and early 1950s. This case has strong overtones of the 'community abandonment' thesis of deindustrialization: the NCB oversaw strategic divestment as part of its policy of diverting resources to other coalfields within the context of manpower shortages.[23] Friction centred on the NCB's plans for reorganization, which incorporated large-scale migration from the declining Lanarkshire coalfields to new and redeveloped collieries in central and eastern Scotland. These priorities were in tension with expectations of community stability that contributed to pressure for industrial investment to replace lost coal employment. The Shotts experience was a formative period for the moral economy which conditioned the more sensitive management of redundancy and transfer in future closures.

Tension between the NCB's economic priorities and community cohesion was summed up by the NUMSA president, Abe Moffat, in 1950. When responding to the proposed closure of Baton colliery he stated: 'the Board should realise that they were not discussing a mining engineer's

---

[21] TNA, Coal 89/103/2A, 986, Meeting of the Board, 1969.

[22] TNA, Coal 89/103/28A, J. B. Burton to D. J. Nisbet, Memorandum: Cardowan, 9 March 1971.

[23] B. Bluestone and B. Harrison, *The Deindustrialization of America: Plant Closings, Community Abandonment, and the Dismantling of Basic Industry* (New York, 1982), pp. 19–20.

opinion but the social life of a mining village'.[24] This outlook solidified within the NUMSA during the eastern periphery's contraction, with future closures negotiated with the objective of alternative employment being provided within commuting distance. The NCB's strategy centred on encouraging labour mobility. This was signified when W. Drylie, the Area industrial relations officer, advised 'local men, particularly young men', to seek transfers within Scotland during the closure of Loganlea colliery in 1949.[25]

Board officials were articulating the dominant understanding of employment policy in the terms of the wartime coalition government's 1944 white paper that formed the cornerstone of subsequent practice: a state commitment to 'high and stable' employment underpinned by workforce mobility obligations.[26] However, the Board's migration schemes were undermined by regional policy concerns, while reluctance to leave Shotts also deterred workers from relocating. H. S. Phillips, the Board of Trade's researcher for Scotland, communicated a sense of policymaker social responsibility in August 1948, when he argued that 'the transfer method is only an additional and short-term method of reducing male unemployment'. These conclusions related to social circumstances: 'quite a high proportion of unemployed persons are not prepared, or able, to move more than a short distance'. Phillips argued that the situation in Shotts justified a 'take work to the workers' policy.[27] Two years later, a Board of Trade research team studied Shotts and concluded that nevertheless limited emigration was desirable. Some success had already been attained in attracting light industry and further developments were necessary 'to preserve and balance the community on a smaller scale'.[28] These concerns had entered NCB and civil service practices in the management of closures by the late 1960s. Correspondence referred to 'hard core' pockets of unemployment in areas to be affected by closures in 1967 and 1968. Both NCB and Ministry of Power officials sought solutions to defuse politically sensitive situations in

---

[24] NRS, CB 222/14/1/21A, Notes of proceedings between the Scottish Divisional Coal Board and the National Union of Mineworkers Scottish Area (NUMSA) regarding the proposed closure of Baton colliery held at 58 Palmerston Place, Edinburgh, Monday 8 May 1950.

[25] NRS, CB 295/14/1, Loganlea CCC: minutes of special meeting, 1 Feb. 1949.

[26] M. Roberts, 'Annotated copy of *Employment policy* (1944)', *Margaret Thatcher Foundation* <http://fc95d419f4478b3b6e5f3f71d0fe2b653c4f00f32175760e96e7.r87.cf1.rackcdn.com/2312B65342E04F2B8107131C635023BD.pdf> [accessed 22 Nov. 2019].

[27] NRS, SEP 4/762, H. S. Phillips, Research studies: geographical movement of labour, 9 Aug. 1948.

[28] NRS, SEP 4/762, Research Section Board of Trade (Scotland) note, Geographical movement of labour, 4 Aug. 1950.

Lanarkshire through the provision of employment within travel distances of the homes of redundant miners rather than opt for encouraging migration as they had in earlier years.[29]

Policymakers' assumption of responsibility for providing communities with employment shaped the workings of the moral economy. Their actions were indicative of the double movement's pressures, which were exemplified by Moffat's juxtaposing community concerns with maximizing productivity. Steps to stabilize labour markets in contracting coalfields, and the provision of colliery transfers within travel distance, hampered the NCB's restructuring. The Board's system of trade union consultation, which included advance warning of closure by placing pits in 'jeopardy' status, created obstacles to inter-coalfield migration. This was compounded by tight labour market conditions and regional policy commitments that directed investment towards declining coalfields. Local employment opportunities encouraged young miners to leave the industry when threatened with jeopardy status. This was confirmed by two formal schemes that operated across the Welsh, Scottish and Northern English coalfields. Only 14,974 transfers of miners and former miners took place between 1962 and 1971. By comparison, 678,000 workers left the coal industry between 1957 and 1967, with the bulk concentrated in the transferring areas.[30]

The policy of linked closures and migration to other coalfields was broadly a failure. Movement was even limited among younger men, with many refusing to transfer. There was a high rate of immobility at Shotts pits, while others cited family in gainful employment and unwillingness to leave the community. At Hillhouserigg in 1949, as at Baton the previous year, there was an anticipated low take-up of transfers. In the face of another 105 redundancies within Shotts, the NUM branch president and secretary argued that one third of workers were too old to transfer, while another third 'will be unwilling to uproot family ties' and leave the locality. They noted that 1,050 Shotts men had been made redundant through recent pit closures and raised concerns for the future of the area.[31] This evidence concurs with Heughan's research from Shotts in the late 1940s, which concluded that the 'deep roots' that most of the population had within a 'self-sufficient, independent community' stimulated reluctance to relocate.[32]

---

[29] TNA, POWE 52/305, J. W. Anderson note, 'Kennox colliery', 28 Jan. 1968; Coal 31/130, R. B. Marshall, Departmental Secretary, note to the Ministry of Power, 5 July 1967.

[30] Ashworth, *British Coal*, pp. 260–2.

[31] NRS, CB 321/14/1, W. Moore, secretary and J. N. Watson, president of no. 121 Baton Branch NUM, Central East Area, to the NCB, 26 Feb. 1949.

[32] H. E. Heughan, *Pit Closures at Shotts and the Migration of Miners* (Edinburgh, 1953), pp. 12, 56.

Records from Shotts collieries reveal resistance to inter-coalfield transfers strongly correlated with community ties and an aversion to disruption. Following the closure of Baton in 1950, miners were offered transfers to the Ayrshire, Lothian and Fife coalfields, yet under a fifth of the 226 men judged eligible transferred to these areas. More men either refused offers for transfer (twenty-eight), or transferred and then subsequently returned (fifteen). This may have been conditioned by the lack of consultation at this stage, with managers instructed that at redundancy interviews, 'the workman should be made a firm offer of employment and not asked if he is prepared to consider it'.[33] The low take-up rate came despite the NUMSA's support for coalfield reconstruction. During the late 1940s and early 1950s, the NUM was committed to trying to make the nationalized industry a success and accepted some closures. This was most straightforward in cases where geological rather than economic reasoning was put forward by the NCB. In 1949, James McKendrick, the NUM Area secretary and a long-standing CPGB activist, accepted that, 'they [the trade unions] could not make out a case for the continuation of Chapel mine', as it became apparent that much of its reserves were unworkable due to flooding.[34] At the Area's annual conference in 1948, Abe Moffat referred to the gains that the union had made under nationalization, and stated that these were tied to industrial reorganization:

> We cannot possibly expect to retain the reforms and make further improvements in miners' conditions unless we are prepared to give wholehearted support to modernization and concentration. We have accepted the principle of modernization and concentration with a view to securing the best conditions possible for our members.[35]

This indicates some parallel with John L. Lewis's strategy in the United Mine Workers of America during the 1940s and 1950s. Under Lewis's leadership, the union agreed contracts with large firms that accepted significant closures and major job losses within less productive regions in return for guarantees of security and better pay and conditions.[36] However,

---

[33] NRS, CB 222/14/1, Offers of employment to redundant workmen, 13 June 1950.

[34] NRS, CB 295/14/1/1F, Minutes of meeting of Chapel Mine Consultative Committee meeting held in Bothwell Office 2 Feb. 1949.

[35] National Mining Museum Scotland archives, Newtongrange, Midlothian (NMMS), NUMSA, Minutes of Executive Committee and Special Conferences, 8 July 1946 to 11 June 1947, p. 499.

[36] G. Wilson, '"Our chronic and desperate situation": anthracite communities and the emergence of redevelopment policy in Pennsylvania and the United States, 1945–1965', in *De-industrialization: Social, Cultural and Political Aspects*, ed. Bert Altena and Marcel van

the major distinction was the context of nationalization which promised a greater social restraint on economic pressure in the construction of a more durable industrial order. For Moffat, joint regulation and consultation was fundamental. When these conditions were seen to have been violated, relations between the NUM and NCB soured. This was evident at Broomside, which closed during 1948, the same year as NUMSA leaders urged conference delegates to accept restructuring. Having provided seventy houses for transferees in Fife, the NCB began redundancies before closure was agreed. The NUMSA objected to this at Scottish Divisional level. Moffat argued these developments went against stipulation as closure had not yet been sanctioned by the trade unions.[37]

There were parallels at Baton in 1950, when eighty-six workers were laid off upon closure. Due to the presence of redundancies before the beginning of consultation, the NUM refused to be party to the closure.[38] Baton's closure saw grievances over the closure process combine with concerns over the availability of facework for transferees and the treatment of elderly and injured workers. During the consultation process, Moffat stated that it was 'not sufficient to say that the workmen would receive twenty-six weeks' redundancy pay and then be forgotten about'.[39] Unease over local responses to the rundown of mining in Shotts was apparent in the Scottish Division's worry over 'press reactions' to the closure of Hillhouserigg during 1950.[40] The Board objected to accusations it was determined to shut Shotts pits. Economic priorities justified redundancies and maintained efficiency. Closures were due to financial losses and the greater productivity of manpower if it was transferred elsewhere. However, the NCB later appeared to be more accepting of social arguments. In 1954, when Kingshill 1 was reorganized, and manpower reduced, the Area production manager recognized that there were 'very few jobs in the area' due to the impact of 'fairly heavy redundancies'.[41]

---

der Linden (Cambridge, 2002), pp. 137–58, at pp. 140–1.

[37] NRS, CB 483/24/1/6A, Divisional Consultative Committee point 127, Broomside colliery, 1948.

[38] NRS, CB 222/14/1/47A, A. Moffatt, NUM, Edinburgh to L. E. Bourke, NCB, Edinburgh, 20 Oct. 1948.

[39] NRS, CB 222/14/1/29A, Note of proceedings between the Scottish Divisional Coal Board and the NUMSA regarding the proposed closure of Baton colliery held at 58 Palmerston Place, Edinburgh on Thursday 18 May 1950.

[40] NRS, CB 321/14/1/14a, Divisional Board, closure of Hillhouserigg colliery-item 153 (1949).

[41] NRS, CB 334/19/2/2, Special meeting of the Consultative Committee held on 26 Jan. 1954.

Moral economy concerns gained traction over time as the NCB demonstrated a greater willingness to meet the responsibilities expected of it by the NUM and within mining communities. In 1955, the chair of the Scottish Division, Ronald Parker, acknowledged 'a community life in Shotts which the Board did not want to see disappear'.[42] Kingshill 2 was gradually rundown during the early 1960s and extensive transfers were provided.[43] In attempts to reach profitability, Kingshill 3 was maintained through several loss-making years after 1970. Consultation procedures were abandoned in 1973 before final closure took place in 1974. A renewed, embedded, social democratization of the NCB was visible in the management of Kingshill 3's eventual closure. Local transfers were provided for all miners under fifty-five years of age to relevant positions according to skill levels, while redundancy payments were made to elderly and disabled workers. Transfers were achieved in part through offers of early retirement for men aged sixty-two and above at the receiving pits, Cardowan and Bedlay, which avoided involuntary redundancy.[44]

For the most part, migration was turned down in favour of extended travel-to-work distances, either to larger pits such as Cardowan and Bedlay or to employment in other industries. This was exemplified by the procedure adopted at Kingshill 3, which broadly maintained patterns of residency but enlarged the locale, stretching the moral economy understanding of local employment. Travel distances grew incrementally through waves of closures. Peter Downie's employment trajectory epitomized these changes. He started working in his local village pit of Glentore in Greengairs, North Lanarkshire, during the 1950s, before transferring to the adjacent Gartshore 9/11, and then the somewhat longer commute of around nine miles to Bedlay. Following the closure of Bedlay in 1982, Peter transferred further afield, to Polkemmet in West Lothian, seventeen miles from Greengairs. Peter ended his working life at the Solsgirth mine, part of the Longannet complex in Clackmannanshire, which incurred a daily commute of nearly thirty miles across central Scotland, amounting to a round trip of almost sixty miles.[45] These developments corresponded to a broader Scottish pattern of increased travel-to-work distances and rising car ownership within traditional mining

[42] NRS, CB 410/14/1/57A, Notes of proceedings between representatives of the board and representatives of the NUM at 58 Palmerston Place Edinburgh, Friday 21 Oct. 1955.

[43] NRS, CB 334/19/2, Transport arrangements for workmen transferring to Polkemmet (1963); Kingshill no. 2, 2nd Stage (1963).

[44] NRS, CB 334/19/3, Minutes of a special consultative committee meeting held at Allanton Welfare, 30 May 1974.

[45] Moodiesburn focus group.

villages. Localized community connections and familial patterns at work and home were disrupted by the dispersal this created, as well as the rising time spent on journeys.[46] But the improved pay and security provided by the nationalized industry increased the value of colliery employment. Restructuring also encouraged workforce solidarities within cosmopolitan collieries, which were understood as valuable economic assets paid for at the cost of the closure of smaller older units.

The contested moral economy

The Scottish industrial modernization agenda was most visible in the Lanarkshire coalfield between the mid 1950s and the late 1960s. Coalfield contraction was accompanied by the expansion of assembly manufacturing sectors. In the southern area of the Lanarkshire coalfield, the rundown of mining employment was broadly managed within policymakers' moral economy obligations. Economic closures were still objected to, and the NCB rarely met expectations entirely. As in Lanarkshire's eastern periphery, travel-to-work journey extensions redefined conceptions of what was considered local employment. However, over this period, employment within travel distance was generally provided upon closure. This included the transfer of miners to pits in the Lanarkshire and Ayrshire coalfields, as well as expanding employment opportunities in alternative industries.

The southern area included relatively large population centres such as Lanark and Lesmahagow, as well as smaller pit villages such as Coalburn and Douglas Water, which were dedicated mining settlements. The rundown of employment was more gradual than in the eastern periphery. To some extent, this resulted from the more profitable status of the seams within this area, which was not at the forefront of the initial closures during the late 1940s and early 1950s. The southern extremity of this coalfield acted as a receiving area for miners in the late 1940s. Alex Clark, Jessie's husband, transferred from Larkhall to Douglas Castle colliery in 1948 and was allocated a council house in Rigside.[47] However, the later maintenance of mining was also indicative of its relative geographical isolation. NCB employment was maintained due to the absence of alternatives.

A social commitment to the preservation of employment within areas which overwhelmingly depended on coal mining competed with financial imperatives in NCB decision making. Proposals for development at

---

[46] D. Wight, *Workers Not Wasters: Masculine Respectability, Consumption and Unemployment in Central Scotland: a Community Study* (Edinburgh, 1993), p. 25.

[47] A. Clark, 'Personal experience from a lifetime in the communist and labour movements', *Scottish Labour History Review*, x (1996–7), 9–11, at p. 11.

Douglas Castle colliery through investing just over £460,000 were put forward by the West Central Area in 1957. This would have enabled access to 3.9 million tonnes of proven reserves, and an estimated total of over 8 million, but the suggestion was rejected by Scottish Division officials. The area projected that the investment would increase output from 270 to 700 tonnes per day with 510 men employed at an output per manshift of 27.56 tonnes. Instead, a lower cost investment of around £90,000 was adopted which maintained employment levels at 392 men.[48] The pit's maintenance was partly justified by concern over the fate of the villages around it. Recognizing moral economy imperatives, the Scottish Divisional Board's deputy reconstruction director justified this investment in a loss-making pit by arguing that there was a 'need to retain the colliery from the manpower and social aspect'.[49]

In 1957, a Scottish Division official speculated 'that closure of the [Douglas Castle] colliery would be something of a disaster for the local community at Douglas Village and Douglas West, and a considerable number of dependable miners would probably be lost to the mining industry'.[50] These reservations over closure centred on the nature of the village and the unwillingness to move to the Ayrshire coalfield which paralleled the earlier experience in Shotts. Douglas was described as, 'a contented homogeneous community with strong local attachment and very unlikely to accept, without opposition, any move to close the colliery or any suggestion to move the bulk of the miners to Barony or Killoch'.[51] Commuting to these collieries entailed a daily round trip of around fifty miles from Douglas. There were tensions within management surrounding responsibilities towards areas experiencing colliery job losses. In this case, a sense of moral economy obligation was embedded at the level of the Scottish Divisional Board. The Scottish Division was done away with in a centralizing reorganization ten years later, further empowering London headquarters, which was more socially distant from Scottish collieries.[52]

---

[48] NRS, CB 280/30/1, Douglas Castle colliery: Rankin and Wilson mines reorganisation (1957).

[49] NRS, CB 280/30/1/1A, Thomson, deputy reconstruction director, NCB, Edinburgh, to W. I. Finnie, deputy area production manager, NCB Lugar Works, Cumnock, 6 Dec. 1957.

[50] NRS, CB 280/30/1, Report on Douglas Castle colliery project, Scottish Division, 6 Dec. 1957.

[51] NRS, CB 280/30/1, Report on Douglas Castle colliery project, Scottish Division, 6 Dec. 1957.

[52] R. Halliday, *The Disappearing Scottish Colliery: a Personal View of Some Aspects of Scotland's Coal Industry since Nationalisation* (Edinburgh, 1990), p. 107.

Opposition to closure emerged in the form of the Douglas Defence Committee. The committee wrote to the Conservative minister of fuel and power, Percy Mills, before the colliery's final closure in January 1959. Their letter described the body as having recently headed 'a torch-lit demonstration through the streets of Douglas', before a large meeting was held at the local Miners' Welfare. Protests were bolstered by two key contingencies that shaped the moral economy: the area's dependency upon coal mining, and perceptions the NCB had given a 'clear understanding that the pit had a long life ahead', which had stimulated local authority investment in housing and amenities.[53] Ultimately, 139 men were made redundant, only six of whom were over sixty-five.[54] Protests against the onset of mid twentieth-century closures in single-industry localities with distinct patterns of cultural life also took place elsewhere in Europe and in North America. Occitan mining communities in Decazville, western France, similarly formed local protest committees against the rundown of their collieries and transfer to the more productive coalfields in eastern France, underlining the threat closures posed to communal cultural bonds.[55] Miners in Appalachia formed motorcades that used car lights to create spectacles in darkness, with some parallel to the torch demonstration in Douglas. These protests were aimed at non-union coal companies that threatened the status of swiftly contracting unionized mines during the 1960s.[56]

The NCB felt that contraction would release manpower for other industries. Officials reprised the 1944 white paper's emphases on occupational as well as geographical labour mobility. An internal paper noted it was 'likely that many of the miners would seek work outside the mining industry near to places where the younger members of their families work'. Glasgow, Hamilton and Motherwell were listed as possibilities.[57] The oral testimonies demonstrate that pit closures led to extended commuting distances. Gilbert Dobby served an engineering apprenticeship at Auchlochan 9 in Coalburn. He recalled that after it closed: 'Well there wisnae, ah don't think there was say an awful lot left the village, but they'd to go further for work the likes ae

---

[53] TNA, POWE 37/481/26, J. McCartney and R. Scott, Douglas, to minister of fuel and power, London, 'Douglas defence committee', 4 Jan. 1959.

[54] NRS, CB 280/30/1, Douglas Castle colliery: summary of manpower as at 6/1/59.

[55] D. Reid, *The Miners of Decazeville: a Genealogy of Deindustrialization* (Cambridge, Mass., 1985), pp. 189–201.

[56] A. Portelli, *They Say in Harlan County: an Oral History* (Oxford, 2010), p. 273.

[57] NRS, CB 280/30/1, Report on Douglas Castle colliery project, Scottish Division, 6 Dec. 1957.

Ravenscraig [steelworks] and things like that'.[58] Workers from areas outwith travelling distance of Ayrshire pits experienced redundancies. Geographical remoteness meant that when Auchlochan 9 closed in 1968, NCB officials noted that 'work within daily travelling distance can be offered only to a very few men but jobs will be available to men willing to move'. Only ten of the 340 workers were able to transfer initially.[59] Opposition was greatest in cases where closure took place on economic grounds and moral economy transfer expectations were not fulfilled. At Auldton in 1963, objections were raised to an economic closure without adequate plans for transfer. G. Stobbs, the NUM district secretary, argued that this contravened established customs: 'as alternative employment could not be offered to the men, he felt that production should be continued in all places w[h]ere coal was available'. Workforce objections were heightened because the jobs that were available did not meet the expectation of a transfer to the same grade. Faceworkers, shotfirers and deputies were all affected.[60] Four months later, upon final closure, seventy-one of the mine's relatively small workforce of seventy-four were transferred to other pits, but only at the expense of a very much enlarged travel distance for the vast majority (sixty-four), who went to Killoch.[61]

Extended travel distances stretched the moral economy's conception of local employment and contributed towards mounting objections to the removal of industrial employment without a perceived adequate replacement. Similar demands to those in the Shotts area were raised in relation to the rundown of the pits in the Douglas Water area. These concerns gained some redress from policymakers. The Scottish Economic Planning Council attempted to link closures with vacancies at new sinkings in other coalfields, including transfers to England. However, the need for 'major developments' in advance factory provision in North Lanarkshire was also recognized.[62] This partly reflected pressure emanating from local political organizations. The Douglas Water Cooperative Society responded to the continuation of the closure programme through proposals for industrial developments to take place locally given the likely closure of Auchlocan 9. In a 1965 letter to the Labour MP for Lanark, Judith Hart,

---

[58] Gilbert Dobby, interview with author, Coalburn Miners' Welfare, 11 Feb. 2014.

[59] NRS, CB 210/14/3, NCB colliery closures: Scotland, 21 March 1968.

[60] NRS, CB 210/25/1/12A, Minutes of meeting held at Auldton Mine, 12 March 1963.

[61] NRS, CB 210/25/1/7B, Notice of closure, concentration or reorganisation, 30 Sept. 1963.

[62] NRS, SEP 4/17/56/20, Scottish Economic Planning Council note, Implications of pit closures for economic planning, Dec. 1965.

Cooperative representatives asked, 'why should the [Labour] government not make some effort to take steps to prepare our people in some other job or employment during the transitory period to the closing of the very doubtful life of the colliery?' Referencing the example of East Kilbride, the letter proposed that a new town should be built on the A74 adjacent to Rigside or Lanark. If that could not be granted, 'at least an industrial estate' was judged necessary 'to revitalize the upper ward of Lanarkshire'.[63] This effort was successful, at least indirectly, as Larkhall was later marked for development. Work at Cardowan and Bedlay, around thirty miles to the north, as well as opencast developments that were closer by, maintained employment in the area, albeit within a redefined and enlarged locale.

Concern over falling coal employment grew as the weight of closures and local opposition mounted through the political as well as industrial links of the Scottish coal lobby. Hart stood alongside the NUMSA and NUM's general secretary, Lawrence Daly, and the NUMSA president, Michael McGahey, in opposing the NCB's proposed closure of Kennox during proceedings over 1968 and 1969. Their terms underlined objection to pit closures on economic grounds as the future of the industry was called into question. Incremental divestment's cumulative impact was evident in the union's pronounced opposition to closures and reassertion of the moral economy. The NUM took the Kennox closure to a national appeal in London, which consolidated the sense that the NUMSA were defending national as well as local interests. McGahey argued in terms resonant with the double movement that 'the Board had already upset the social structure of what had been a prolific mining area' through many closures. Daly also emphasized the threat to community cohesion by calling for the pit to be 'kept open until the Government had arranged to introduce measures to alleviate the serious social consequences which would arise from closure'.[64] The NCB conceded extending the colliery's life to allow the installation of chain conveyors. Despite early fears that the improved performance was inadequate and that closure 'appears inevitable', the pit was later withdrawn from jeopardy status.[65] The final closure of Kennox in 1972 related to the

---

[63] NRS, SEP 4/17/56/20, Scottish Economic Planning Council note, Implications of pit closures for economic planning, Dec. 1965; SEP 4/17/56/23, Douglas Water Cooperative Society to J. Hart, 15 Dec. 1965.

[64] NRS, CB 327/14/1/37B, Note of meeting between representatives of the NCB and NUM held at 3.30pm on Wednesday 29 Jan. 1969, in the board Room, Hobart House.

[65] NRS, CB 327/14/1/52A, T. D. M. Scrimgeour, Scottish Southern Area NCB, to W. Kerr Fraser, Regional Development Division, St. Andrew's House, Edinburgh, 1 May 1969; CB 207/14/5/52A, J. M. Trail, Scottish Southern Area NCB, to K. S. Jeffries, NCB, London, 30 Dec. 1970.

erosion of the pit's reserve. Trade unions accepted this without contesting formal jeopardy procedures. This demonstrates miners' understanding of the Board's obligation to secure employment through the continued mining of coal reserves.[66]

The renegotiated moral economy

The moral economy was renegotiated through a series of confrontations between the NCB and the NUM over the late 1960s and early 1970s. Opposition to economic closures became more pronounced within the NUMSA following Michael McGahey's election as area president in 1967. Industrial relations tensions developed in response to the impact of incremental closures, and in opposition to the diminution of miners' pay and conditions. A shift in NCB policy took place during the early 1970s that reasserted the moral economy obligation towards maintaining employment. The analysis in the final sections of this chapter focuses on the management of closures in the northern core, which encompasses the established mining and industrial area of North Lanarkshire. Under nationalization, there had been significant investment in the area, including the redevelopment of Cardowan and Bedlay as well as major expenditure at Gartshore 9/11 and Wester Auchengeich.[67] The availability of transfers mitigated resistance, but there was evidence of grievances related to the coordinated rundown of pits in the area to provide manpower for Bedlay and Cardowan. This contributed to a sense of instability within the industry as perceptions grew that the NCB's promises of economic security were being broken.

Closures in the northern core were predominantly justified on economic grounds. They were contested by NUM representatives who embodied the protective countermovement to the NCB's market logic by asserting the Board's obligation to preserve collieries which the Board had previously stated were economically viable units. The NUM was supported by local politicians. Margaret Herbison, the Labour MP for North Lanarkshire, wrote to Ronald Parker, opposing the closure of Auchengeich in 1965. Her concerns echoed those of the workforce regarding the NCB's promises of economic security given that a colliery ranked as Grade A and considered to have a long-term future in 1962 was now being closed. A further grievance was that men faced a considerable loss of earnings through transfers. Herbison stated she was 'willing to help in any way possible to try to keep this colliery' and called for investment, referring favourably to the

---

[66] NRS, CB 207/14/5/92A, Kennox colliery note, 22 May 1972.

[67] M. Oglethorpe, 'The Scottish coal mining industry since 1945', *Scottish Business and Industrial History*, xxvi (2011), 77–98, at p. 82.

NUM's suggestion to develop the main coal area.[68] Parker's reply sought to placate these concerns by emphasizing that closure took place within moral economy parameters. Only nine men were to be made redundant and pieceworkers would be found suitable positions through transfer.[69]

The NUM responded to the proposed closure of Wester Auchengeich in 1967 by compiling an extensive engineering report arguing for its redevelopment. This demonstrated concern over the mounting termination of what had been earlier understood to be economically viable units. Union engineers concluded that long-term possibilities depended on six months of development work to reach the Cadder area, which could be profitably mined. Their report detailed tension between moral economy obligations and cost control priorities within different levels of management: 'some [unnamed] Board officials' supported further investment, but the Scottish North Area production manager was opposed.[70] Developments at Wester Auchengeich had parallels with the experience of Michael colliery in Fife, which was closed in 1967 following a major fire but, as in Lanarkshire, NCB officials favouring continued operations was ruled out by superiors.[71] These cases confirm the moral economy's basis in localized communities. Support from colliery management were overruled by central investment priorities, displaying the declining autonomy afforded to Scottish NCB operations during the 1960s as the industry felt intensifying cost control pressures.

The connection between closures and industrial relations tension is vividly described in NCB correspondence and the minutes of CCC meetings. Mineworkers' rising frustrations centred on objections relating to the legitimacy of consultation and perceptions that the NCB falsely promised employment stability to transferees. These factors were apparent at the closure of Gartshore 9/11 in 1968. Plans for the closure of the colliery and Wester Auchengeich were published in the NCB's *Coal News* and the Glasgow *Evening Times*. In a letter to the Scottish North Area's director, D. J. Skidmore, McGahey argued that this breached protocol and encouraged animosity between workers and management: 'no information or statements should be issued publicly before the miners and other workmen at the pit

---

[68] NRS, CB 207/24/1/30A, M. Herbison, House of Commons, London to R. Parker, NCB, Edinburgh, subject: proposed closure of Auchengeich Colliery, 6 March 1965.

[69] NRS, CB 207/24/1, R. Parker, NCB, Edinburgh to M. Herbison, House of Commons, London, 19 March 1965.

[70] NRS, CB 207/14/3, L. Johnston, Report on Wester Auchengeich coal reserves (1967).

[71] J. Phillips, 'Deindustrialization and the moral economy of the Scottish coalfields, 1947 to 1991', *International Labor and Working-Class*, History, lxxxiv (2013), 99–115, at p. 106.

are informed of the actual position'.[72] The cumulative effect of closures was apparent in the mistrust towards the NCB which workforce representatives articulated at CCC meetings. At Gartshore 9/11, it was alleged that transfers to the colliery following the closure of Boglea in 1962 had contributed to overmanning the pit and depressed its financial performance.[73]

Previous area management promises of ten and twenty years of economically productive life were mentioned during closure procedures at both Gartshore 9/11 and Wester Auchengeich, indicating that these pits were understood as having been secured through investment that accompanied earlier closures. Doubt had been cast upon the transfer policy by Gartshore men in 1966 with reference to the fact that forty-nine men had recently been moved to Wester Auchengeich, which an NUM representative, reflecting on his belief the pit would soon close, stated had 'no future'. This was borne out when closure proceedings began the following year.[74] Discontent over Gartshore 9/11's closures related to perceptions of the future of the industry in the area. Among sixty men who refused transfer, one miner, commenting on recent closures, stated that 'the pits are finished', while another simply said, 'I've been through too many closures'. Others highlighted the fact that in five years four nearby pits, an NCB workshop, and Wester Auchengeich's coking plant, had been closed.[75] These feelings were collectively articulated at a local level. Kilsyth town council wrote to the chair of the NCB, Alf Robens, to protest at Gartshore 9/11's closure and alleged that the Board was 'deliberately re-deploying large numbers of men into Bedlay to force a closure on economic grounds within three years'. They further protested, in concurrence with the transferees' grievances over grading, that these miners were 'not gainfully employed'.[76] It is notable that grievances were directed towards Robens and not Scottish NCB officials. This was another indication of the centralization associated with the 1967 reorganization, which enhanced the role of the NCB's UK headquarters in scheduling pit closures.[77]

---

[72] NRS, CB 300/14/2, M. McGahey, NUMSA, Edinburgh to D. J. Skidmore, NCB, Alloa, 9 Jan. 1968.

[73] NRS, CB 300/14/2, Minutes of special CCC Meeting held in Grayshill office at 1.30am, on Wednesday 25 Aug. 1966.

[74] NRS, CB 300/14/2, Minutes of special CCC Meeting, Wednesday 25 Aug. 1966; Wester Auchengeich CCC: Minutes of meeting held in manager's office, Wednesday 27 Dec. 1967; CB 207/14/4, Minutes of special CCC at Wester Auchengeich colliery, 9 Jan. 1968.

[75] NRS, CB 300/14/1, Manpower branch, NCB, Glasgow to legal department, NCB, Edinburgh, memorandum: Background notes: what happened at Gartshore 9/11 closure? (1968).

[76] NRS, CB 300/14/1, Manpower branch, NCB, Glasgow to legal department, NCB, Edinburgh (1968).

[77] A. Robens, *Ten Year Stint* (London, 1972), pp. 117–23.

Perceptions of distance between mining communities and the Coal Board's senior management were accentuated by a clash of worldviews. The divide centred on the productionist priorities of the NCB and the moral economy expectations of mining communities, which limited the extent to which the generation of miners represented by McGahey and Daly viewed nationalization as a social advancement. Tommy Canavan, who was an NUM representative at Cardowan, recalled McGahey claiming that the NCB had continued with the priorities and personnel of the private industry, stating that 'the management just changed their jerseys. They went fae the Bairds, Scottish Steel and aw these different private companies and became managers'. However, Canavan qualified this by claiming the NCB 'worked two ways', and that despite its economic priorities it demonstrated a social understanding and commitment to mining communities at a local level through its support for community activities at Miners' Welfares.[78] Willie Doolan, who was an active communist and trade unionist at Cardowan, described the 'hated ... hierarchy' of the NCB as 'enemies of the working class' who retained a privileged and distant position. Nonetheless, their activities were constrained by public ownership, which perhaps reflected some colliery managers' commitment to moral economy obligations: 'I'm not trying to paint the picture that all colliery managers were vicious towards the miners because that wasn't the case. They were under instructions obviously, the National Coal Board 1947'.[79]

There was a significant turning point in coalfield contraction during the late 1960s and early 1970s, which coincided with the diminishment of alternative employment opportunities. Awareness of economic insecurity was increased by the publication of the government's *Fuel Policy* white paper in November 1967. This forecast that British coal usage would decline from around 175 million tonnes in 1966 to 120 million tonnes by 1975, and that employment in the industry would decline at an even faster rate. Job losses would be highest in 'peripheral' coalfields, including Scotland.[80] In Lanarkshire's northern core, these projections correlated with local experience, where miners were asked to accept pit closures in return for transfers. The NCB gave assurances of security on each occasion, yet planning documents suggest an awareness of future closures. This extended to the possible closure of Bedlay and Cardowan soon after the

---

[78] Tommy Canavan, interview with author, residence, Kilsyth, 19 Feb. 2014.

[79] Willie Doolan, interview with author, The Pivot Community Centre, Moodiesburn, 12 March 2014.

[80] Phillips, 'Deindustrialization', pp. 107–8; *Fuel Policy* (Parl. Papers 1967 [Cmnd. 3438]), pp. 27–32.

remaining pits in the northern core had been closed.[81] The fact these closures did not take place is indicative of both the strength of workforce and community opposition, but also reflects market changes which benefited these collieries' positions.[82] A more favourable position for coal in relation to power generation was provided by the quadrupling of oil prices over 1973–4, following the beginning of the Arab-Israeli war. This created political uncertainty over the viability of supplies.[83] Grievances over wages and closures, which culminated in the industrial action of the late 1960s and early 1970s, were underpinned by coal's strengthened energy market position.

Rising tensions within the mining industry slowed contraction. Large unofficial strikes in 1969 and 1970 led to official action over wages during 1972 and 1974.[84] K. S. Jeffries, the NCB's secretary in London, sent a letter to area officials in December 1971 specifying that 'in view of the fluid industrial relations situation', jeopardy meetings were to be suspended. The instruction applied to economic closures, with 'exhaustion closures' to be resolved through contact with headquarters.[85] Richard Hyman concluded in 1974 that 'coal miners have been able to argue that their work is dangerous and unpleasant and also that it has a new and strategic importance in economic life, as justification for substantial improvement in the incomes hierarchy'.[86] A focus group of retired Lanarkshire miners found that this perspective resonated with them. In 2014, the participants remembered that five decades earlier they had compared their conditions to men who worked in the Rootes car factory at Linwood, which opened in 1963 and was subsequently taken over by Chrysler. Peter Downie recalled that thousands of workers left coal mining for such opportunities and that while following Rangers Football Club, he had often met men who had done so:

---

[81] NRS, CB 256/33/2, Cardowan colliery: extract from note of meeting between NCB and NUM in London, 13 Nov. 1969; W. Rowell, director NCB Scottish North Area, Alloa, to G. Teel, agent, NUM, Edinburgh, 10 June 1971.

[82] NMMS, FC/3/2/3/2, National appeal meeting, Cardowan colliery, 16 Aug. 1983, Appendix 1: special extended CCC meeting held in the Parochial Hall, Stepps, Friday 13 May 1983, p. 9.

[83] C. R. Schenk, *International Economic Relations Since 1945* (London, 2011), p. 56.

[84] A. Taylor, *The NUM and British Politics*, ii: *1969–1995* (Aldershot, 2006), pp. 31–6.

[85] NRS, CB 207/14/5/67A, K. S. Jefferies, secretary, NCB, London to Area officials, Colliery closures, 22 Dec. 1971.

[86] R. Hyman, 'Inequality, ideology and industrial relations', *British Journal of Industrial Relations*, ii (1974), 171–91, at p. 189.

> We were standin at Ibrox in 1966 and '68. Linwood was open. Linwood. The Imp. The making of the Imp motorcar. We were goin from our village in Greengairs in to watch the Rangers playing football and we're standing with boys that were working in Linwood, working in the motor factory. And they were coming home with thirty-five pound, and our wages was twenty-two fifty. That was the difference for leaving the pit to go to industry.[87]

The relative decline of miners' wages vis-à-vis those of other manual workers was a major component of the discontent which fuelled miners' industrial action in the early 1970s. Linwood played a seminal role in this. Chrysler resolved a dispute in February 1972, while the miners were on strike, by granting a 14 per cent annual pay rise. This took basic weekly wages to thirty-seven pounds, which was nine pounds more than the NUM's basic claim.[88] Reflecting on this disparity, the 1972 and 1974 strikes were recalled in terms of restoring economic justice, which accorded the miners their strategic industrial status and demanding physical labour deserved:

> Billy Maxwell: There were two strikes, '72 and '74 we won both of them. I think the power-loading wage went up.
>
> Peter Downie: It was thirty-five, up to thirty-five Billy. But we'd nothing before that.
>
> Billy Maxwell: And then we got a good wage in 1974. We won that second strike in 1974 and it was like a good wage. We'd beat the government at that time. Heath had to chuck it up.[89]

A settlement was reached in the 1974 dispute through the incoming Labour government's *Plan for Coal*. It 'held out the prospect of a stable environment' by cementing a place for coal within a mixed energy policy, overturning the position of the 1967 white paper. This included a pledge to secure employment through a long-term commitment to new sinking extending into the 1980s.[90] Colliery closures resultantly slowed down. The experience of energy market fluctuations between the 1960s and 1970s, especially accelerated closures due to the displacement of coal by oil, had a long-term impact on the NUMSA's perspective. It reinforced a rejection of liberalized market approaches and reinforced an understanding based on

---

[87] Moodiesburn focus group.

[88] J. Phillips, *The Industrial Politics of Devolution: Scotland in the 1960s and 1970s* (Manchester, 2008), p. 123.

[89] Moodiesburn focus group.

[90] TNA, Coal 31/159, Department of Energy note, Energy policy framework and forecasts, July 1974.

an energy policy approach that utilized national resources endowments to sustain workforce and community integrity. In 1979, McGahey moved a resolution at the Scottish Trades Union Congress (STUC) conference that praised the work of the Labour government in committing to expanding coal production and employment and making moves towards a more planned energy policy. However, he also warned of the experience of the recent past, reflecting on how reliance on imported oil had dislocated the UK economy and contributed to inflation:

> I remember in this Congress many years ago when my predecessors – Abe and Alex Moffat – warned in the 1950s and 60s that this country would pay a heavy price for the contraction of the coal industry, that oil would not always be plentiful and not always cheap. We have paid that price.[91]

The moral economy under threat

Deep coal mining ceased in the northern core during the early 1980s. The striking differences in the responses to the closure of Bedlay in 1982 and Cardowan in 1983 were due to the broad adherence to the moral economy at Bedlay and a clear transgression of its customs by the NCB at Cardowan. Bedlay was the last moral economy closure in Scotland, with all those that followed being marked by managerial hostility to union consultation and workforce opposition. The colliery had traditions of collaboration between managers and anti-communist trade unionists. Cardowan contrasted with Bedlay. It was a large cosmopolitan colliery and was a stronghold of politicized left-wing trade unionism.[92] Descriptions of these distinct ideological alignments were present in the oral history interviews. For instance, Pat Egan, an NUM youth delegate at Bedlay, recalled:

> Cardowan was always quite a militant pit. Bedlay wisnae, and Bedlay was run by *pause* Cardowan's mainly a communist pit and Bedlay a lot a Catholic groups, the Knights of St. Columba, all these kindae organizations. It wis probably Knights of St. Columba. They used tae say if you wanted overtime at Bedlay go for a pint at the Knights on Saturday night or a Friday night and you'd ask 'how much is that?' Most ae the management were all in the Knights of St. Columba or the Masonic Lodge ... Union and management wis pretty much what would be termed right wing noo. Cardowan was always left wing.[93]

[91] STUC, *Annual Report 1978–1979*, lxxxii (1979), 500–2.

[92] A. Perchard, *The Mine Management Professions in the Twentieth-Century Scottish Coal Mining Industry* (New York, 2007), p. 373; J. McIlroy and A. Campbell, 'Beyond Betteshanger: Order 1305 in the Scottish coalfields during the Second World War part 2: the Cardowan story', *Historical Studies in Industrial Relations*, xvi (2003), 39–80, at pp. 46, 66.

[93] Pat Egan, interview.

These distinctions were not the fundamental cause of the differing responses to closure. It was the difference in the treatment of closure by the NCB, through their relative adherence to the moral economy at Bedlay and clear breach of it at Cardowan, which was fundamental in determining the stance taken by the NUM and within the communities affected. In the case of Bedlay, the closure was less controversial as it took place on geological rather than economic grounds. Extensive consultation and discussions with all unions were spread over 'several months,' while a 'joint examination of all possible areas of reserves' took place with the involvement of the union's mining engineers. Closure was agreed due to 'insurmountable geological conditions'.[94]

After consultation was complete, transfers were arranged to 'neighbouring collieries'. This stretched the understanding of local employment further than previous Lanarkshire transfers. Receiving collieries included Polkemmet and extended across central Scotland to Fife pits. However, the 640 men from Bedlay were absorbed by 200 redundancies, 440 transfers and 100 redundancies within other collieries. Although the NCB undertook these costs, its accountants objected to a net loss of £3.3 million through redundancy, transfer and pension payments. They were most concerned by taking on 340 extra workers without increasing production and prescribed a solution of an additional 340 redundancies.[95] This attitude indicates that commitments to further development through the *Plan for Coal*, upon which the renegotiated moral economy rested, were increasingly being undermined by financial priorities. The accountants' perspective was shaped by the Coal Industry Act 1980, introduced by the first Thatcher government, which projected an end to subsidy by 1984 and effectively scrapped the long-term investment framework of the *Plan for Coal*. The legislation was 'drawn so as to make it almost impossible to operate the industry in the way it had been operating until then'.[96] These major changes to the political economy of energy policy disembedded coal employment from communitarian routines and asserted the primacy of financial priorities.

Cardowan's closure originated in the assertion of market logic across the nationalized industries through British Steel's use of coal imports

---

[94] NRS, CB 223/14/3, P. M. Moullin, NCB, London, to P. McPake, Bedlay, Glenboig, 30 Nov. 1981.

[95] NRS, CB 223/14/3, Area chief accountant to area director, memorandum: Closure of Bedlay, 4 Nov. 1981.

[96] Ashworth, *British Coal*, p. 352.

to supply Ravenscraig.[97] Nicky Wilson was an electrician and acted as a Scottish Colliery, Enginemen, Boilermen and Tradesmen's Association (SCEBTA) delegate at Cardowan. In 2014, he remained angry at the loss of the Ravenscraig market for the pit's 'high grade coking coal'. The NCB were then obliged to 'mix it wi a lotta rubbish' for lower value power station use.[98] Cardowan's closure transgressed the moral economy's expectations of consultation with trade union representatives, while financial compensation was retained in return for the acceptance of closure, transfer and redundancy. Albert Wheeler, the NCB's Scottish Area director, made this clear at a meeting during 1983 when he stated that 'he wanted the opinion of the 1,090 men employed at the colliery and not just the few who attended branch meetings'. In place of negotiation with trade union representatives, he made an 'offer' to individual workers. Wheeler enticed miners to eschew collectivist imperatives with redundancy payments that incorporated a lump sum payment of up to £20,000 and pensions of up to £100 a week for men over the age of fifty. Protected earnings and transfer allowances were promised to younger workers.[99]

The NUM's principal objections related to the NCB taking steps to close the pit before any intimation of closure procedure was made towards trade unions. By the time of the trade union's appeal, which saw the NUM supported by the National Association of Colliery Overmen, Deputies and Shotfirers in opposing closure, 300 men had already left Cardowan. Both Michael McGahey and Arthur Scargill, the NUM's president, argued this breached procedure. Transfers were used to undermine workforce solidarity and collective agreements across the Scottish coalfields, leading to far greater intra-workforce disputes than the tensions unearthed at previous transfers over access to coalface positions. Grievances centred on the undermining of joint regulation, with strikes following the entrance of unnegotiated transfers from Cardowan to Pokemmet in West Lothian and Frances and Bogside in Fife.[100] The most serious moral economy transgression took place at Polmaise in Stirlingshire. Cardowan men were transferred in June 1983, while the pit was undergoing reconstruction. This breached promises to local miners, who had been assured of first refusal on employment at

---

[97] Ashworth, *British Coal*, p. 401.

[98] Nicky Wilson, interview with author, John Macintyre Building, University of Glasgow, 10 Feb. 2014.

[99] NMMS, FC/3/2/3/2, National appeal meeting, Cardowan colliery, 16 Aug. 1983, Appendix 1: special extended CCC meeting held in the Parochial Hall, Stepps.

[100] NMMS, FC/3/2/3/2, Background brief for national appeal meeting on Cardowan colliery, 1983.

the redeveloped pit, and led to the NUM branch pursuing a policy of non-cooperation with unnegotiated transferees. These developments contributed to a lockout at the pit.[101]

The tendency for collective memories of social conflict to involve a 'cluster of tales, symbols, legends and imaginary reconstructions' is evident in recollections of Cardowan's closure where boundaries between singular events are porous.[102] Willie Doolan recollected the closure as an anticipation of the divisive policies deployed by the NCB during the 1984–5 miners' strike:

> The Coal Board were offering, and I'm going back to 1983. They were offering fifteen-hundred pounds to an individual to transfer tae another pit … And it wis the same kindae tactic the Coal Board used during the miners' strike, '84–85. They were offering vast amounts of money to people who'd been out on strike for the best part of a year … All taxpayers' money by the way![103]

In August 1983, the workforce at Cardowan rejected industrial action against the closure. Just under 40 per cent voted to strike following several months of demoralization as the NCB disregarded consultation with the cooperation of a minority within the workforce who accepted transfers and redundancy.[104] Memories of the build-up to the final defeat of the NUM's appeal the following month emphasized that Wheeler's methods of management diverged from the nationalized industry's traditions. Tommy Canavan recollected that when Wheeler visited Cardowan to explain the closure to the workforce he was met with physical violence. The confrontation centred on the workforce's and community's understanding of the colliery as a communal resource and source of employment as opposed to the property of senior management to be disposed of according to the logic of profitability:

> Wheeler came to Cardowan. Previous to that Kinneil [in West Lothian] went on strike [against closure]. They had a sit-in doon the pit. Cardowan was a gas tank it wis full aw methane gas … He was aw 'this is a gas tank we will no be responsible if anything happens in this pit or anything'. And he went on and

---

[101] T. Brotherstone, 'Energy workers against Thatcherite neoliberalism: Scottish coal miners and North Sea offshore workers: revisiting the class struggle in the UK in the 1980s', *International Journal on Strikes and Social Conflict*, i (2013), 135–54, at p. 144.

[102] A. Portelli, *The Death of Luigi Trastulli and Other Stories: Form and Meaning in Oral History* (Albany, N.Y., 1991), pp. 1–2.

[103] Willie Doolan, interview 2014.

[104] TNA, Coal 89/103, Bill Magee, press release: Cardowan ballot results, NCB press office, Edinburgh, 26 Aug. 1983.

on. And it finished up, which wasnae a nice thing tae see. One of the guys went up and knocked him up. He just went up and banjoed him. I'm no sayin that was right, but he brought it upon himself cause he was nothin but a bully. He was sayin 'yous are wearin ma uniforms', a boiler suit! As if we were some he was the commander of some big army or something. It's ma, his uniform! It was his pit! It wisnae oor pit, it was his!'[105]

Nicky Wilson described opposition to Wheeler's visit in terms of a mobilization from across the settlements which depended upon employment at Cardowan: 'you had men in fae various groups ae workforce, and aw the families from Cardowan and roond aboot came. And he couldnae get oot'. Wheeler was unable to leave using the main exit and was eventually escorted from the premises by police. The protestors physically blocking Wheeler's exit symbolized a rejection of the NCB's financial priorities in favour of a moral economy claim on the colliery and the employment it provided. Both testimonies were strongly shaped by the requirements of 'composure': the telling of a congruent life-story, which is filtered by contemporary social and political sensibilities.[106] Tommy Canavan, articulating himself through the coalfield cultural circuit's emphasis on the legacy of struggle, underlined the militancy of the Cardowan workforce in their opposition to Wheeler. Nicky Wilson, perhaps reflecting his role as the NUM's Scottish president, argued that opposition to Wheeler was more comical than violent: 'Somebody stuck an ice cream on his head right enough, that's aboot as much damage as he got!' It was Wheeler's intransigence and arrogance in lecturing the workforce and their families, and his insistence upon attempting to use the main exit, which was behind the incident.[107]

The acrimony over Cardowan's closure was focused on rising unemployment and accelerating deindustrialization. Pat Egan remembered that in the early 1980s there was a major rundown of industrial employment around his home in Twechar, North Lanarkshire: two foundries, a brickworks, Bedlay and Cardowan closed. There were also redundancies at the Burroughs electronics factory in Cumbernauld, which finally shut in 1987. By the mid 1980s, 'that was basically it for big employers in the locality'.[108] The incremental growth of unemployment and the lack of opportunities for work in other industries increased opposition to closure.

---

[105] Tommy Canavan, interview. Kinneil colliery in West Lothian was closed in Dec. 1982. The closure was unsuccessfully opposed through a stay-down strike.

[106] P. Summerfield, 'Culture and composure: creating narratives of the gendered self in oral history interviews', *Cultural and Social History*, i (2004), 65–93, at pp. 74, 81.

[107] Nicky Wilson, interview.

[108] Pat Egan, interview.

McGahey drew attention to the fact that Cardowan was the last colliery in Lanarkshire at the pit's national appeal. Moral economy objections to closures were further bolstered by the rejection of the uneconomic status of the colliery: Cardowan was not a 'clapped out pit'. Unlike Bedlay, Cardowan had large workable reserves with eight years of immediately accessible coal, and long-term development prospects for thirty-five years more work. J. Varley, a representative from the NUM's white-collar federate, the Colliery Officials and Staff Association (COSA), starkly commented that it was 'ludicrous that such large reserves should be sterilised for purely political reasons'.[109]

The view that Cardowan's closure was willed by the Conservative government and their agents within the NCB was strengthened by the treatment of the pit's manager, John Frame. Social obligations were felt more keenly by the lower rungs of management who were often embedded with their workforce and communal life in coalfield settlements. This led to them being more conducive to moral economy customs. Cost control imperatives from officials at area and headquarters level became irreconcilable with commitments to the workforce in the context of the NCB's abandonment of the moral economy. Market logic and aggressive anti-trade union policies were imposed by new figures, including Ian MacGregor and Albert Wheeler, while officials such as John Frame who adhered to moral economy principles were removed.[110] In the view of SCEBTA's president, Abe Moffat, the son of the NUM president with the same name, Frame was 'virtually forced' to retire, and 'was leaving the industry because he disagreed with the closure and knew that the miners had a good case'.[111] During 2014, Willie Doolan recalled Frame, who he described as a 'devout Christian[,] … telling the union privately, "fight for the retention ae your pit because you have millions of tonnes of reserves of coal there, you have a bright future, good coal"'.[112] In similar terms, Tommy Canavan stated, 'Even the manager knew the pit shouldnae shut. But it was laid doon by the Coal Board, by Wheeler, MacGregor, the whole lot ae them'.[113] In their determined pursuit of closure, the NCB ceased any

---

[109] NRS, CB 256/14/1, memorandum: Cardowan colliery national appeal meeting minutes, 19 Sept. 1983.

[110] NRS, CB 256/33/2, Cardowan colliery: extract from note of meeting between NCB and NUM in London, 13 Nov. 1969.

[111] TNA, Coal 89/103, M. Gostwick, 'Miners win unanimous backing', *Morning Star*, 26 May 1983.

[112] Willie Doolan, interview 2014.

[113] Tommy Canavan, interview.

e of adherence to moral economy responsibilities. Rather than tion and a negotiated procedure of closure and subsequent transfer, under Wheeler, the Board's Scottish management sought to divide the workforce and break community opposition.

The closure of Polkemmet in 1985 confirmed that the NCB was wilfully disposing of valuable assets and employment. Both Seumas Milne's investigative journalism and oral testimonies from managers within the NCB's Scottish Area have confirmed that Albert Wheeler was personally responsible for ordering that managerial staff desist from pumping and maintenance activities during the 1984–5 strike. This resulted in the pit flooding. Final closure followed shortly after the dispute finished. A crisis had been created by a return-to-work effort that involved the NCB collaborating with police in enrolling six of the pit's workforce to cross picket lines. The NUM resultantly withdrew safety cover.[114] Polkemmet's workforce included transferees from Bedlay who commuted from Lanarkshire to West Lothian. These events were recollected as a clear transgression of the moral economy within several testimonies. Gilbert Dobby, who worked as an engineer at the pit, stated it 'wis closed wi a lie. It wis supposed to be flooded durin the miners' strike. Now, ah cannae say it wisnae flooded, because it wis flooded. But it wisnae flooded'. He explained that although on strike engineers continued to provide safety cover to maintain the pits intact but that:

> One day we hears the pits flooded. And we couldnae understand why it was flooded. But it was. You could put it as they 'persuaded' people. I cannae say any more than that cause I don't know one a hundred per cent, but I'm ninety-nine per cent sure I cannae say but they were persuaded tae go back. Now the electrical work the manager ae the pit y'know, the manager, he couldnae interfere personally, physically, personally wi that. Okay. So, they managed tae get, from what I understand, doon the pit tae the pit bottom. Then there was a slight incline, you maybe walk forty yards, fifty yards, on a very slight uphill incline cause that'll be the very first coal tae work on. Other seams were lower and it wis a more steep downward trend tae get tae where the other coal seams were. Now some o these coalfaces had a lot ae water in them, so the water got pumped tae the pit bottom. There wis a big pit reservoir built and water got pumped in there. You'd a huge pump at the pit bottom that pumped the water intae there. As I say, they managed to persuade this electrician back tae his work by what means I don't know *laughs* And he wis told 'switch that pump off, the pit bottom pump', so he did what he wis told, switched it off. So, the

---

[114] S. Milne, *The Enemy Within: the Secret War against the Miners* (London, 2004), pp. 319–21; A. Perchard and J. Phillips, 'Transgressing the moral economy: Wheelerism and management of the nationalised coal industry in Scotland', *Contemporary British History*, xxv (2011), 387–405, at p. 398.

> water's no gettin pumped oot from the pit bottom oot the pit but the water's still getting pumped from the lower workings intae there. So, yes, there was a flood in the pit bottom cause I wis doon the pit after it. I spoke to the mines rescue, there wis a guy in it that I worked beside, and he worked beside my father as well, who was in the mines rescue, and they went doon to check things oot because ae this. And, I wis talkin tae him and I says 'how bad is it?', 'aw, three weeks' he says 'it'll be back in full production'. I says, 'so it's no flooded?'. 'Naw,' he says, 'a little bit of water in the facelines but they're no bad cause the pump's been pumpin the water oot anyway'. But the water had only gone so far up the incline at the pit bottom it didnae go over and run back where it'd come fae. And that wis the excuse they used for closin the pit.[115]

Similar allegations of intentional damage were put forward by Peter Downie. He explained that an expensive development, including assets valued at an estimated £300 million, suddenly ended with no adequate explanation. The equipment was in his view intentionally left at the pit which was deliberately flooded by management:[116]

> We had a development working and it was within a hundred feet of being complete. And there were miles and miles of coal going away up by Forth and up the Forth Hills and in tae, which all went opencast eventually. There were big machines set doon that pit before the strike and they never turned a wheel. Never turned a wheel. And they know I'm telling the truth. I was there when I seen them going in. The machines that were costin millions of pound to go in to production. The washer they had in Polkemmet was outdated. It was completely gutted out, renewed. And the strike came on. It never turned a wheel. It never turned a wheel. But they continued, the man that was daein the installation. He continued workin during the strike when we were idle. We're fightin for conditions in the pit and this firm's coming in and getting millions of pounds! For engineering, keeping engineering going, we had men that could have showed them how to build machines and put machines in. Cost millions. And when the pit shut they just did like that and shut the pit. They just said, 'it's finished, it's finished'.[117]

Peter Downie's memories emphasized the growing involvement of private contractors within the nationalized industry which expanded markedly during the 1980s. Their profitmaking undermined the solidity of strike action and constituted a waste of resources by the NCB. His analysis of the closure rested on the broader process of deindustrialization, the decline of

---

[115] Gilbert Dobby, interview.

[116] J. Phillips, *Collieries, Communities and the Miners' Strike in Scotland, 1984–5* (Manchester, 2012), p. 153.

[117] Moodiesburn focus group.

the steel industry's demand for coking coal, as well as the greater assertion of market principles within it. As with Cardowan, Polkemmet had previously supplied Ravenscraig, and similarly its economic viability was hampered by the shift towards imported coal.[118] The accelerated closures of modernized pits confirmed the NCB's abandonment of the moral economy. During and after the miners' strike, the competing dimensions of the double movement clashed over the governance of the industry. The NCB's market logic overcame the NUM's protective countermovement. A decisive victory for the Board resolved the contested ownership of collieries between the Board and coalfield communities: they were the property of NCB management to be dispensed with according to financial priorities.

The nationalized coal industry was subject to the double movement's competing logics of financial performance and productionist pressures on the one hand and the workforce and coalfield community's claims to colliery employment on the other. Moral economy customs evolved as methods by which closures could be legitimated through dialogue with trade union representatives and the provision of collective economic security, usually via job transfers. Closures in Lanarkshire's eastern periphery during the late 1940s and early 1950s were formative to the moral economy's customs. These experiences reinforced the importance attached to maintaining the integrity of small-scale territorial communities in the offer of job transfers at future closures. During the 1950s, 1960s and 1970s miners affected by closures were generally not expected to uproot themselves and their families to retain colliery employment. But as demonstrated by the transfers that followed closure in the southern area, and later at Bedlay, this was only maintained through greatly extended commuting distances which loosened relationships between localities and workplaces. Moral economy expectations were not static, and the operation of colliery closures was continually contested and renegotiated. The incremental logic of closures in Lanarkshire's northern core contributed to a growing sense of economic instability among miners, which stimulated industrial action during the early 1970s. Although nominally fought over wages, these disputes were attached to a broader agenda for dignified and secure employment that was partially realized through the *Plan for Coal*'s guarantee of investment and comparative employment stability. During the 1980s, the Board repudiated its moral economy obligations to pursue a pronounced assertion of managerial prerogative instead of dialogue and the logic of financial costs instead of industry planning. Objections to colliery closures in the

---

[118] S. Fothergill, 'The new alliance of mining areas', in *Restructuring the Local Economy*, ed. M. Geddes and J. Benington (Exeter, 1992), pp. 51–77, at p. 57.

lead up to the 1984–5 strike, and afterwards, coalesced around the radical restructuring of industry governance from dialogue with trade unions towards individualized 'offers' to workers affected by closures. Managerial transgression of moral economy sensibilities was heightened by the sense that the Board was wilfully disposing of valuable public assets which had been paid for at the cost of earlier closures.

3. Communities: 'it was pretty good' in restructured locales

Community is a central feature within cultural representations of British coal mining and deindustrialization. Prominent cinematic examples of this include the depiction in *Billy Elliot* of the hardship caused by the 1984–5 miners' strike in a Durham pit village and the story of a Yorkshire colliery band's struggles through the last major wave of pit closures during the early 1990s dramatized in *Brassed Off*.[1] Twentieth-century scholarship similarly emphasized how occupational identities were consolidated through the strong homogeneity of mining settlements. The 1956 *Coal is Our Life* study of 'Ashton' (Featherstone) in West Yorkshire detailed the close link between work and social life in a village which continued to serve a nineteenth-century colliery.[2] Stuart MacIntyre's later historical account of interwar radicalism in West Fife and the Rhondda Valley similarly focused on single-industry localities.[3] Despite the character of these depictions, the Scottish coalfields experienced extensive reconstruction during the second half of the twentieth century that significantly reshaped relationships between work and residence. Localized communities were reconstructed through the building of public housing to replace miners' rows, which were demolished in slum clearance efforts. The concentration of production in larger 'super pits' and increasing commuting distances discussed in the last chapter loosened traditional ties between mining settlements and collieries.

This chapter utilizes oral testimonies, primarily drawn from the Lanarkshire coalfield, to understand the meanings attached to the long experience of coalfield reconstruction. It emphasizes that the phase of reconstruction between the 1940s and 1960s involved the loss of older bonds but also included significant material improvements and the making of new industrial communities. Accelerated deindustrialization and private housebuilding from the 1980s was more dislocating. It disrupted working-

---

[1] J. Arnold, '"Like being on death row": Britain and the end of coal *c*. 1970 to the present', *Contemporary British History*, xxxii (2018), 1–32, at pp. 8–9.

[2] N. Dennis, F. Henriques and C. Slaughter, *Coal is Our Life: an Analysis of a Yorkshire Mining Community* (London, 1956).

[3] S. Macintyre, *Little Moscows: Communism and Working-Class Militancy in Inter-War Britain* (London, 1980).

'Communities: 'it was pretty good' in restructured locales', in E. Gibbs, *Coal Country: The Meaning and Memory of Deindustrialization in Postwar Scotland* (London, 2021), pp. 91–118. License: CC-BY-NC-ND 4.0.

class consciousness constructed around shared investments in occupational identities and public housing tenures. The testimonies are appraised through the lens of critical nostalgia. They demonstrate criticisms of the social conservatism that often characterized coalfield communities during the mid and late twentieth century, while also noting the tangible and intangible cost of deindustrialization. Brendan Moohan summarized this by explaining that 'it was pretty good' growing up within the associational life of the mining industry, and that its social infrastructure provided fulfilment for himself and his family in the Lothians between the 1960s and early 1980s.[4] Finally, the chapter investigates religious sectarianism, demonstrating both the reconstructive value and reflective nature of oral history. The testimonies reveal that ethno-religious identities remained a diminished but nevertheless present source of political and social division between the 1940s and 1990s.

Scottish coalfield communities developed around towns and villages with a historical basis in the mining industry. Community structures strongly intersected with class solidarities which were founded on a shared attachment to male manual industrial employment and public housing tenancies which typically characterized Scottish coalfield settlements between the 1940s and 1980s.[5] Multiple identities have evolved and endured in the coalfields through complex interactions. Different elements of identity relating to locale, class and ethno-religious allegiances, have been emphasized in different contexts and in response to varying social and economic pressures. Institutions, leisure activities and labour movement political affiliations were instrumental in shaping the cultural 'mediation' of social relations and the formation of working-class consciousness in the coalfields.[6] Expressions of class-consciousness developed in spaces that furnished active associational lives: the workplace, public space and social institutions such as Miners' Welfares. Deindustrialization was not only a threat to industrial workers' employment, it also undermined identities invested in occupation and place. These experiences also shaped national allegiances. Alan Little's broadcast ahead of the 2014 referendum on Scottish independence empathetically described growing support for devolution and then independence within formerly industrial areas of Scotland. He recalled that four decades earlier, 'the British state dug coal. It milled steel. It built ships'. British identities had dissipated through privatization and the

---

[4] Brendan Moohan, interview with author, residence, Livingston, 5 Feb. 2015.
[5] J. Phillips, *Scottish Coal Miners in the Twentieth Century* (Edinburgh, 2019), p. 62.
[6] R. Williams, *Culture and Materialism: Selected Essays* (London, 2005), pp. 32–4.

long-term effects of workplace closures.[7] Links to both Scottish and British national identities were embedded by connections to nationalized industries as well as through the labour and trade union movement. These included the distinct Scottish identity of the NUMSA but also its integration with unitary British structures.

The oral testimonies indicate a strong symbiosis between class and community. Dialogue was structured around questioning what it meant to be part of a mining community. Connections formed around workplaces, neighbourhoods, and collective social activities were highlighted. Examples of both the closeness of community bonds constructed on this basis, and their exclusivity through barriers to inclusion, were vividly discussed by the interviewees decades later. Michael McMahon was raised in Newarthill, North Lanarkshire, where he worked alongside his uncle and father in the American-owned Terex heavy vehicle-manufacturing factory. Both of Michael's grandfathers had been miners. His response to being asked the character of Newarthill was to emphasize its homogeneity and comment that 'They were very working class'.[8] The association between class and community is confirmed in a negative sense by Mary Spence's experience. Mary's narrative confirms that social boundaries limited community involvement despite geographical proximity and familial connections. She returned to Bantyre, South Lanarkshire, in 1959, aged twelve, with her father, who was from a mining background but had entered the civil service and risen to a relatively senior rank. Mary described the gulf which this created between her father and his family, who lived in impoverished conditions owing to her grandmother having to care for both her grandfather and uncle, who had been disabled by work in coal mining:

> The people, my grandparents and aunt, they were so side-lined. They were like people on an island. They were away from the centre of things. They were no longer part of a community to the same extent with immediate neighbours, you could feel the isolation … We lived from '61 onwards in Blantyre, round the corner from them, round the corner, round another corner, about a couple of hundred yards away. My grandmother was able-bodied. Never visited us. She was never nasty to us. But my father thought he was going to move back into bosom with his family. He was wrong. It didn't happen. There was a barrier … They were in absolute poverty. Very unfair. And we had a very nice house. They were still living in their council house. And my grandmother was very proud. And she thought, well this is just me speculating: I've done my job I've brought

---

[7] A. Little, A. 'Scotland's decision', *BBC News*, 4 Sept. 2014 <https://www.bbc.co.uk/news/special/2014/newsspec_8699/> [accessed 1 Dec. 2019]. [accessed 1 Dec. 2019].

[8] Michael McMahon, interview with author, constituency office, Bellshill, 21 Feb. 2014.

up my children now they must get on with it. She belonged in her world, and she was in that world, and we were in ours.[9]

Mary's comments underline the major social distinction between middle-class owner-occupiers and working-class public housing tenants. They also emphasize the growing isolation felt in some mining communities during the 1960s through the impact of mounting colliery closures. Mary's later relocation to the new town of East Kilbride clearly maps social mobility onto the geographies of economic development in mid twentieth-century Scotland. Jessie Clark also recalled the retention of sharp class divisions into the period of the nationalized industry, as well as the often painful process of community breakup inherent within social reconstruction. Her memories highlight the importance of the connection between workplaces and community life, which continued under the public ownership. Jessie commented that nationalization 'wasn't just plain sailing, you still had that manager you know'. Despite the promise of 'more democracy' within the National Coal Board's (NCB's) collieries, there were evident elements of continuity in employer power. A member of colliery management, who had continued in his post from the private industry, 'prevented us getting a house for a wee while'. She speculated this was the result of her husband's reputation as a 'troublemaker' given his communist and trade union affiliations. The NCB hierarchy, or at least this individual, used the Board's control of Scottish Special Housing Association (SSHA) houses at Rigside, in South Lanarkshire, to exercise power over the workforce in a manner reminiscent of paternalist private employers who acted as landlords.[10]

Jessie's memories exemplify Little's reflections on the British state's direct presence in the coalfields. Her reflections on the eventual move to new housing combined happy memories of a major material improvement with mourning for the community life in the demolished miners' rows:

> It was a big change for me because I'd been living in a room in somebody's house when I got married at first, you know. And to go tae a house where I had three bedrooms and a bathroom and running water. Fantastic, you know! It was good to have a bath, because that was the one thing about living in the miners' rows. You didn't have a bath and unlike the people in cities we didn't have the baths that we could go to, you know, next door to the steamie ... But I must say, when it came to friendship, comradeship, the village, the old village, was a hundred per cent. Second to none. And even the move up to a mile way to these new houses, it was different. It was different. It was really, it wasn't as

[9] Mary Spence, interview with author, The Terraces café, Olympia shopping centre, East Kilbride, 11 Aug. 2014.

[10] Jessie Clark, interview with author, residence, Broddock, 22 March 2014.

close a community you know. Literally close I mean, miners' rows, when you think about it.[11]

These comments on proximity emphasize the degree to which public and private spheres overlapped in coalfield communities. The community formed a restricted public with membership defined on class, occupational and within the pit itself, gender terms. However, the nominally private family sphere had traditionally been inhibited by physical closeness and shared amenities and social lives. Jessie's interview underlined the self-organized nature of community activities through Douglas Water's Miners' Welfare: tennis courts, bowling greens, a pipe and silver band as well as dances. At a focus group in Shotts, similar recollections to Jessie's about the neighbourly closeness of miners' rows were also accompanied by references to leisure activities. During the interwar period this had also included illegal gambling syndicates organized around 'safe houses' in the dense miners' rows, as well as more formal public events. Bill Paris, who grew up within a Shotts mining family and worked in local collieries, recalled that:

> It was actually quite a busy town and there was many local amenities for everyone. Well, the Miners' Welfare existed at that time and I can remember there was swimming baths and a library and various others. I would say, y'know, associations, clubs built round aboot it. There was tennis courts, bowling green. Snooker, there was a billiards club and also there was the junior football clubs. Football was a strong very strong thing in the area because there were junior clubs, amateur clubs, juvenile clubs and they were they say probably quite a lot at least half a dozen, maybe even more. And all sorts ae other pastimes associated. You had the dog track as well, doon there.[12]

As in Jessie Clark's narrative, the focus group in Shotts recalled that the community was disrupted by pit closures and the growth of commuting to work:

> Ella Muir: Would it be fair to say if you've got these people changing under the umbrella if you moved onto different things would it do anything to the community the feeling in the community?
>
> Bobby Flemming: It certainly fragmented. In the mining there was a common strand running through everything, whereas going to all different industries it certainly fragmented
>
> Betty Turnwood: The more people went out to different jobs the less people were all the same.

---

[11] Jessie Clark, interview.
[12] Shotts focus group, Nithsdale Sheltered Housing Complex, Shotts, 4 March 2014.

Willie Hamilton: The community as you say fragmented, they werenae so close as what they were.[13]

Memories of community fragmentation associated with intertwined residency and employment shifts emphasize the growth of privacy. Socially and geographically mobile individuals entered new routines that led their employment and social lives to be less connected with their neighbours and work colleagues. Yet alongside these developments it is quite clear that there were elements of continuity. Willie Hamilton played for Shotts Vics junior football team, which had a historic connection with both local miners and steelworkers. He mentioned there had been an ongoing presence of amateur teams in Shotts which had only declined in recent years. Willie also recalled that these activities formed connections within Southfield pit where men from Fauldhouse in West Lothian and Carluke in South Lanarkshire bonded with miners from Shotts by playing sports together.[14]

The remaking of coalfield communities was positively affected by housing developments. Housing schemes constructed after the Second World War retained close forms of workplace and neighbourhood connections, forming the basis of new communities. Margaret Wegg's family moved into an SSHA house in Cardowan village during 1948, when she was seven years old. Her father worked at Cardowan colliery and had previously travelled from Feriegair near Hamilton, but was offered a house in the village through a friend, a 'union man'. Reconstruction was marked by the continuation of coal industry involvement in housing and the extension of joint regulation within the nationalized industry beyond the workplace. Margaret fondly recalled social life in the village which revolved around connections to the colliery. She referred to miners as 'the salt of the earth', reflecting on the generosity on display in regular collections being held at Cardowan colliery, which supported a range of initiatives including a pensioners' club that she is presently a member of. Margaret's testimonies also underlined the importance of neighbourly connections with 'mining families' several times, which were characterized by friendliness and shared social activities:

> When the snow was on the ground, we used to sledge down the street and straight into the field, you know. You had to duck under the wire, but we used to you know or we used to go into the field and play. I mean, we used to play out in the street and our parents used to. Well, ma mum and Gladys stayed across the road. It was all mining families. And my mum and them used to go out in the street and play rounders. We'd play tennis, you know. They played

[13] Shotts focus group.
[14] Marian and Willie Hamilton, interview, residence, Shotts, 14 March 2014.

with the kids. No the first time the knock had come to the door, 'is your mum coming out to play?'[15]

Billy Ferns recollected that social life in Bishopbriggs also linked the workplace and residence. Billy worked alongside his father at Cardowan colliery and moved to the area during the mid 1960s. He estimated there were up to forty mining families who lived in NCB houses nearby. The men and their families socialized, indicating the establishment of a new coal community: 'We used to go oot to the miners' club on a Saturday up in Kirkintilloch they were aw there wi their wives. I used tae go tae sometimes Twechar or Kilsyth and you'd meet all the boys there Saturday night with their wives. I knew them all well'.[16]

There is a clear distinction between these descriptions of earlier social reconstruction and the period of private house building and intensified deindustrialization up to major final closures during the 1980s and 1990s. These developments contributed to the rise of commuter towns and the decline of relatively autonomous industrial settlements with interlinked residency, work and social life patterns. Intensified deindustrialization disembedded the economy from communities, defamiliarizing towns and villages and weakening historic attachments to associations at a national level. Siobhan McMahon, Michael's daughter, remembered growing up in Bellshill, North Lanarkshire, during the 1980s where an 'industrial community' was marked by a crossover of family and neighbourhood connections. Siobhan's conception of an industrial community confirmed a view of the working class being defined by a common culture and sense of social standing. She felt that families of miners, steelworkers, engineers and welders, including her father and both her grandfathers, had a stake in the 'shared struggle the community had gone through'. This was juxtaposed with the more private, middle-class and white-collar character of the 'fancy new houses' where lawyers and accountants reside. Siobhan emphasized the impact that the town's increasing reliance on commuting and relatively low-paid service sector jobs has had in disrupting traditional community patterns and the essence of what she saw as having defined life in Bellshill:

> We've changed in that the jobs aren't coming to the area anymore. And when they do come, it's what I would, it's not, it's not the same types of jobs, not the skilled jobs that you required to keep people. We've got lots o big retail jobs. Tesco coming's great, and it is great. But our town centre's decimated because it's charity shops, it's bookies, it's pubs. What does that say to young people,

---

[15] Margaret Wegg, interview with author, residence, Stepps, 17 Nov. 2014.
[16] Billy Ferns, interview with author, residence, Bishopbriggs, 17 March 2014.

y'know? What job do you get there? And when you're trying to better yourself at uni then you don't come back to Bellshill. You're not going to get a job in Bellshill. You go somewhere else. So, people aren't remaining in Bellshill. They're seeing it as a town to build nice new houses in, absolutely, because it's half way between Glasgow and Edinburgh so it gets you along the motorway. That was never what it was supposed to be about. That was never what Bellshill was.[17]

Siobhan McMahon's comment 'that was never what it was supposed to be about', indicates a clear mental synonymy between Bellshill and an 'industrial community'. Bellshill's transformation into a suburban commuter town is a challenge to its very essence and has disrupted a strongly felt sense of belonging.

Duncan Macleod's father and both his grandfathers were Lanarkshire miners before his father migrated to Derbyshire during the late 1940s after colliery closures in in Carluke, South Lanarkshire. This migration experience exemplifies the unitary nature of the nationalized industry which connected miners across the British coalfields. Duncan's perspective on Carluke's shift towards becoming a dormitory town paralleled Siobhan's appraisal of Bellshill. The population has increased in size while signs of the area's mining legacy have been removed:

Carluke's gone from say about eight thousand population to about fourteen thousand. They've built about two thousand new houses in the last thirty, forty, years. These'll be the oldest of the new ones if you like. These are about forty years old. I work in my grandson's primary school, and if you talk to the people, let's call them incomers for want of a better word, they're not aware of the mining history of this area. And yet, if you come up even twenty years ago, which was thirty years after the pit closed, you couldn't fail to be aware that it was a mining area simply cause of the bings everywhere. And if you come in from Airdrie, from the Airdrie direction coming along the A73, you could see it from as far as the land was flat. It was almost like an artificial mountain.[18]

Duncan emphasized the demise of regular community events such as the annual miners' gala day, which had continued after the local pits had closed. The interlinked changes in housing tenure and occupational structure disrupted the conceptions of class and community that had pervaded for four decades. In 1981, Motherwell, Wishaw, Bellshill and Coatbridge in North Lanarkshire all had council house tenancy rates of

[17] Siobhan McMahon, interview with author, Central Scotland Regional List MSPs Office, Coatbridge, 28 March 2014.

[18] Duncan and Marian Macleod, interview with author, residence, Carluke, 1 March 2014. Bing is a Scots term for slagheap.

over eighty per cent.[19] This was reflected in the overwhelming majority of the oral history respondents who took part in this study having lived in public-sector housing. Jennifer McCarrey grew up in Mossend – to the east of Bellshill – during the 1970s and 1980s. Her parents were both active trade unionists. Jennifer's father was the convenor for non-manual workers at the Ravenscraig steelworks, where her grandfather had also worked. She recollected that the economic restructuring of the 1980s was interpreted through a linked community and class consciousness. This reinforced identification with the Labour party, and affirmed the social division based on housing tenancies which Mary Spence referred to:

> You've got to remember our existence was very different. Like I had never even been in a bought house tul I was like a teenager. Nobody I knew even lived in a bought house! Everybody lived in a scheme, in a council house. I was saying the other day I never met a Tory tul I was fourteen. One of them was leafleting up in a scheme in Bellshill. I remember looking at him as if to say you look just like us. It was that uniformity of the political identity in my community. It was the safest Labour seat in Scotland, North Lanarkshire.[20]

These comments resonate with Hassan's and Shaw's view of post-1945 'Labour Scotland' as having been strengthened by the 'institutional pillars' of public housing, local government and trade unionism. Jennifer's parents' trade unionism and her involvement as a young Labour party activist are indicative of those connections. These networks sustained a 'Labour state which extended far into the lives of communities in a way unimaginable now', linking the workplace, housing and political representation at local, Scottish and UK levels.[21] Jennifer's recollections also affirm that working-class identity was associated with the growing electoral divergence between Scotland and UK, which accompanied the British state's withdrawal from industry and the abandonment of moral economy obligations. During the last four decades of the twentieth century, electoral support for Labour in Scotland was generally higher than its UK-wide performance, while Conservative governments were secured through English parliamentary representation. Growing divergences stimulated demands for devolution, which became especially pronounced as deindustrialization intensified during the 1980s and 1990s.[22]

---

[19] W. W. Knox, *Industrial Nation: Work, Culture and Society in Scotland, 1800–present* (Edinburgh, 1999), p. 262.

[20] Jennifer McCarey, interview with author, iCafé, Woodlands, Glasgow, 9 Oct. 2014.

[21] G. Hassan and E. Shaw, *The Strange Death of Labour Scotland* (Edinburgh, 2012), pp. 5–7.

[22] J. Phillips, V. Wright and J. Tomlinson, 'Deindustrialization, the Linwood car plant

Nostalgia and critical nostalgia: 'it was pretty good'

In former industrial localities, nostalgia parallels 'a mourning process'. It is characterized by grieving for lost social connections, cultural activities and occupational identities.[23] A latent critique of present circumstances is implicated within nostalgia.[24] These sentiments are communicated in the form of 'broad brushed contrasts' between a past defined by community and a present whose main feature is its erosion.[25] Feelings of loss within deindustrialized areas are often characterized by a 'smokestack nostalgia', which emphasizes sensual experiences of industrial activities.[26] Marian Hamilton, whose grandfather and husband were miners while her father had been an iron moulder, fondly recalled that in Shotts, 'You used to get up and suddenly you heard the boots in the morning the tramp, tramp, tramp. And that was folk going to the pits and going tae the ironworks cause it was a big works as well'.[27] Jennifer McCarrey remembered that the sound of both the Lanarkshire and Ravenscraig steelworks were defining features of life in adjacent Mossend. She found their absence displacing upon moving to Birmingham in the mid 1980s where she worked as a paid trade union organizer:

> I didn't sleep very well. And I realized it's because I couldn't hear the clanging of the steel at night. Because that was constant through ma whole life, you would hear the clanging of the steel at night at Clydesdale and wherever. You always heard it. And I noticed it wasn't here anymore! It was bizarre it was like, what is that? Only when I went home I realized it was the steel! It sounds ridiculous now, but it was absolutely true.[28]

Affinity with industrial activities was entrenched by the infrastructure of community life built around them. Mick McGahey, the son of Michael McGahey, the NUMSA president, referred to coal mining as providing

---

and Scotland's political divergence from England in the 1960s and 1970s', *Twentieth Century British History*, xxx (2019), 399–423 .

[23] T. Strangleman, '"Smokestack nostalgia," "ruin porn" or working-class obituary: the role and meaning of deindustrial representation', *International Labor and Working-Class History*, lxxxiv (2013), 23–37, at p. 28.

[24] B. Jones, *The Working Class in Mid-Twentieth Century England* (Manchester, 2012), pp. 187–8.

[25] R. Samuel, *Theatres of Memory: Past and Present in Contemporary Culture* (London, 2012), p. 6.

[26] S. High, *Industrial Sunset: the Making of North America's Rustbelt* (Toronto, 2003), p. 50.

[27] Marian and Willie Hamilton, interview.

[28] Jennifer McCarey, interview.

the basis for a 'social fabric' of cultural activities and social connections.[29] Annual gala days were mentioned in several testimonies aside from Duncan Macleod's considered above. They played a key role in several respondents' recollections of community life. Rhona Wilkinson's grandfather worked at Woodmuir and then Polkemmet colliery. She had fond childhood memories of the annual celebration in Breich, West Lothian:

> Everybody all went to the miners' gala day. We always had our gala day, all that sort of stuff. It was a big day, even although Breich couldnae afford shows or anythin. I think one year we got a coconut shy. That would be aboot it! But it was just like races over the park and a cauld mince pie. Fauldhouse always seemed grand cause they had a Tunnock's box at their gala day. But we just get a cauld pie and a German biscuit or something like that. That'd be it![30]

Rhona's emphasis on material deprivation alongside social cohesion concurs with the assessment of community apparent in memories of interwar miners' rows. A similar view of social connections spurred by impoverishment was given by Barbara Goldie, who was from a mining family in Cambuslang, and went on to marry a steelworker. She stated that mining families 'had nothing then, we were aw in the same boat'.[31] Sam Purdie recalled the objections to the depopulation of Glenbuck, a village of traditional miners' houses in East Ayrshire. In 1954, Glenbuck's population were rehoused in modern council housing in the larger adjacent settlement of Muirkirk:

> It was a tremendous place. No gas, no electricity, no water in most of the houses. There was a communal water pump at the end of the street where you went to fill your pails. When you entered Glenbuck at first, here's a row of miners' houses, twenty-six, one pump at the end. So, the housewives had to go to the end, get the pail and carry it for anything they needed to do. No cooking facilities except the fire, the open range. That was Glenbuck. I'll guarantee that Burns would not have seen much difference in the place. Sewage didn't exist. Open middens. The toilets were just a hole in a board … Oh, Glenbuck people were desperate to stay. We had the impression that you could have built the houses in Glenbuck.[32]

---

[29] Mick McGahey, interview with author, Royal Edinburgh Hospital, Edinburgh, 31 March 2014.

[30] Rhona Wilkinson, interview with author, residence, Fauldhouse, 7 Nov. 2014. A German biscuit is a variety of sweet biscuit popular in Scotland. It is more commonly known as an Empire biscuit.

[31] Barbara Goldie and Margaret Keena, interview with author, Whitehall Bowling Club, Cambuslang, 8 Dec. 2014.

[32] Sam Purdie, interview with author, UWS Hamilton campus, 3 May 2018.

Sam's insistence on the maintenance of a Glenbuck identity over half a century later is matched by his work with other former residents to develop heritage activities and memorials on the former site of the village. The most prominent commemoration has been for the highly successful Glenbuck Cherrypickers football club, which produced a multitude of high-achieving professional players, including Bill Shankly. Shankly went on to manage Liverpool but retained an iconoclastic socialist philosophy which he attributed to his experience of growing up in Glenbuck and working in local collieries. In 2019, Sam was involved in unveiling a memorial in Shankly's partially restored house. It was attended by Liverpool fans involved in the 'Spirit of Shankly' supporters' union.[33] Similar activities are also visible in lost Lanarkshire mining villages. While trade unions were broadly accepting of pit closures during the late 1950s, concerns were raised about Hamilton Palace shutting in 1959 due to 'the grave social consequence this closure will have on the village of Bothwellhaugh'.[34] The village was subsequently demolished in 1965, yet thirteen years later an annual commemoration, 'Palais Day' was inaugurated and it subsequently became an annual event headed by a committee of former residents. The committee continues to maintain memorials and are archiving memories of the village and colliery.[35]

Within the oral testimonies, the dialogue was shaped by a strong sense of passing on memories for posterity as the industrial era receded from human memory.[36] Respondents identified me as somebody who had grown up following the end of large-scale industrial employment in Scotland. I was known to the interviewees as a University of Glasgow student or University of the West of Scotland lecturer originally from Edinburgh, which contributed to perceptions of both geographical and social distance. This perhaps heightened contrasts between past and present. In the case of Alan Blades, it also emboldened claims of the solidarity engendered through neighbourhood connections in the mining village of Greengairs, North

---

[33] D. Kay, 'Liverpool legend Bill Shankly's spirit rekindled in village he first kicked a ball', *Liverpool Echo*, 3 Sept. 2019 <https://www.liverpoolecho.co.uk/sport/football/football-news/liverpool-most-important-date-history-16843384> [accessed 2 Dec. 2019].

[34] National Records of Scotland, Edinburgh (NRS), Coal Board (CB) 313/14/1/16A, D. Kelly, Scottish Colliery, Enginemen, Boilermen and Tradesmen's Association (SCEBTA), to R. W. Parker, NCB, Edinburgh, 20 March 1959.

[35] S. Swarbrick, 'Breathing fresh life into the story of forgotten Lanarkshire mining village Bothwellhaugh', *Herald*, 11 Feb. 2017 <http://www.heraldscotland.com/life_style/pictures/15084711.display/> [accessed 2 Dec. 2019]; 'Bothwellhaugh Ex-Residents Committee', *Bothwellhaugh* <http://www.bothwellhaugh.com/> [accessed 2 Dec. 2019].

[36] A. Portelli, 'What makes oral history different', in *The Oral History Reader*, ed. R. Perks and A. Thomson (Oxford, 2006), pp. 32–42, at p. 40.

Lanarkshire, through a contrast with my own background: 'You're a tight-knit community, y'know what I mean? It's not like growin up in Leith!' He later specified that he felt the 'village mentality' of Greengairs was particular to small-scale settlements, arguing that differing social attitudes, especially a tendency towards suspicion and to treat others as strangers, were visible in the comparatively large town of Airdrie where he now resides.[37]

Assessments of industrial nostalgia require a 'generous critical cultural reading' of working-class memories.[38] Recollections of social life in industrial settlements are not limited to crude longing for the past, but instead demonstrate forms of critical nostalgia. Nostalgia can facilitate recognition of improvements as well as regression associated with deindustrialization and connected social changes. Critical nostalgia's 'radical imagination' is a productive means to articulate criticisms of the past as well as the present.[39] Oral history interview participants voiced critiques of hierarchical industrial communities but also reflected on the losses associated with workplace closure and community fragmentation. Generational distinctions encouraged critical considerations of the changes associated with deindustrialization. Alan Blades, perhaps influenced by my presence as a representative of a generation which had matured after most major industrial closures had taken place, and my status as a university researcher, argued that economic changes had presented opportunities for some young men as well as social dislocation for others:

> Boys and then obviously their sons arenae getting in the local pit, so you've got the young ones aw comin through that were expected tae work in the pits they'd to go elsewhere and look for the jobs. Y'know what I mean? ... It'd maybe be good for some ae them cause some ae them would probably say 'I'll need to be good I'll need to start gettin into ma education', y'know. And go to uni and college and aw that.[40]

A critical outlook was visible towards the culture within mining communities in some testimonies. Brendan Moohan grew up in Musselburgh, East Lothian. His grandfather was a communist who was blacklisted out of Lanarkshire collieries during the interwar period. Brendan's grandfather subsequently found employment in the Midlothian coalfield where Brendan and his father also later worked. Brendan's comments

---

[37] Alan Blades, interview with author, resident, Airdrie, 26 Feb. 2014.

[38] Strangleman, '"Smokestack"', pp. 25–28.

[39] A. Bonnett, *Left in the Past: Radicalism and the Politics of Nostalgia* (New York, 2010), pp. 1–3.

[40] Alan Blades, interview.

indicated his feelings of ambiguity towards the organized associational life that characterized mining communities:

> There was something about that mining lifestyle that mining communities had that was a little bit conservative with a small c, and if you were adventurous it could be quite restrictive. And there was a clear hierarchy to it as well. You know, you had the guys who were on the committee, and the guys from the union. There was a very definite kind of structure to it.[41]

However, indicating the facets of critical nostalgia, Brendan's criticisms were qualified by reflections on the loss of 'social cohesion' and the 'form of socialism' that associational life encouraged. He highlighted the occupational identity and activities of Miners' Welfares, which anchored a pride and consciousness that has been eroded through deindustrialization. Brendan stated there had been a transition from an active community with a unique social life that celebrated its role in the mining industry to one increasingly based on passive receivership of media:

> But the miners' clubs again, the other kind of socialistic element to it was in your community. You're gonnae celebrate your community, and we're gonna have this gala day every year. The high point of the year in the summer. And we'll have all these kinda sporting activities, all that kind of stuff. It would all centre round the club, but it would be something whereby the community celebrated itself. I have to say in a very hierarchical way, with a 'gala queen' and all the rest of it. But it was something that was for working-class people, accessible for working-class people and being bold enough to celebrate who they are and in their own community. I think that was a good thing. And you know, nowadays if people are more recipient of things to celebrate, i.e. they'll watch the World Cup, they'll celebrate whoever wins, the Olympics, something that's put on their TV screen. Whereas in those events you had all ages having their races, their singing competitions, their boxing competitions, the garden competition was often judged then. So, there was a variety of things and it was celebrating no further than the boundaries of the village. It was pretty good.[42]

Brendan's sentiment that growing up in a mining community 'was pretty good' contains definitive elements of critical nostalgia in noting the past's detractions but nevertheless asserting positive comparisons with present circumstances. His narrative articulated criticisms of social conservatism. Nevertheless, in contrasting mining communities' social life with less mobilizing and politically conscious contemporary routines he felt they had developed a culture imbued with activities and sensibilities of social and

---

[41] Brendan Moohan, interview.
[42] Brendan Moohan, interview.

political value. Brendan's perspective is redolent of Mark Fisher's critique of the transition from 'engagement to spectatorship' which has taken place under Britain's neoliberal political and economic transformation. Fisher singled out the politicized assault on mining communities during and after the 1984–5 miners' strike as a significant moment in consolidating the perspective that there was no alternative to an individualistic and consumption-orientated society.[43]

Memories of mining accidents exemplified the tension between recollections of hardship and collectivity which characterize coalfield memories. Scott McCallum remembered the 1982 gas explosion at Cardowan colliery where his father and brother worked. To Scott, who was only eight years old at the time, the explosion imbued an awareness of the dangers associated with the industry. Yet he also recalled that the sense of community in Cardowan village, where his family lived, was strengthened by its stance in the face of adversity. The explosion fortunately led to no deaths but did have a casualty toll of twenty-five, with seven men having to be stretchered to hospital with burns:[44]

> That brought everybody quite close, there was lots of services it was broadcast all over the world. We had family all over the world, Australia and Canada, saying they'd seen it on the news and checking everybody was okay. And then for years after it you still seen people that was affected. One man, just mentally disturbed in his head, he could remember bits of it. It was a close-knit community. Everybody knew everybody.[45]

Margaret Wegg worked in the pit's canteen at the time. The accident served as both a memory of the danger miners endured but also as a pertinent example of solidarity, symbolized by the efforts of the workers who were off shift, to assist with the rescue:

> That was a bad one, that was, you know. But even then, that's when you knew there was a community. I mean, people ma dad's age was retired, but the minute the word went out that that had happened all the old miners were right down to the pit to see what they could do, you know. Could they go down to see, help with the rescue ... We just got ready and that was it. We were down, the canteen staff were all there working you know. It didn't matter whether you were off duty or on duty or that you were down, you know. Oh, there was a

---

[43] M. Fisher, *Capitalist Realism: is There No Alternative?* (Winchester, 2009), p. 5.

[44] Hansard, *Parliamentary Debates*, xvi, cc889–91 (27 Jan. 1982), Cardowan Colliery (Accident) <http://hansard.millbanksystems.com/commons/1982/jan/27/cardowan-colliery-accident> [accessed 2 Dec. 2019].

[45] Scott McCallum, interview with author, The Counting House, Dundee, 22 Feb. 2014.

bad. That was when John O'Rourke got brain damage. As I say, that was when we knew there was a community.[46]

Willie Doolan, who was the NUM's workmen's inspector at Cardowan, recalled the incident in harrowing terms. Having seen colleagues leave for hospital on stretchers with their burned skin peeling off their bodies, he was tasked with completing an inspection of the site with Her Majesty's Inspectorate of Mines: 'the minute you walked in you could smell burning flesh. It was horrible. Horrible … It was a thing that will stay with me for the rest of ma life'.[47] Willie's memories demonstrate the important links between government and local communities through the nationalized industry. These were especially present during mining accidents. He recalled undertaking similar social responsibilities during the twilight years of deep coal mining in Scotland. When he was an NUM delegate at Castlebridge colliery during the 1990s, Willie collaborated with the pit management during an unsuccessful attempt to rescue a man who was trapped underground. Willie was responsible for informing the man's wife of his death. The woman was comforted by her father who had also worked in the industry.[48] These memories indicate the strong demarcation of gender roles within coalfield communities. Male pit managers and union representatives took responsibility for managing operations while mines rescue volunteers led underground efforts. Women acted as auxiliaries above ground and, along with children, anxiously awaited news from an alien underground environment.

These themes were apparent in reflection on other major incidents of injury and death within collieries, especially the Auchengeich disaster of 1959 in which forty-seven men died following an underground fire that resulted from mechanical failures in a large fan.[49] This event's significance transmitted through the cultural circuit of coalfield memory. During 2014, Siobhan McMahon recollected her maternal grandmother, a miner's wife, recently telling her about the experience of the disaster and the role that women played in coal mining families. Siobhan retold her grandmother's memories of the emotions felt by miners' wives and mothers. These included a sense of terror at the incident and the combination of relief and guilt, with thought for other families who had not been so fortunate, when they learned that their sons and husbands had survived:

[46] Margaret Wegg, interview.

[47] Willie Doolan, interview with author, The Pivot Community Centre, Moodiesburn, 14 June 2019.

[48] Willie Doolan, interview, 2019.

[49] E. Gibbs and J. Phillips, 'Remembering Auchengeich: the largest fatal accident in Scottish coal in the nationalised era', *Scottish Labour History*, liv (2019), 47–57, at p. 49.

She said, 'it had been on the news and one of the other wives had come up and said "y'know there's a disaster at Auchengeich?"' So, all the females got their children and waited at Bellshill Cross waiting on the bus coming back to see if their partner would get off the bus. How harrowing that must have been. To wait and to see if your partner was coming off a bus, because they had no other means of communication, you know, at that time. And she said, y'know obviously that was a worried time for her and the support. It was the female. Just the image of that, standing for hours at a bus stop waiting on that and holding each other together. I mean how strong those women must have been.[50]

These experiences parallel developments in South Wales, where 'blood on the coal ... generated a moral claim on the local mine'.[51] Disasters served as pivotal events in framing conceptions of what it meant to be from a mining community, emphasizing collective solidarity in the face of industrial dangers. This embedded moral economy arguments within the sense that a community that had sacrificed for the industry had a right to economic security. Striking miners constructed a memorial for the Auchengeich disaster during the 1984–5 strike. It was erected by men affiliated to the Cardowan colliery strike centre that organized strikers from the recently closed pit. The main feature was the Cardowan colliery winding wheel which was 'unofficially' donated by pit management. An annual commemoration for the Auchengeich disaster began during the strike and now acts as source of community continuity. Willie Doolan was only aged four at the time of the disaster, but his father was present as a mines rescue volunteer. He remembers the event and its impact on his schoolmates who had lost fathers and brothers. Willie was involved in the establishment of the memorial as well as its subsequent expansion, including the erection of a bronze statue designed by Kevin McKenna, for which the Auchengeich Miners' Memorial Committee raised £35,000.[52] The attendees at the commemoration extend beyond the immediate areas of Lanarkshire which served Auchengeich colliery:

> I don't know whether I said to you about it before or not Ewan, but we have every year we have an annual memorial service to commemorate the miners that lost their lives in the Auchengeich pit disaster. I mean the fiftieth anniversary was in 2009, and we could command three thousand people at that memorial service. But it wasn't only the families of ex-miners from this community. We had people from as far as the Lothians coming through. We had people from

---

[50] Siobhan McMahon, interview.

[51] A. J. Richards, *Miners on Strike: Class Solidarity and Division in Britain* (Oxford, 1996), p. 22.

[52] Willie Doolan, interview, 2019; Phillips and Gibbs, 'Auchengeich', p. 52.

Figure 3.1. Auchengeich Mining Disaster Memorial, Moodiesburn, ©John O'Hara (2019).

down south coming up tae share with this community once again the sorrow and the sadness that we had undergot due to that disaster that happened in oor pit.[53]

Willie's presentation of the annual commemoration indicates the connections and tensions between occupation, community, class and nation. The reconstruction of the coal industry in Scotland encouraged the construction of a national coalfield community, embodied by the presence of former miners from other coalfields at the annual commemorations. Donations from Longannet miners assisted the memorial's expansion after the return to work in 1985, underlining the links between miners across Scotland.[54] These connections developed through pit transfers and deindustrialization. They were encouraged by the political culture within the Scottish coalfields, especially the activities of Communist Party of Great Britain (CPGB) members and their allies within the NUMSA. There is an extent of critical nostalgia in the annual memorial service, especially through

[53] Willie Doolan, interview with author, The Pivot Community Centre, Moodiesburn, 12 March 2014.

[54] Willie Doolan, interview, 2019.

memories of miners' successive struggles for improved safety conditions. The Auchengeich disaster took place within the context of significant long-term health and safety improvements. Death rates in the Scottish coal industry approximately halved between their nadir in the interwar years and the early years of nationalization. But there was a slight rise over the late 1950s and early 1960s. More complex mechanical systems, including mechanized ventilation and haulage systems, contributed to renewed dangers. These were augmented by pressures in the context of accelerating colliery closures and competition with oil and nuclear fuels.[55]

Nicky Wilson addressed the 2019 memorial service in his capacity as the NUM Scotland president. He commemorated the lives lost at Auchengeich with reference to his union's long fight for improved health and safety conditions, including the campaign for the issuing of the self-rescuer breathing masks to miners. This was ultimately achieved following another disaster at the Michael colliery in Fife when nine men perished following an underground fire in 1967. Michael was Scotland's penultimate mining disaster.[56] After the death of five men following a roof collapse at Seafield colliery in 1973 there were no further accidents with multiple fatalities in the Scottish coalfields.[57] Mick Hogg summarized his conception of Auchengeich's place in the long history of the Scottish coalfield during his contribution to the 2018 memorial service. He addressed the rally as a fraternal delegate from the Rail and Maritime Transport union, for whom Hogg works as a paid organizer, and as a Midlothian miner who was victimized following his arrest during the 1984–5 strike. Hogg referred to this conflict as part of the long struggle of miners for safety and dignity which originated in resisting 'feudal bondage' during the eighteenth century.[58] However, the outlooks of Willie Doolan, Nicky Wilson, Mick Hogg and their comrades were not universal. Identities and affiliations were complex and renegotiated through evolving historical circumstances.

Sectarian divisions

The history of the Scottish coalfields intersects industrial conflict with religious sectarian and anti-trade union employment practices. These were pivotal in shaping personal identities, communal affiliations and the social and territorial borders of communities. Struggles over the distribution of resources, especially industrial employment, are central to understanding

[55] Phillips, *Scottish Coal Miners*, pp. 88–117.
[56] Observervation notes from Auchengeich colliery memorial service, 15 Sept. 2019.
[57] Phillips, *Scottish Coal Miners*, p. 98.
[58] Observeration notes from Auchengeich colliery memorial service, 16 Sept. 2018.

the continued significance of sectarianism during the latter half of the twentieth century. Kelly defined sectarianism as 'a social setting in which systematic discrimination affects the life chances of a religious group, and within which religious affiliation stands for much more than theological belief'.[59] This is an appropriate basis on which to construct an analysis in the context of the Scottish coalfields. Sectarian trends were especially concentrated in Lanarkshire where different Irish ethnic backgrounds, Catholic and Ulster–Protestant, intersected with residence, work patterns, and political affiliations.

Sectarian practices figure significantly within the oral testimonies collected for this study, especially as they relate to the private industry. Jessie Clark recalled that in the South Lanarkshire mining village of Douglas Water, her father, a blacklisted trade unionist, felt 'the members of the Masonic Lodge were the ones that always got the work, you know. And that was a fact of life in the village that I lived in'. Jessie's father had rejected such a path, breaking with his father's affiliation in favour of socialist politics through the Independent Labour party and later the CPGB.[60] But sectarian connections retained some bearing on colliery employment into the nationalized period. Pat Egan's memories of the influential role played by a Catholic fraternity, the Knights of St. Columba, at Bedlay colliery, are demonstrative of the pit's social embedding through strong links between workers and management.[61] These practices were informed by a defensive and divisive mentality, which protected access to premium employment and promotion for those of a particular ethnic background and religious–political affiliation.

Pat's contentions about Bedlay are corroborated by the memories of John Hamilton who was originally from Lesmehagow in South Lanarkshire. John had worked alongside his father at Ponfiegh colliery, adjacent to Lesmehagow, but took up employment at Bedlay during the early 1980s. His recollections also confirm that sectarian affiliations were embedded in other collieries. Before Bedlay closed, John transferred to Polkemmet in West Lothian. In contrast to Bedlay, it had a Protestant loyalist character:

> I'm of the Protestant religion. I worked at the Bedlay and it was, the majority was Catholic religion. Big time. So you couldnae even talk aboot Glasgow Rangers when you were doon the pit. You'd just to watch what you were saying when you were saying it! *laughs* So, when that closed, I got transferred to Polkemmet. And in Polkemmet they've got pictures o the Queen in every

[59] E. Kelly, 'Review essay: sectarianism, bigotry and ethnicity – the gulf in understanding', *Scottish Affairs*, i (2005), 106–17, at pp. 109–10.
[60] Jessie Clark, interview.
[61] Pat Egan, interview with author, Fife College, Glenrothes, 5 Feb. 2014.

corner you can think ae ... But you were accepted nae matter where you came fae, didnae matter to who you were working wi. No. That was okay, as long as you were daein your job and aw that. There was never any trouble.[62]

John's eagerness to stress that sectarianism did not contribute towards serious divisions in the workforce is indicative of elements of composure, especially the influence of the coalfield cultural circuit's emphasis on workforce and community togetherness. Mick McGahey recollected that both Catholic and Protestant factions were active within mining communities and had a presence within the NUMSA: 'You had the Communist party, you had the Labour party, you'd have the Catholic Action, the Masonic Lodge you know. You'd these major factions that were competing with one another'.[63] In an interview recorded during the late 1990s, Mick's father recalled growing up amid the divide between Catholics and Protestants in Cambuslang, South Lanarkshire, and claimed to have been rejected by both sides as a communist. However, the former NUMSA president also underlined the distinction between weekends characterized by the Old Firm football rivalry and the return to work on Monday when miners were reunited.[64] This was indicative of both the NUM's universal presence within the nationalized industry and shared investment in the moral economy. In Pat Egan's view, sectarianism was historically stimulated by employers who used religious differences to divide the workforce. But like John Hamilton and Mick McGahey, he underlined that these had declined during the second half of the twentieth century:

> I know Twechar, was a mining village, and the next village up is Croy, which I think is predominantly, in fact it wis aw Catholic. It wis one hundred per cent Catholic. And they used to play baith villages cause it was the same mine-owner. It wis Bairds who used to play them aff each other – 'they're producing much more than yous are' – that kinda stuff and he just played them aff each other. I think it was a big part in the twenties, thirties, forties, even fifties, ah would say.[65]

The private industry's legacy of fostering social control through sectarian division has been noted in recent research. In Coatbridge, the journalist Peter Geoghegan was informed by Jim MacDonald, a former steelworker

---

[62] John Hamilton, interview with author, South Lanarkshire Integrated Children's Services office, Larkhall, 26 Apr. 2016.

[63] Mick McGahey, interview.

[64] Busby, Stein and Shankly: the football men' *Arena*, episode one, H. McIlvanney, BBC, UK, originally broadcast on 28 Mar. 1997, 55 mins.

[65] Pat Egan, interview.

and an Orange Lodge veteran of fifty years standing, that Bairds had given land to construct both Catholic and Protestant churches as well as an Orange Hall.[66] MacDonald's interpretation of local history is consistent with Alan Campbell's account. He notes that, 'to very varying degrees, the Scottish mining communities were fractured by ethnicity and religion', emphasizing this was prominent within Lanarkshire. These divisions were maintained by the activities of secret organizations including the Knights of St. Columba and the Orange Order, which had a 'mass membership' among Protestants from Ulster backgrounds. Into the interwar period there were regular major clashes at parades and other events such as local football matches. Orange organizations were utilized by employers to pursue policies of 'class collaboration', often clashing with communists who were understood as threats to a Protestant-led social order and allies of Irish republicans.[67]

However, the Catholic Church itself also possessed a powerful anti-communist influence. Pat Egan's description of Catholic organization at Bedlay colliery was part of a wider prolonged effort to retain a basis within working-class communities, and to undermine the CPGB. This involved the mobilization of groups working under the banner of 'Catholic Action' within the labour movement and led to clashes over emotive subjects such as birth control, divorce, and support for the Soviet Union and the secularist Republicans during the Spanish Civil War.[68] Jennifer McCarey's awareness of family and community history in Newarthill, North Lanarkshire, exemplified this religious and political divide, and the power of the Catholic hierarchy:

> My grandpa had been in the Communist Party. We were brought up with that, it was a real identity in that you know. There was some families that had that identity and they were ashamed of it. And they hid it because they had been victimized by the Catholic Church in the community. Some of them, their fathers, had been thrown out the parish, the whole of the family. And they were embarrassed and ashamed aboot that, so they didn't really talk about it. It was only later on when I asked about some of the characters, who the people were, that I kind of realized that there were others that weren't proud of it.[69]

Jennifer's recollections accord with other memories from the testimonies in noting the sustained presence of Catholic-infused anti-Communism in

---

[66] P. Geoghegan, *The People's Referendum: Why Scotland will Never be the Same Again* (Glasgow, 2015), pp. 39–40.

[67] A. Campbell, *The Scottish Miners, 1874–1939*, i: *Industry, Work and Community* (Aldershot, 2000) pp. 317–27, 342–6.

[68] G. Walker, 'The Orange Order in Scotland between the wars', *International Review of Social History*, xxxvii (1992), 177–206, at p. 199.

[69] Jennifer McCarey, interview.

the Scottish coalfields after the Second World War. Sam Purdie recalled the NUM delegate at Cairnhill mine in Ayrshire during the early 1960s was Ed Donagher, a prominent local Catholic Action activist who opposed Abe and Alex Moffat and the communist orthodoxies that prevailed within the NUMSA under their presidencies.[70] John Brannan's father had been a coal miner but he took up employment at the Caterpillar tractor factory in Tannochside during the mid 1960s and went on to become the engineering union convenor at the plant, eventually leading the 103-day occupation against its closure during 1987. Brannan remembered that the previous generation of shop steward leaders at the plant had included Tom Dougan, who was also the chair of the local Catholic Action group. Dougan approached Brannan to act as an informant. Brannan's refusal was to the chagrin of his father, but he saw this as the only consistent way to act as an anti-sectarian trade unionist:

> The boy came to me and he says, 'how are you doing John?' I said 'not bad', he said, 'my name's Tom Dougan'. I said 'how are you doing Tam?' 'I'm the chairman of the Catholic Action group.' I said, 'very good. What do you want?' 'I'm a bit concerned about the Communist infiltration in the Caterpillar. If you dae know who they are could you tell me their names.' I said 'No problem, what are you wanting?' He said 'members of the Communist Party, fellow travellers,' as they called them at that time. 'Anybody at aw'. 'Have you got a pen?' 'Oh aye' I says, 'John Brannan'. He says, 'I think I've made a mistake'. I says, 'who telt you to come to me?' 'Oh I couldane tell you that'. So, two weeks later I'm over at my mother's hoose, and my father says to me 'you embarrassed me'. He said, 'did Tom Dougan talk to you?' I says, 'did you tell Tom Dougan to talk to me?' He said, 'well I did say talk to oor John'. I said, 'well you made a mistake n aw dad'. About a year or so later there was a job going. Tom Dougan should have got the job and he didnae get it. The next time I goes up to Tom and I says 'why are you not challenging that job?' He says, 'I don't expect to get it. I need to go to you anyway to take the case up'. I said, 'I already took it up because you deserve to get the job'. I says, 'don't think because you did that makes a difference to your union membership'. I says, 'Tom, that was in the days of your Catholic Men's Society'. And I did the same thing with the Masons one time. I did exactly the same thing. Obviously, somebody thought for some reason I wasn't a Catholic for lack of a better world.[71]

John's memories confirm the continuity of a sectarian influence in Scottish labour movement politics, including within assembly plants operated by multinationals' subsidiaries. Yet they also indicate sectarian practices had considerably diminished when contrasted with earlier in the century,

---

[70] Sam Purdie, interview.

[71] John Brannan, interview with author, UWS Hamilton campus, 21 Feb. 2017.

especially where trade union representatives were committed to opposing them. Jennifer McCarey's memories also indicated a continuation not only in the Catholic Church's political power but also in historical geographical distinctions between Catholic and Protestant communities of the sort Pat Egan recalled within North Lanarkshire. However, these were also complicated by ethnic distinctions within Catholic communities between Lithuanians, Italians and those of Irish backgrounds:

> If you get in a taxi the first thing they'll ask you if you're going to Mossend they'll ask you 'are you Lithie, an Italian or a Tim?' Cause that was the only three options, so that was the standard joke, you had to be one of those three. There was Protestants that lived there. Mr and Mrs Scott were Protestants that lived next door to us. Some of the other people on the street were Protestants. There was a Catholic identity in Mossend, that's true, but I mean there was a Protestant primary school right in the middle of it. But much more of a Catholic community than compared to Bellshill.[72]

Michael McMahon similarly recollected this area was marked by the form of sectarian 'micro-geography' that Campbell identified as dividing communities within bordering locales in the Lanarkshire coalfields:[73]

> You ask anyone now and they'll tell you Mossend is still seen as a Catholic place, Carfin is still the Catholic village, New Stevenson is the Protestant village, Holytown is a Protestant village. That still exists, no doubt about that. If you actually check now, you'll see that the make-up of those villages is much different from what it would have been in the days when the pits were there. I mean there was also the fact that Terex in Newhouse was seen as the Protestant factory and Caterpillar was seen as the Catholic factory.[74]

Michael's narrative indicates that manufacturing inward investment involved an extension of the geographies that characterized the Lanarkshire coalfield during the late nineteenth and early twentieth centuries. Continuities in the presence of sectarian affiliations within the nationalized coal industry, of a diminished form, also demonstrate that local autonomy facilitated persistence in industrial relations practices from the private era. Broadly, the pattern of community and industrial relations fits Knox's characterization of sectarianism being increasingly marginalized by the demise of major Scottish Protestant employers and paternalistic apprenticeship systems as well as the consolidation of trade union strength.

---

[72] Jennifer McCarey, interview.

[73] A. Campbell, 'Exploring miners' militancy, 1889–1966', *Historical Studies in Industrial Relations*, vii (1999), 147–64, at p. 158.

[74] Michael McMahon, interview.

Retention was visible though in secluded 'close-knit industrial communities' where sectarian norms remained more powerfully inchoated into social life, including the workplace.[75] As at Bedlay and Polkemmet, industrial relations were the product of core management expectations being filtered through local traditions. Michael McMahon's memories of workplace organization at the Terex factory reveal how plant management were able to employ sectarian industrial relations practices despite the intent of American owners:[76]

> Terex opened up in the early 1950s, and I was told this by one of the American bosses because I was a shop steward in the place ... He'd come across and he was talking to management and the foremen and all the rest of it, and it struck him how open they were in their pride in the fact the factory was predominantly Protestant. And so, he asked the management to check what the make-up of the factory was and then he asked them to check what the make-up of the local community was. And it didn't match up. So, he insisted that the management do something about that. And the company was expanding, the reason why he was over was cause the company was expanding. They built an extension onto the factory and the manager insisted that there had to be a higher proportion of Catholics hired to make the balance of the workshop reflect better the local community.
>
> All the new recruits were all hired ontae one shift. What you ended up with was a shift that was predominantly Catholic and a shift that was predominantly Protestant ... When it came to the end of oor apprenticeship we were told tae well, we were invited to ask what shift did we want to go ontae. And it was made pretty clear to us that we would be expected to go onto the shift that suited the religion best y'know. And most of the Catholic apprenticeships went onto one shift and most of the Protestant apprentices went onto another shift. And that was in the late 1970s, early 1980s, that kind of stuff was still going on. And there was one shop steward who ran an election campaign on the basis that he thought there was too many Catholic shop stewards and he wanted to get elected to redress the balance and make sure there was less Catholics.
>
> When the company closed down in 1984, what they decided to do was close the company down and then rehire those that would make the company efficient, i.e. those that were trained in more than one skill. As apprentices, we were quite high up on the list of desirables because we could work all the machines because we were trained on them all. Whereas a lot of guys had come in and only worked on one machine. Thirty years and every day only working on one

---

[75] Knox, *Industrial Nation*, p. 269.

[76] I. R. Paterson, 'The pulpit and the ballot box: Catholic assimilation and the decline of church influence', in *Scotland's Shame?: Bigotry and Sectarianism in Modern Scotland*, ed. T. M. Devine (Edinburgh, 2000), pp. 219–30, at p. 220.

machine. So, when we were getting hired back most of the apprentices were getting hired back but the guy who was hiring them back was a very senior individual in the Orange Order and he was making sure he was selecting the ones he didn't want to come back as well. And it was noticed a lot of the people that weren't getting re-recruited, re-hired, were the Catholics and it just reverted to the way it had been. And there was a lot of religious tension built up around that time.[77]

Within a subsidiary where senior management made commitments to opposing sectarianism, divisive practices were fostered by local level managers and trade union representatives. These accommodated senior management directives but ensured the workforce remained demarked by religious denomination into the 1980s. Sectarian distinctions were to some extent embedded within at least part of the inward investment assembly goods sector. Upon closure, as industrial employment became a scarcer resource, sectarian affiliations were strengthened through their deployment as a mechanism to lay claim to and ration well-paid engineering jobs.

Associations formed through shared ethno-religious affiliations also continued within Lanarkshire coal mining into the 1980s. Alan Blades grew up in the village of Greengairs, adjacent to Airdrie, which is a characteristic example of the small settlements Knox identified with the continuity of sectarianism. Orange connections provided a common link between workers and managers. Alan emphasized, however, that the closure of local collieries meant there were also strong links between Catholic and Protestant miners. His testimony corresponds to the requirements of composure, as well as the dialogical nature of oral history interviews, by emphasizing that sectarianism was limited to 'banter' and comparing it to the Edinburgh football rivalry he assumed I was familiar with:

> I grew up wi ma gaffer. I was in the Orange bands wi ma gaffer, y'know what I mean. We were a wee Orange village, so I was in the bands wi aw the boys I worked wi y'know … But see, the difference is, a lot ae folk think that because yer in the Orange bands you hated the Catholics. Some ae ma best mates are Catholics. Oh aye, I grew up wi some guys, but they accepted it cause they grew up knowin that we were aw in the bands and oor Dads were aw in the bands, y'know. Ye have yer bit ae banter y'know. Ye Orange bee y'Fenian bee, y' effin bee and aw that. We aw liked each other, y'know. It was a bit of banter. They went tae Celtic games on a Saturday and we went tae Rangers games, know what I'm saying? A lot ae Catholics worked in the pits wi us as well, especially as you moved out ae Moodiesburn and Croy and that. They're big Catholic villages. Still get on great wi aw they boys. Banter, that's aw it is, a wee bit ae

---

[77] Michael McMahon, interview.

banterin, y'know. Celtic Ranger, y'know how it goes. You have Hearts and Hibs. *laughs*

These recollections chime with Wight's study of 'Cauldmoss', an anonymized coal and steel town also located within Central Scotland. Wight detailed the importance attached to an annual Orange march during the mid 1980s. Although only attended by a small minority of dedicated members of the local Lodge, the march was greeted by a few hundred spectators, who flew Union flags, brandished Rangers Football Club scarves, and cheered as the band played while the march passed the town's Catholic church.[78]

Yet the nationalized coal industry and the operation of moral economy prerogatives in the management of pit closures minimized divisions which markedly declined from their scale and legitimated status within the private industry. While trade union organization and Labour party politics created semblances of unity, they also facilitated the management of division. This is visible in the continuation of residential divides under public housing and in workplace practices.[79] These were relative rather than absolute and diminished during the second half of the twentieth century. Willie Hamilton recollected the existence of prohibitive religious divisions in the Shotts area within new public housing developments that prevented one of his friends from marrying his partner. Reflecting the more open attitudes of his generational cohort who matured during the mid twentieth century, Willie felt these distinctions had dissipated over time:

> People from different areas had moved y'know it took quite a while to integrate and then you had the religious problem ... My mate courted quite seriously, but they never got married cause he was a Protestant and she was Catholic. It was their parents. It was quite restrictive for a lot of young couples.[80]

Sectarianism was a fading but present source of division in the Scottish coalfields into the 1980s. Although not marked by the same extent of employer-instigated fractiousness or street violence of earlier in the century, after 1945, ethnic and religious ties continued to forge important bonds of identity and difference. The example of Terex shows that when industrial employment again became scarce in the 1980s it was contested on sectarian lines in some instances. The continued strength of sectarian identities was also confirmed in the 1994 Monklands East by-election. As

---

[78] D. Wight, *Workers Not Wasters: Masculine Respectability, Consumption and Unemployment in Central Scotland: a Community Study* (Edinburgh, 1993), pp. 54–6.

[79] G. Walker, 'Sectarian tensions in Scotland: social and cultural dynamics and the politics of perception', in Devine, *Scotland's Shame?*, pp. 125–34, at pp. 127–31.

[80] Marian and Willie Hamilton, interview.

at Terex ten years before, it was the distribution of economic resources, in this case local authority spending and employment, which was disputed. The election was marred by accusations that the Labour-led Monklands District Council heavily favoured the largely Catholic town of Coatbridge over predominantly Protestant Airdrie. In her victory speech, the Labour candidate, Helen Liddell, accused the Scottish National Party (SNP) of playing the 'Orange card' after her party's majority fell from over 15,000 to 1,660.[81] The rise in support for the SNP was also indicative of prolonged disillusionment with the British state in communities where its industrial presence had diminished through major coal and steel closures and privatization during the 1980s and early 1990s.

Recollections of sectarianism demonstrate the critical lens with which coalfield communities are remembered and the value of oral testimonies in reconstructing otherwise unrecorded events. Memories of the industrial past are filtered through a critical nostalgia which located the positive dimensions of associational life and links between workplaces and neighbourhoods that were lost through deindustrialization. This was a long and complex process. The new communities constructed through public house-building along with the nationalized industry's concentration on super pits and inward investment were fondly remembered: 'it was pretty good' in restructured locales that offered improved material conditions and the appearance of stability. However, a potent ambivalence towards mining's occupational dangers is visible in recollections of disasters as well as the patriarchal social conservatism and sectarianism that were embedded in community identities and rivalries. Localized coalfield communities were founded on the collectivism engendered by shared experiences of adversity. But links to a Scottish national coalfield community were reinforced by both the extension of travel to work distances under nationalization and collective struggles for improved conditions. Deindustrialization has not entirely ended associations between coal mining experiences and localities in the Scottish coalfield. The commemorations of the Auchengeich disaster and the depopulated pit villages of Bothwellhaugh and Glenbuck demonstrates that industrial culture's influence extends beyond the industrial era.

---

[81] Paterson, 'Pulpit', p. 224; J. Aldridge, 'Labour to act over Monklands scandal', *Independent*, 2 July 1994 <http://www.independent.co.uk/news/uk/labour-to-act-over-monklands-council-scandal-mp-to-heal-wounds-caused-by-allegations-against-local-authority-john-arlidge-reports-1410968.html> [accessed 2 Dec. 2019].

4. Gendered experiences

Experiences of job losses and workplace closures were strongly dictated by gender: dislocated male workers dominate collective memories of deindustrialization, especially in coalfield settings. This focus obscures a longer set of changes that developed across the second half of the twentieth century. Gender relations were altered by increasing women's employment and major changes within the industrial workforce from the 1940s, before sustained contraction and intensified deindustrialization during the 1980s and 1990s. These developments were not just economically important, as the rising cultural status of women's work challenged family and community hierarchies. Scottish coalfield areas remained distinguished by highly patriarchal patterns of employment into the mid twentieth century, but this was significantly altered by rising married women's employment in the welfare state and a diversified economic structure (see table 1.2). The relative working-class prosperity experienced between the 1940s and 1970s was often secured through the paid employment of both parents in nuclear families. In the accelerated deindustrialization that followed, women's new-found status and economic security was undermined alongside the occupational identities and social standing of their brothers, husbands and fathers.

This chapter begins by exploring the operation of gendered social boundaries within the Scottish coalfields. It locates their origins in women's exclusion from underground colliery work during the first half of the nineteenth century and the reinforcement of the 'breadwinner' wage ideology, but emphasizes that these outlooks were always an idealized picture of a more complex reality. Boundaries between private and public spheres were always blurred and women participated in economic life out of necessity, especially in contexts of mass male unemployment and poverty earnings. The second and third sections compare how masculinity and femininity were reshaped by deindustrialization. Colliery closures and redundancy undermined the capacity of both older and younger men to meet their breadwinner obligations and challenged their traditional status as household and community leaders. Women gained opportunities through the extension of employment openings in industrial restructuring and then from service sector expansion. However, these gains were tempered by labour market precarity and shouldering both household and employment responsibilities in the context of renegotiated, but nevertheless, patriarchal, gender relations.

'Gendered experiences', in E. Gibbs, *Coal Country: The Meaning and Memory of Deindustrialization in Postwar Scotland* (London, 2021), pp. 119–53. License: CC-BY-NC-ND 4.0.

Scottish coalfield life was strongly characterized by gendered social boundaries, often marked by physical and geographical barriers. Broadly, they mapped onto the familiar feminine private domestic space and the masculine domination of the public realm. These were never absolute, always contested, and continually renegotiated. Nevertheless, distinct spheres were vital to forming an idealized understanding of gender roles which remained a powerful influence into the late twentieth century, even as its material foundations in gendered divisions of labour were eroded. Male claims to a household breadwinner status were at the centre of this understanding. Masculinity has been defined as 'a configuration of practice within a system of gender relations', which must be understood as 'an aspect of a larger structure'.[1] Hegemonic conceptions of masculinity are shaped by other aspects of social relations and prevalent ideological assumptions. Distinct political standpoints were important in shaping competing conceptions of middle and working-class masculinity. In the case of the latter, trade union wage demands merged with paternalist family ideology to assert the male worker as producer and provider.[2]

Coal miners' commitment to a breadwinner wage was legitimated through nineteenth-century public morality and concerns over the employment of women in dangerous and potentially character compromising underground work. The 1842 Mines Act banned women and boys under the age of ten from working below the colliery surface. It was the product of campaigning by miners and their representatives, but also the efforts of aristocratic politicians and state officials.[3] Polanyi viewed such a heterogeneous 'tangle of interests' as typical of countermovement efforts to embed the operation of industrial capitalism into societal norms.[4] From 1842 onwards, underground coal mining in Britain was therefore male only. Collieries were pivotal to the development of masculinities within localities which often overwhelmingly depended on coal mining for employment. Male domination of public spaces encouraged a highly gendered understanding of class identities. These were performed through intra as well as inter-class plurality of masculinities, especially through the presence of 'rough' as well as 'respectable' working-class masculinities. An upright self-image

---

[1] R. W. Connell, *Masculinities* (Cambridge, 2005), pp. 67, 83.

[2] M. Roper and J. Tosh, 'Historians and the politics of masculinity', in *Men and Masculinities: Critical Concepts in Sociology*, i: *Politics and Power*, ed. S. M. Whitehead (Oxford, 2006), pp. 79–99, at p. 88.

[3] C. Mills, *Regulating Health and Safety in the British Mining Industries, 1800–1914* (Farnham, 2010), pp. 56–66.

[4] K. Polanyi, *The Great Transformation: the Political and Economic Origins of Our Time* (Boston, Mass., 2001), p. 101.

was constructed in part on the judgment of 'failures' often accorded to alcoholics and other men who did not live up to expectations of familial stability or suitable employment. This continued into the late twentieth century. During the 1984–5 strike, strikebreakers were condemned in highly gendered terms. Supporters of the strike portrayed them as 'wasters': poor workers and bad husbands who lacked the gumption to join the struggle for jobs.[5]

An association between mining identities and respectable masculinity was communicated in the oral testimonies. Gilbert Dobby followed his grandfather and father into the industry during 1961. He remembered feeling pride in preserving a family tradition and contributing to the household by handing over his pay packet to his parents after he started work as an apprentice electrician at Arlochan 9 colliery in Coalburn, South Lanarkshire:

> Ma first wage was five pounds and fifteen shillings, which was a lot ae money. Once everything was taken off, I went home wi four pounds and ten shillings. And I remember you felt really proud handin your pay packet over to your Mum sealed, no open. And she counted the money out, took the four pounds and handed me the ten shillings and I thought I was one rich boy! *both laugh* That wis enough money for me tae take a girl to the pictures in the next village, pay her bus fare, pay her in, everything, and still have money left. A lot ae boys went tae the factories in Larkhall, and they were maybe more than a pound a week less than what I was makin.[6]

Gilbert's memories confirm that a pay packet was a 'symbol of power,' which underpinned his household responsibilities and allowed him to court young women. Colliery employment was embedded through familial obligation and community membership. The social status associated with being a contributor to the family was confirmed through the 'respectable' hallmark of Gilbert handing a sealed packet to his mother, who was responsible for the household budget, and being given 'pocket money' of ten shillings from his wage.[7] Gilbert's emphasis on the relative affluence afforded by his wages indicates the changing context of coal employment during the 1960s and 1970s. Gilbert left Coalburn for the prospect of higher

---

[5] D. Morgan, 'Class and masculinity', in *Handbook of Studies on Men and Masculinities*, ed. M. S. Kimmel, J. Hearn and R. W. Connell (London, 2005), pp. 165–77, at pp. 168–72; J. Phillips, *Scottish Coal Miners in the Twentieth Century* (Edinburgh, 2019), p. 254.

[6] Gilbert Dobby, interview with author, Coalburn Miners' Welfare, 11 Feb. 2014.

[7] R. Johnston and A. McIvor, 'Dangerous work, hard men and broken bodies: masculinity in the Clydeside heavy industries, c. 1930–1970s', *Labour History Review*, lxix (2004), 135–51, at pp. 142–3.

earnings in Nottinghamshire when Auchlochan 9 closed during the late 1960s. Jimmy Hood made a similar journey, transferring from the same pit. As a young mining engineer, he perceived transferring as the chance to make 'mega money' at highly productive pits in the English Midlands.[8] The development of earning maximization orientations towards coal employment concerned National Coal Board (NCB) industrial relations personnel. During 1973, a paper was circulated around Scottish officials which analysed changes in sources of social status. As the industry was concentrated in larger depersonalized 'cosmopolitan' collieries less embedded in localized communities, younger miners increasingly engaged in 'conspicuous consumption', while living more private family centred lifestyles.[9] This represents another incarnation of the double movement, which combines gendered perspectives with social standing. Unlike during the 1840s, in the 1970s it was miners' employers who were concerned that their workforce was pursuing a market orientation which undermined the norms upon which the industry depended. Their answer was to attempt to preserve the social incentives that mining had traditionally provided, including the 'male status' of colliery employment.[10]

The effects of increased material expectation should not be exaggerated in terms of the erosion of social structures and associated gendered divisions. During the late twentieth century, the Scottish coalfields contained 'whole areas of life where only men congregated'.[11] Large heavy industrial workplaces and associated trade union activism as well as other key locations of social interaction such as social clubs, pubs and football matches were relatively gender exclusive. The maintenance of gendered spaces within the coalfields was evident from the testimonies of former miners, and not just in Scotland. Peter Mansell-Mullen recalled commencing his work in colliery management in Nottinghamshire during the early 1950s. His wife, who was also undertaking management training, was treated with suspicion by the male workforce. This was rooted in traditions and superstition over women's presence underground:

---

[8] Jimmy Hood, interview with author, South Lanarkshire Council offices, Lanark, 4 Apr. 2014.

[9] TNA, Coal 101/488, J. C. H. Mellanby Lee, Operational Research Executive (Scotland), 'A paper for consideration by the Scottish Area Monday and Friday Absence committee', 8 Aug. 1973.

[10] J. Arnold '"That rather sinful city of London": the coal miner, the city and the country in the British cultural imagination, c. 1969–2014', *Urban History*, xlvii (2019), 292–310.

[11] R. Horrocks, *Masculinity in Crisis: Myths, Fantasies and Realities* (Basingstoke, 1994), p. 167.

Women were treated with considerable reserve. There weren't many in there except as secretaries and so on ... My wife, actually, was a junior management trainee and she went underground. And there was a fire shortly afterwards. And the two were supposed to be connected. So, there was a good deal of prejudices wandering around, or what we now think of as prejudices.[12]

Margaret Wegg recalled hostility towards women in paid work in Cardowan village to the east of Glasgow. Margaret was intermittently employed between leaving secondary school during the mid 1950s until Cardowan colliery closed during 1983, when she lost her job in the pit's canteen. Her wages were necessary to maintain the household. Notably, however, Margaret also underlined that her own employment was organized around that of her husband Jerry, who worked at Cardowan, and her children. During the 1980s, Margaret's status as a working mother and miner's wife earned her the chagrin of a neighbour who felt that by working, she set a negative example for his own wife:

> The miners were very chauvinistic. *laughs* The wife was for the house, you know. In fact, I used to get in to trouble off one chap because he said I was putting ideas in his wife's head. But I had always worked ... Well after oor Dale got to school age, I went back out to work, you know. Because you had to work to keep our house. Well I say, I started work after I was married and that for luxuries, but it got that it was in. The money was in the house and it was to keep the house, you know. But as I say, I've always worked. Always worked at different jobs. I trained as a shorthand typist bookkeeper which I did up tul after I was married, but then you had to take jobs. You took jobs that suited your family circumstances, you know what I mean. That if I could go out [to] any work while the kids were at school. But I had to be in the house when they came back you know. My mum brought ma brother and I up, why should she bring my kids up? You know, that's up to me to bring ma kids up. That's how I say, up tul they got to the age when they were old enough you know. But by then you could only get the jobs you know, you still took jobs whatever you could get. Well I've worked in a clothing factory, I've worked in the bottling plant, I've worked in shops, I worked in the pit canteen as I say you just took, took what job you could to suit in with the family you know.[13]

Margaret's life story indicates that women's wages were often vital to obtaining economic security despite breadwinner wage aspirations. This sits within a long historical continuity, but there were distinct regional and temporal variations within Scotland.[14] The 'metal working and mining towns

---

[12] Peter Mansell-Mullen, interview with author, residence, Strathaven, 3 Oct. 2014.
[13] Margaret Wegg, interview with author, residence, Stepps, 17 Nov. 2014.
[14] A. Hughes, *Gender and Political Identities in Scotland, 1919–1939* (Edinburgh, 2010), p. 19.

and villages of Fife and Lanarkshire' often lacked significant labour market opportunities for women until the mid twentieth century. These can be counterpoised with both textile areas where female work, often of a skilled nature, was widely available, and the 'diversified female labour market' of larger cities such as Glasgow.[15] Jessie Clark recollected women's employment in the textile factories of Lanark, South Lanarkshire, indicating a longer history of women's industrial employment even in such areas, which is also recorded in table 1.2.[16] Within the Shotts focus group, memories of women who worked on the surface of collieries sorting coal under the private industry were also present.[17] However, Margaret's memories of the expanded employment opportunities available after the Second World War, and the Wegg household's pronounced dependency on her formal employment, also confirms the impact of significant labour market changes discussed in greater detail below.

Coal mining's gender exclusivity continued to shape attitudes towards work and employment during the late 1980s. Billy Ferns recalled that during 1988, when he was in his early fifties, he unilaterally chose to take redundancy from the Solsgirth mine in the Longannet complex after being approached by Willie Doolan, who was a colliery union official. He had begun working at Solsgirth after being transferred following the closure of Cardowan in 1983. Billy justified not consulting his wife because she had never worked underground. In his view, she did not have the necessary experience to provide judgment on a decision which went on to shape both of their lives:

> Willie says, 'go and draw your cheque before you put your clothes on your going for an interview there's men getting out'. He says, 'they're getting rid of men'. So, I went in to see them. So, they says, 'Well right, gonna tell you what you're gonna get if you're interested'. In fact, they told 'what you want to do is take that away with you and go home and speak it over with your wife and see what she thinks aboot it'. I says, 'excuse me I'm no going home to thingmae with ma wife'. I said, 'where do I sign?'. And the two of them looked at one another and they said, 'to sign?' I said 'the reason I'm going to sign rather than talk it over with my wife. My wife never put pit boots on in her life, she never worked doon the pit', 'ah but do y'know?' I said 'ma wife'll be fine with ma judgment', and she was to this day. [I] signed it and that was it, never looked back, never looked back after it.[18]

---

[15] A. McIvor, 'Women and work in twentieth century Scotland', in *People and Society in Scotland*, iii: *1914–1990*, ed. T. Dickson and J. H. Treble (Edinburgh, 1992), pp. 138–73, at p. 144.

[16] Jessie Clark, interview with author, residence, Broddock, 22 March 2014.

[17] Shotts focus group, Nithsdale Sheltered Housing Complex, Shotts, 4 March 2014.

[18] Billy Ferns, interview with author, residence, Bishopbriggs, 17 March 2014.

Billy achieved composure through an appeal to the breadwinner model. The insecurity of mining employment contributed to this situation. When faced with an offer that could be withdrawn, Billy saw it as his obligation alone to swiftly negotiate the scenario to a conclusion which offered his family economic security. With strong parallels to Walkerdine and Jimenez's study of 'Steeltown' in the South Wales valleys, in the Scottish coalfields, 'masculinity was produced out of a certain distance from femininity' through the demarcation of places, spaces and social roles.[19] Rhona Wilkinson recalled that gender distinctions contributed towards feeling removed from her family's coal mining connections in Breich, West Lothian. She remembered collecting racing pigeons for her grandad when they returned from training flights and being 'fascinated' by the bird's homing abilities, but 'being a girl I wasn't really allowed into the doocot [enclosure]'. Gendered exclusion extended more directly into coal mining itself, with Rhona unable to join her male relatives in trips to the colliery, and potentially in hearing details about her grandfather's underground experiences:

> My granda worked Woodmuir … Never spoke about it though. No, none of them did. None of them spoke aboot it. But it really, I was kind of excluded cause as a girl, and I'm the only girl. So, I grew up with like male relatives, boy cousins, boy brother, you know, that sort of stuff. They used to get to go for showers at the pit heid and all that sort of stuff, which obviously I couldnae be part of.[20]

Duncan Macleod's testimony confirmed Rhona's inclination that boys were more readily socialized into mining culture. Duncan's father was trained as a colliery blacksmith in Lanarkshire, but later became a police officer in Derbyshire after an inter-colliery transfer. Nevertheless, as a young boy during the 1950s, Duncan was also introduced to the world of mining through pit baths:

> I grew up among miners and the pits were working in Derbyshire at the time. I mean, I used to go down to the pits with my Dad. We used to go down and use the pit baths. Although my Dad was a policeman, he knew a lot of miners. Because my Dad was a blacksmith, he used to go down to use the tools to make me a swing or whatever he happened to be making.[21]

---

[19] V. Walkerdine and L. Jimenez, *Gender, Work and Community after De-Industrialization: a Psychosocial Approach to Affect* (Basingstoke, 2012), p. 174.

[20] Rhona Wilkinson, interview with author, residence, Fauldhouse, 7 Nov. 2014.

[21] Duncan and Marian Macleod, interview with author, residence, Carluke, 1 March 2014.

Ian Hogarth provided another vantage on the highly gendered colliery environment. He explained the precautions that had to be taken before a Workers Educational Association study group took part in an underground visit at Cardowan, specifying in relation to the pit's baths and lockers that 'We were not going to let ladies loose in the baths cause the idea of proprietary among the boys just didn't exist ... You weren't going to let thirty women loose. You might do it today, most certainly not in 1956'. This unusual instance of cross-gender contact underground proved formative for Ian, who led the visit as a young managerial employee. His future wife, Muriel, met him for the first time when she attended the tour.[22] Sam Purdie's testimony affirmed the gendered dynamics of Scottish heavy industry. He elaborated on his experience as a manager in Clydeside shipyards during the 1970s in terms that underlined the tough masculine attitudes he had learned in Ayrshire collieries during the previous two decades: 'Ma background in the pits helped. If an academic had gone into that, or somebody who had worked as a personnel manager in a ladies' hosiery factory had gone in and tried to do that, they wouldn't have succeeded'.[23] Sam's memories also revealed facets of a 'hard man' culture in relation to experiences of death and disaster. As a young colliery engineer, Sam discovered that loss of life in the industry was 'regular', and explained that masculine expectations led miners to internalize their grief and carry on with their jobs. Sam contrasted these routine deaths with the collective grief associated with the death of seventeen men in an underground explosion at Kames colliery on 19 November 1957. However, it was still characterized by the expectation of a normalized return to work:

> Kames pit killed about one man a year. It was regular. We'd usually take the day off and go to the funeral. Kames disaster, that was different, we took about three days to bury them. And I suppose [there was] no such thing as counsellors, but we had the community spirit. And as they used to say, men don't greet. And so, we just all went back.[24]

Willie Doolan recollected that his father, who was a rescue worker at the Auchengeich disaster two years later, similarly never spoke about his experience. He generalized that 'mining was a dangerous game, a dangerous industry to work in. Deaths were pretty commonplace within the industry. Fatalities were a common thing. But I think you could say it was a trait

---

[22] Ian Hogarth, interview with author, National Mining Museum, Newtongrange, 28 Aug. 2014.

[23] Sam Purdie, interview with author, UWS Hamilton campus, 3 May 2018.

[24] Sam Purdie, interview.

of the miners nationally that they never spoke about these things'.[25] These attitudes are perhaps indicative of the 'silences' oral historians have identified as common in instances of individual and collective trauma.[26] They also preserved a masculine sphere from which women and children were excluded, and secured an occupational identity anchored in masculinity. In their influential mid 1950s study of a West Yorkshire mining village, *Coal is our Life*, Dennis et al detailed that miners exhibited 'a pride in the fact that they are real men who work hard for their living, and without whom nothing in society could function'.[27] Similar combinations of occupation, class, and nation and gender consciousness developed in the Scottish coalfields. Mick McGahey, who worked as a surface worker at Bilston Glen colliery in Midlothian during the 1970s and 1980s, explained the key role that miners had played in developing British industry and supporting the economy at major points of national distress. These feelings were intertwined with and embedded in kinship and family connections to the industry. In Mick's case these included his father, Michael, who was the NUMSA president from 1967 to 1987 and grandfather, Jimmy, a blacklisted miner and founder member of the Communist Party of Great Britain (CPGB), who was jailed for his involvement in the 1926 general strike and miners' lockout. This lineage bolstered the moral economy status of colliery employment:

> Miners took a pride in the fact that they were producing coal which kept Britain going. You know, the reality is that you know we talk aboot, you know, the great wars: World War One, World War Two, these wars were kept going because miners produced coal that produced the steel, that made the guns and bullets and aw the rest ae it ... So, they seen themselves as being an integral part ae the economy. Mining kept the wheels ae industry turning and we knew that. And that's why they talk about coal being king and so on. But it was, it was aw aboot that, and it was aboot places for jobs for their kids, their families.[28]

Within the Scottish experience of deindustrialization, these factors were crucial in the articulation of opposition to contraction and closures. They were not only economic threats. Closures and redundancy undermined strongly held collective identities and the self-image of male industrial

---

[25] Willie Doolan, interview with author, The Pivot Community Centre, Moodiesburn, 14 June 2019.

[26] L. Passerini, 'Work ideology and consensus under Italian fascism', *History Workshop*, viii (1979), 82–108, at p. 91.

[27] N. Dennis, F. Henriques and C. Slaughter, *Coal is Our Life: an Analysis of a Yorkshire Mining Community* (London, 1956), p. 33.

[28] Mick McGahey, interview with author, Royal Edinburgh Hospital, Edinburgh, 31 March 2014.

workers. This was evident during the Upper Clyde Shipbuilders (UCS) work-in of 1971–2 when its public leader, Jimmy Reid, overtly associated shipbuilding with constructions of masculinity by stating, 'We don't only build ships on the Clyde, we build men'. Reid's other infamous uttering from the dispute, that there would be 'no bevvying' at the work-in, demonstrates how industrial identities overlapped with notions of respectability.[29]

The work-in was characterized by a rhetorical fusion of class, nation and masculinity which underlined the morality of the argument against closure. At a meeting of shop stewards from across Britain, Reid asserted that by occupying the yards, Clydeside shipbuilders had 'reasserted the dignity of working men'. Their objective was to 'establish that they've got rights, and they've got commitments and privileges and principles, and they are going to utilize their ability and capacity to resist these measures, to fight and to unite around them their brothers and sisters'.[30] The work-in was a struggle for dignity that implicated industrial employment into the heart of communal identities. In the terms of the double movement, it was a defence of the socially embedded industrial economy against liberalizing market imperatives. Reid titled his autobiography *Reflections of a Clyde-Built Man*, which indicates that the imagery of shipbuilders he espoused during the UCS dispute connected to a strongly held sense of self.[31] This outlook received widespread support across Scotland during the work-in because it had a deep-seated basis within the experience of industrial communities and chimed with the imagery of a distinctive Scottish 'industrial nation'.[32]

Miners and their leaders also commonly presented themselves as self-sacrificing male workers. In 1957, Alex Moffat, the NUMSA's vice-president, and future president, contributed to a discussion about industrial diversification and employment for disabled miners by underlining 'the social responsibility of the government to ensure that the work was found for those [disabled] men so that they could become part of an active working community'. Moffat was speaking to a resolution submitted from Newton colliery. Michael McGahey had moved the resolution on behalf of his branch by arguing that continued employment and retraining was the 'minimum responsibility accepted by the NCB towards these men who

---

[29] Johnston and McIvor, 'Dangerous work', pp. 138–43.

[30] J. Foster and C. Woolfson, 'How workers on the Clyde gained the capacity for class struggle: the Upper Clyde Shipbuilders' work-in, 1971–2', in *British Trade Unions and Industrial Politics*, ii: *the High Tide of Trade Unionism, 1964–1979*, ed. J. McIlroy, N. Fishman and A. Campbell (Aldershot, 1999), pp. 297–325, at p. 306.

[31] J. Reid, *Reflections of a Clyde-Built Man* (London, 1976).

[32] J. Foster and C. Woolfson, *The Politics of the UCS Work-In: Class Alliances and the Right to Work* (London, 1986), pp. 322–5.

have sacrificed their health in the mining industry'.[33] The moral economy was reinforced by an obligation towards men who had put themselves at harm for the collective good.[34] This rhetoric stood miners in parallel with other occupational groups such as soldiers who are depicted as masculine heroes facing dangers in the national interest.

Stories of workplace strife and heroism have often been reprised in narratives of class struggle which depict cohorts of coal miners in a similar manner to generations of military combatants.[35] During the mid 1970s, the Scottish nationalist poet, T. S. Law, exhibited this trend. Law was the son of a Lanarkshire coal miner. In light of the significant victories achieved through industrial action in 1972 and 1974, he lauded the 'men of a different temper' who 'began to impose their ideas of independence upon the structures of coercion, intimidation and violence' following the expansion of coal mining during the nineteenth century and through the major industrial struggles of the twentieth.[36] Participant comments on the 1984–5 strike replicated Law's frame of reference. Billy Ferns recollected that he was 'sent' to picket in Nottinghamshire along with four other Kirkintilloch miners with whom he shared a strong bond and comradeship. This collective strength saw them through the 'terrible' events of police violence at Hunterston ore terminal in Ayrshire and at Orgreave coking plant in Yorkshire. In both cases, Billy and his comrades joined other Scottish miners on early morning journeys to attend mass pickets. A sense of occupational distinction was affirmed through Billy's exclusion of non-combatants from violent elements of the strike. He recalled instructing a non-miner regular at Kirkintilloch Miners' Welfare that 'you cannae get involved in this' after the man began to confront off-duty police officers who were goading Billy.[37] These memories indicate that while British coal strikes were not characterized by the same levels of violence as American miners' disputes, they also instigated a sense of battle experience informed by masculine sensibilities.[38]

---

[33] National Mining Museum Scotland archives, Newtongrange, Midlothian (NMMS), National Union of Mineworkers Scottish Area (NUMSA), Minutes of Executive Committee and Special Conferences from 18 June 1956 to 5 to 7 June 1957, pp. 725, 814.

[34] A. McIvor and R. Johnston, *Miners' Lung: a History of Dust Disease in British Coal Mining* (Aldershot, 2006), p. 63.

[35] D. Nettleingham, 'Canonical generations and the British left: narrative construction of the British miners' strike, 1984–85', *Sociology*, li (2017), 850–64, at p. 852.

[36] T. S. Law, 'A Wilson memorial', *New Edinburgh Review*, xxxii (1976), 22–8, at p. 25.

[37] Billy Ferns, interview.

[38] S. N. Horwood, *Strikebreaking and Intimidation: Mercenaries and Masculinity in Twentieth-Century America* (Chapel Hill, N.C., 2002), p. 115.

Billy's description of more routine activities during the 1972 and 1974 wages strikes allude to the continuation of a gendered division of labour. He described miners from Bishopbriggs in East Dunbartonshire searching for firewood and raiding a coal ree in Ayrshire to provide fuel for mining families and elderly people who relied on NCB supplies.[39] Billy's memories parallel the earlier 1926 miners' lockout when women ran soup kitchens and men risked arrest gathering fuel and rustling livestock.[40] Yet the enforcement of these spheres was never absolute. Women's economic activities punctured the breadwinner ideal out of economic necessity, especially in times of industrial conflict or male mass unemployment. Young girls were 'caught up in the survival practices' of coalfield households from an early age.[41] Jessie Clark explained that after her father was blacklisted following his involvement in the 1926 miners' lockout, she used to help her mother pick coal from bings in Douglas Water, South Lanarkshire:

> I don't know if you've ever heard about people in mining villages picking the coal off the bings? Right well that happened in oor village as well, and it happened usually wi the guys that were [unemployed]. I mean my mother did it and I went with her, you know? And it was partly a walk. It was a family sort ae outing as it were! *laughs* And you picked the coal as well, and that would be was mainly at the time when ma father wasn't working. Although you got coal a bit cheaper, you still couldnae afford to buy it.[42]

Jessie described a complex arrangement whereby the Coltness Iron Company would defer rent charges while miners were unemployed, but then impose double payments when they returned to work. These practices created a strong interdependency typical of paternalistic links between coal companies and communities.[43] Paternalism consolidated itself through bridging the gap of the double movement. The breadwinner model was protected by measures which constrictively embedded colliery employment into family and community routines, partly by creating bonds of obligation that extended outside the immediacy of hourly wage earnings. Over the

---

[39] Billy Ferns, interview. Ree is a Scots word for a store.

[40] S. Bruley, 'The politics of food: gender, family, community and collective feeding in South Wales in the general strike and miners' lockout of 1926', *Twentieth Century British History*, xviii (2007), 54–77, at pp. 66–7.

[41] H. Barron, 'Women of the Durham coalfield and their reaction to the 1926 miners' lockout', *Historical Studies in Industrial Relations*, xxii (2006), 53–83, at p. 73.

[42] Jessie Clark, interview. Bing is a Scots term for slagheap.

[43] P. Ackers, 'On paternalism: seven observations on the uses and abuses of the concept in industrial relations, past and present', *Historical Studies in Industrial Relations*, v (1998), 173–93, at p. 178.

second half of the twentieth century, sustained changes to the industrial structure of the Scottish coalfields led to significant alterations as married women's employment increased and gained cultural recognition. However, the expectation that men would continue to act as the prime household provider continued and supported exclusionary practices which sustained a form of masculinity strongly characterized by male-only or male-dominated social spaces. As deindustrialization intensified, it posed a growing threat to male identities.

Male responses to deindustrialization

Deindustrialization posed a major challenge to masculine sensibilities within coalfield communities, especially as it eroded men's occupational identities and breadwinner status. These effects accelerated in the environment of mass unemployment that developed as workplace closures and job losses intensified during the 1980s. Labour market trajectories separated generations of men, differentiating older men who withdrew from employment from younger men who left industries for service employment, or never found work in industrial workplaces as they had anticipated. In John Slaven's case, entering the labour market as opportunities dried up entailed being sent on the Youth Training Scheme while claiming unemployment benefits before he eventually left Lanarkshire to work on the railway in London. John dubbed this 'the most blatant attempt at manufacturing the unemployment figures that I've ever seen'.[44] For the older generation facing industrial contraction in the 1980s, it was the social democratic environment which had characterized the years between 1945 and the late 1970s that informed their perspective. These reference points added to the dislocation experienced upon redundancy, and the threat to their established social position. John Slaven summed this up by claiming that most male industrial workers in their fifties and sixties who were made redundant during the 1980s survived but felt deprived of their former status: 'I think there was cultural social poverty but economically people kinda got by. Maybe on a slightly [lower] living standard, so there was a lot ae people that just didnae work again'.[45] John's father was laid off in the mid 1980s after almost three decades of employment at the Caterpillar tractor factory in Tannochside, North Lanarkshire. He was officially retired due to an industrial injury, but this related to losing a hand in an industrial accident two decades previously. These experiences exemplified a broader trend which led to an older generation of men experiencing social redundancy:

[44] John Slaven, interview with author, STUC Building Woodlands, Glasgow, 5 June 2014.
[45] John Slaven, interview.

I think there was a lot ae issues with alcoholism. My dad was very much a physical manifestation. I think people definitely seemed vastly reduced, y'know, because they were just the wrong age to go back to uni, to retrain, or do stuff. They just werenae of that generation. So y'know, there was an awful lot of guys who when I look back on it were quite young guys in their forties and their fifties with big redundancy payments. Maybe with the culture then, their kids grown up, cause you had kids in your twenties then, not in your forties and fifties like now. Kind of, shuffling out the last twenty-five years o their life no working again. Going to the old club, going to the pub, no doing very much to be honest. I think, y'know, I would love to know how many people, men ae Lanarkshire, done that.[46]

Within their distinct generational contexts, John and his father were both incorporated into government attempts to mask the labour market impact of the contraction of industrial employment. John's testimony accords with statistical analysis of labour market restructuring in UK coalfield areas. Official unemployment rates fell between 1985 and 1994, despite over 200,000 redundancies taking place within the coal industry, due to the operation of state incentives to withdraw from the labour market. The relaxation of incapacity benefit requirements, and their adjustment to be more financially rewarding than unemployment benefits, contributed to large-scale male labour market withdrawal, and older men taking early retirement.[47] John's comments are also redolent of Daniel Wight's observations from the coal and steel community of 'Cauldmoss' in central Scotland where he found men 'relying on the masculinity of one's work for one's self-respect'. Male identities were preserved by unemployed men retaining strongly held kinship groups which gathered outside the feminine space of the home. Men who had been made redundant struggled to adapt to the changed environment and attempted to preserve a gendered division of labour. This led to the spectacle of the 'carrier bag brigade' disguising their activities when doing the family shopping which was understood as a woman's activity.[48]

However, these experiences were not universal. Jennifer McCarey recalled that when Ravenscraig steelworks closed, her family 'were all really worried about how he [her father] would take it. He was up for work half past six every day. And even when he went to college, he did the studying even though he knew he would never use the qualification. They only did it to

---

[46] John Slaven, interview.

[47] C. Beatty and S. Fothergill, 'Labour market adjustment in areas of chronic industrial decline: the case of the UK coalfields', *Regional Studies*, xxx (1996), 627–40, at pp. 631–5, 644.

[48] D. Wight, *Workers Not Wasters: Masculine Respectability, Consumption and Unemployment in Central Scotland: a Community Study* (Edinburgh, 1993), pp. 37–47.

get a year's money'. Jennifer's father adapted more happily to his new status upon his retirement at the age of fifty-seven than John Slaven's father. As well as contentedly becoming 'a professional golfer', 'he became someone that looked after the grandweans and it was great'.[49] Male experiences of deindustrialization were varied, even in relative proximity. They often differed according to how far a man could reconstruct their identity in a way that preserved their community and family standing.

Strong feelings of a reduced social status were also visible in testimonies from men of younger generations. Scott McCallum recalled that he had grown up anticipating following his grandfather, father and brother into coal mining. The absence of this opportunity following the closure of Cardowan colliery in 1983, and the rundown of the Scottish coalfield after the miners' strike, had cultural as well as economic costs:

> I probably would have worked in it maself if it hadn't have closed, it's like a family generation thing … Oor kind ae education was to leave school and go and work in the mines. You never really stuck in much. That was what the plan was. You'd leave school and go and work in the coal mines … I feel at the time it felt, like sad because you weren't following in the family's footsteps: your brother worked there, your dad worked there, your granddad worked there. So kindae breaking up a family tradition you can say. But that was the problem. My brother that did work in it sometimes says, 'you don't know cause you didn't work in the pits you had to be there', but I had a good knowledge, more than a lot of people in school, of what it wis like.[50]

Workers in their late thirties and younger had differentiated experiences of intensified deindustrialization from those in their forties and fifties who withdrew from the labour market. Younger men faced the difficulties of raising families while navigating the benefit system and less secure or well-remunerated employment. In the Durham coalfield, former miners attempted to maintain collective support mechanisms that held existing social routines together. These included getting up early to walk dogs together, and the ongoing presence of the Durham Miners' Association.[51] There were parallels in the retention of the retired miners' group in Moodiesburn, North Lanarkshire, who have continued to meet monthly and who acted as a focus group.[52] Displacement was most strongly experienced by men

---

[49] Jennifer McCarey, interview with author, iCafé, Woodlands, Glasgow, 9 Oct. 2014.

[50] Scott McCallum, interview with author, The Counting House, Dundee, 22 Feb. 2014.

[51] T. Strangleman, 'Networks, place and identities in post-industrial communities', *International Journal of Urban and Regional Research*, xxv (2001), 253–66, at pp. 253–7.

[52] Moodiesburn focus group, retired miners' group, The Pivot Community Centre, Moodiesburn, 25 March 2014.

who no longer had a stable counterpoint to the domestic sphere, which was constructed as feminine. In response, associations were maintained and, in some cases, relocated to the street, which replaced the workplace as a site of male bonding. These routines are emblematic of the 'crisis' of masculinity that occurred when declining male workforce participation coincided with retained patriarchal expectations during the late twentieth century.[53] Rhona Wilkinson remembered that younger men made redundant from local collieries reprised their social bonds in public spaces within the West Lothian coalfield and removed themselves from domestic settings:

> There was a change in the villages. Suddenly, there was a corner in Fauldhouse known as Lawrie's corner. And as a kid all the men used to stand at it. But it was older men. But then, there was a younger influx. They would all stand there. But it was a lot of men who got money from the pits aw at one time. They'd never had any money. They had nowhere to go, and they drank it, you know. So, it's like, what do you do when you're so isolated? You've had your identity pulled away from you as well. Not just a livelihood. You know, I remember seeing things like that. That changed.[54]

Alcohol abuse emerged as a response and coping strategy for social displacement across generations. This is confirmed by the memories of both Billy Ferns and Brendan Moohan. Billy remembered the disastrous impact that redundancy had on his work colleagues at Cardowan who were largely in their early fifties. He felt the removal of work contributed to excessive drinking and early deaths:

> When they closed them that was, you, you'd had it. I know quite a few ae the boys at Cardowan had taken their redundancy they were just fifty, fifty-one. Most ae them are all dead. Most ae them were all dead by late fifties, sixties. Well, some ae them just couldnae handle it went to drink and that. Nae job and whatever. Nae lifestyle then.[55]

Brendan Moohan recollected the trauma of being dismissed towards the end of the miners' strike in February 1985 after he was arrested picketing during the summer of 1984. As a young man of twenty, Brendan struggled to cope with losing employment despite winning a subsequent industrial tribunal:

> I won ma tribunal. I was supposed to be reengaged. Although it was the

---

[53] C. Haywood and M. Mac An Ghaill, *Men and Masculinities: Theory, Research and Practice* (Buckingham, 2003), pp. 22–3.

[54] Rhona Wilkinson, interview.

[55] Billy Ferns, interview.

government's industrial court, it wisnae law abiding. So, I was never brought back in. I was given some compensation, which I drank. Though to be fair I didn't realize it at the time, but the impact of the miners' strike had been huge on me personally. And then somebody gives you, you know, eight grand in compensation. You've never seen eight thousand pounds. Whittled it within a year.[56]

Brendan also recalled that he was not alone in this experience, telling the story of a friend who eventually died after succumbing to alcoholism in the aftermath of the strike.[57]

Michael McMahon remembered that at the Terex factory in Holytown in North Lanarkshire, workers struggled with the disembedding of industrial workplaces from established social routines. During the 1980s, the plant moved towards a 'just-in-time' production regime, which disrupted the workplace's culture:

> We were moving towards 'just in time'. We stopped the five-day week. We worked day shift, night shift and then it went to what was called the double day shift. So, you started at six and worked to two, and then the backshift came in and started at two and worked to ten. And there was no nightshift, except for maintenance, or just a handful of people to keep certain machines going through the day. So, I was involved in seeing that change and how that impacted on peoples' lives. A lot of men just couldn't come to terms with working backshift. The old traditional things you would do, go for a pint, and then go for work, or whatever, or finish your work and go for a pint. People had to rethink how they did that. That was some of the issues. It wasn't so much that working two o'clock to ten o'clock was any more arduous than a nightshift. It probably wasn't. In terms of their social life it had a huge impact. That's what they were railing against.[58]

The social disembedding of industrial production through the assertion of market logic was visible in other testimonies in more extreme forms. Alan Blades worked at the short-lived Chunghwa factory in the Eurocentral industrial estate near Airdrie after accepting redundancy from the Longannet complex in 1997. Chungwha lacked the social connections evident in the recollection of earlier Lanarkshire assembly goods factories such as Caterpillar and Terex, which opened during the period of social democratic economic management and relative trade union strength. Alan described the non-union factory as having a prohibitive and even anti-social working atmosphere:

---

[56] Brendan Moohan, interview with author, residence, Livingston, 5 Feb. 2015.
[57] Brendan Moohan, interview notes.
[58] Michael McMahon, interview with author, constituency office, Bellshill, 21 Feb. 2014.

Stricter, stricter, aye. Yer dealin wi boys from Taiwan. They were the managers. They were brung over fae Malaysia and Taiwan. It was a Taiwanese company, and like, y'know. It's hand up tae go for a pee, y'know what I mean? Can I go to the toilet? You're at one bit aw the time. At the pit yer movin aboot, talkin tae boys. Y'know, communicating and that. No camaraderie. Nae banter. Doon the pit there was banter, y'know. You werenae allowed tae talk.[59]

Alan recalled working twelve-hour shifts at weekends without an overtime rate, but mostly reserved his moral economy anger for the 'white elephant' status of the factory which received extensive public funding but failed to deliver the long-term employment the company promised.[60] It was the context of growing economic insecurity across the engineering sector that convinced Michael McMahon to leave Terex and pursue studies at Glasgow Caledonian University instead:

Things were not going well. We were forever on short-time working at the plant. Sometimes it was three-day weeks. Sometimes it was one week on one week off. I just thought, the shipyards were closing. The oil rigs were starting to run out of work. I just thought, where can you go? It was time to change and I went to university to do politics and sociology.[61]

Michael, who had served as a workplace union representative and on the Scottish Trades Union Congress youth committee, had developed skills that assisted him in adapting to a labour market increasingly dominated by the service sector. This was true of several of the interviewees who had been relatively young men when they left or lost industrial employment. In some cases, it was not the experience they had gained in their jobs that furnished them with the necessary skills for future employment. Instead, it was the political culture of large workplaces, principally labour movement activism, which had helped some men develop competences they used in later life. Brendan Moohan, who had been politically active through the Trotskyist Militant tendency within the Labour party in East Lothian, also 'reinvented' himself, partly by deploying the skills he had learned as an activist. Brendan became involved in youth work and went on to study community education at Edinburgh University. He remembered feeling driven towards these steps by his experience of political involvement. Brendan noted that other 'politicals' among his former colleagues had been most likely to later study at university. They subsequently found jobs that utilized their skills as organizers, orators

[59] Alan Blades, interview with author, resident, Airdrie, 26 Feb. 2014.
[60] Alan Blades, interview.
[61] Michael McMahon, interview.

and representatives by becoming teachers, welfare rights officers and elected politicians.[62]

Other former miners found work in the public sector via alternative routes. Mick McGahey was also victimized during the miners' strike and lost his job as a surface worker at Bilston Glen, in Midlothian. He described himself as 'a refugee fae the pits y'know. There was nowhere else fir me to go so I got a job here wi the NHS [as a porter]'. Mick recalled that several former miners had worked alongside him as porters and care assistants at both the Astley Ainslie Hospital in Edinburgh and the Edinburgh Royal, where he was then working and was the Unison convener.[63] To some extent, Mick was able to find solace and continuity in this story by an appeal to the history of 'industrial gypsies' like his grandfather and father who had similarly been blacklisted and had to relocate between collieries. Mick's retained involvement in trade unionism also provided another important source of pride and continuity. However, the finality of his removal from colliery employment created a significant lacuna. Mick claimed he could trace back coal mining through eleven generations in his family tree but remarked three times in the interview that he was from 'the last generation' of Scottish or British coal miners.[64] This sense of removal from a trodden path illuminates how far deindustrialization represented a pivotal challenge to masculine identities in the Scottish coalfields.

'A sense ae opportunity': women's experiences of labour market changes

Women shared the communal experience of deindustrialization with men, but gender dynamics significantly differentiated perspectives. Female workers were among the beneficiaries of growing labour market opportunities and an accompanying rise in social status during economic restructuring after the Second World War. Light manufacturers prioritized attentiveness and dexterity, qualities typically socially constructed as feminine, as opposed to the physical strength demanded in coal mining and steelmaking. Assembly plants also provided a cleaner and safer environment than heavy industries. Women's industrial work developed a moral economy status due to rising family dependence upon female earnings, and the increasing prevalence of married women's employment. This was accompanied by women placing increased value on paid work, which became strongly integrated into personal life-stories and attached to collective narratives of incremental social progress.

---

[62] Brendan Moohan, interview and interview notes.
[63] Mick McGahey, interview.
[64] Mick McGahey, interview.

The *Clyde Valley Regional Plan* viewed women's labour as a necessary and valuable contribution towards postwar industrial diversification.[65] Policymakers in the Scottish Office and the Board of Trade shared this outlook. They expressed significant worries over a shortage of female labour. In 1946, the Board of Trade warned that industrial reconstruction might be hampered by 'indications that in certain areas the danger line was being approached in regard to shortage of women'.[66] This was followed by a warning in 1948 that regions were guilty of 'the overselling of female labour' reserves in order to attract new industrial development.[67] These concerns grew during the 1960s, as regional policy shifted from diversification to economic growth aims following the publication of the Toothill report. This involved a growing reliance on the assembly goods manufacturing sectors that demanded large female workforces.[68]

Women workers' rising importance in policymaking agendas was combined with the retention of highly gendered attitudes and priorities. Forward planning projections continued to classify industries or occupations as 'male' or 'female'.[69] When presenting figures for unemployment in the west of Scotland in 1949, Ministry of Labour officials referred to a 'hard core of unemployment' that contained around 20,000 people. As was customary, the figures were divided by gender and categorized as either 'primary' or 'secondary' labour. The latter incorporated those judged to be on the periphery of labour market involvement. Around two thirds of the 20,000 unemployed workers in question were married women classified as secondary, and officials noted that managers were 'not keen to employ them owing to possible absenteeism'.[70] Despite the promise of expanding married women's employment, civil servants accepted gendered attitudes towards the domestic division of labour to the extent of sharing employers' assumptions that married women's obligations as mothers and wives would prohibit them from making a full commitment to paid work. The retention of this perspective was confirmed in 1952 when Scotland had 'a problem

---

[65] P. Abercrombie and R. H. Matthew, *The Clyde Valley Regional Plan, 1946* (Edinburgh, 1949), p. 77.

[66] National Records of Scotland (NRS), Scottish Economic Planning (SEP), 4/690/48, Minutes of the nineteenth meeting of the research committee, Board of Trade, Glasgow, 6 June 1946.

[67] NRS, SEP 4/784, Progressive statement no. 4, 1948.

[68] NRS, SEP 4/781, Scottish Physical Planning Committee: Population Working Party comparison of estimated labour requirements and estimated population in Scotland for the year 1961–62, 1961.

[69] NRS, SEP/4/784, Progressive Statement no. 4, 1948.

[70] NRS, SEP 4/1199, West of Scotland District unemployment figures, 10 Oct. 1949, p. 9.

of finding an industry employing largely semi-skilled and unskilled men'. This was despite the growth in unemployment between 1951 and 1952 being entirely due to an increase in women's joblessness. Over half of the secondary unemployed labour category, which totalled 27,000, remained married women.[71]

There was a marked, if gradual, shift in policymakers' attitudes, which was discernible by the early 1960s. The Scottish Physical Planning Committee forecast that to achieve satisfactory economic growth rates over 1961–2, women's involvement in the workforce would have to increase, and that this would best be attained by encouragement through higher wages.[72] During the 1970s, analysts viewed the employment of women within the electronics and instrument sectors positively. Inward investment provided employment opportunities in regions formerly dominated by heavy industry where female employment had often been limited. Both academic economists and the Scottish Council (Development and Industry) assessed these developments favourably.[73] The expansion of part-time employment contextualized growing agitation for women's rights and legal changes including the 1970 Equal Pay Act and the 1975 Sex Discrimination Act. Up to the 1970s, this was achieved with a significant concentration in manufacturing. Often such workers were married women, especially those with children, either re-entering employment after childbirth or maintaining paid work.[74] In 1979, East Kilbride Development Corporation underlined the importance of female industrial employment to working-class households. They ruled it 'unacceptable' to view the opportunities that the new town's manufacturing industries provided for women, who commuted from across Lanarkshire, as less valuable than male jobs.[75]

---

[71] NRS, SEP 4/1199, Unemployment in Scotland: Dec 1951–Feb 1952.

[72] NRS, SEP 4/781, Scottish Physical Planning Committee: Population Working Party comparison of estimated labour requirements and estimated population in Scotland for the year 1961–62, 1961.

[73] N. Hood and S. Young, 'US investment in Scotland: aspects of the branch factory syndrome', *Scottish Journal of Political Economy*, xxiv (1976), 279–94, at pp. 281–2; Scottish Council Research Institute, *US Investment in Scotland* (Edinburgh, 1974), pp. 4–5.

[74] C. Wrigley, 'Women in the labour market and the unions', in *British Trade Unions and Industrial Politics*, ii: *the High Tide of Trade Unionism, 1964–1979*, ed. J. McIlroy, N. Fishman and A. Campbell (Aldershot, 1999), pp. 43–69, at pp. 55–7; S. Connolly, 'Women and work since 1970', in *Work and Pay in 20th century Britain*, ed. N. Crafts, I. Gazeley and A. Newell (Oxford, 2007), pp. 142–75, at p. 153.

[75] NRS, SEP 15/437, G. C. Cameron, J. D. McCallum and J. G. L. Adams, The contribution of East Kilbride to local and regional economic development: an economic study conducted for the East Kilbride Development Corporation, East Kilbride, June 1979.

Both the oral testimonies and archival records underline that the social gains and labour market opportunities that women achieved during this period were within the restructuring, rather than elimination, of patriarchal relations and gendered norms and values. During the Shotts focus group, Betty Turnwood recollected that before the Second World War, 'A woman's job was really quite an important job and a heavy job in the house. Not like today when you can just put things in the washing machine'. The introduction of pithead baths, central heating and domestic appliances, alongside the availability of employment within the expanded welfare state, including local hospitals, were remembered as heralding significant improvements.[76] However, it is also apparent that there were severe limitations to the extent of liberation that these developments delivered for women. Marian Hamilton recollected the differences between Shotts and those visible within Windsor, to the west of London. She briefly lived there in the early 1960s so that her husband Willie could take up employment at Ford's factory in Langley. The distinctions she felt centred on the absence of community she had been used to within Shotts but also related to differences in social habits and attitudes:

> No, they werenae so friendly. But then we'd been used tae a village. That was Windsor we moved to. Which, really, well, it was a different way of life really. Women went tae the pubs doon there, and families went. But not here. Women didinae go to the pubs.[77]

Marian's comments are indicative of continuities in distinct patterns of social life within the UK. During the interwar years, a sharp differentiation was evidenced in the rise of working-class commuter households in the south-east of England which broke with the forms of 'kinship' networks that were preserved within industrial communities in Scotland after 1945. Affluent English suburbs developed 'consumption communities' marked by home ownership and increased female employment, including within assembly goods plants. Similar developments only arrived in Scotland after the Second World War and did not necessarily have the same immediate impact on social life.[78] Marian's recollections have parallels with the 'different notions of lifestyles and sexuality', specifically in relation to gender roles, unearthed in the 'discovery' of poverty in Harlan County, Kentucky, by

---

[76] Shotts focus group.

[77] Marian and Willie Hamilton, interview, residence, Shotts, 14 March 2014.

[78] P. Scott, 'The household economy since 1870', in *The Cambridge Economic History of Modern Britain*, ii: *1870 to the Present*, ed. R. Floud, J. Humphries and P. Johnson (Cambridge, 2014), pp. 382–86, at pp. 372–3.

metropolitan American political elites during the 1960s. While Marian's comments are not as drastic a contrast as Mildred Shackleford's recollection that 'we [residents of Harlan] were more like the people in Vietnam than the people in the rest of the country', they share a commonality in experiencing the co-existence of distinct social and cultural temporalities.[79] Marian's memories question the extent to which privatist lifestyles transformed the Scottish coalfields. Into the late twentieth century, traditional social structures remained in place, including the strongly gendered demarcation of public space and socializing.

The qualification of women's relative advancement was compounded by the retention of chauvinistic attitudes. Jennifer McCarey elaborated on this, reflecting on her own experiences within the trade union movement during the 1980s. When interviewed in 2014, Jennifer referenced a recent speech by the TUC general secretary, Frances O'Grady, about the exclusion of women who felt a strong connection to the trade union movement to illuminate her own experience. Jennifer underlined her links to the labour movement, including her father's role as an APEX shop steward at Ravenscraig, which she felt was often ignored by male activists:

> In retrospect, I think people probably were a bit threatened and categorized me as quite an extreme feminist and I was kinda unaware of that at the time. But I think talking to John [Slaven], he was telling me stories about things and I think, god, they really thought I was this powerhouse of feminism, a big woman. And it couldn't have been further from the truth. I was more of a socialist trade union rep than anything else, y'know. Frances O'Grady made a great speech once, and I thought it totally epitomized women in the movement. My generation. And she said she got involved in the trade union cause her family was trade unionists and it's where she wanted to be. It was her movement, and she got involved in it. Whenever she went to meetings people treated her as a visitor, you know. They would be like, 'oh hen you sit down there'. They treated her like she wouldn't understand the movement, or she was in a place that she didn't know, or she didn't understand. They treated her like the odd one out. That was my experience in the trade unions as well. I was trade union to the core. I grew up in a house where people would come chapping at the door and sit and talk to my dad about their problem and he would be taking case notes while I was watching Doctor Who lying on the floor! Trade unionism was all around me.[80]

Margaret Wegg's and Jessie Clark's enjoyment of work in pit canteens underlines the need to appraise women's self-valuation of their working lives.

---

[79] A. Portelli, *They Say in Harlan County: an Oral History* (Oxford, 2010), pp. 283–5.

[80] Jennifer McCarey, interview.

Despite patriarchal assumptions, women not only worked to earn additional income for their families but also gained a sense of identity and validation from their occupations.[81] Unlike her mother, when Margaret married, she continued working to support the household. She commuted to Glasgow city centre from Cardowan village on the city's outskirts and was trained in shorthand typing. Margaret felt her late daughter had advanced further than herself by receiving college education and entering management training with British Telecom. Within her memories, aspects of growing working-class affluence and rising social expectations are also apparent. Margaret's wages afforded 'luxuries' for the household and allowed her own children to remain in education for longer, which developed their capacity for social mobility.[82] These comments indicate a sense of incremental improvement in working-class families' living standards which was partially secured by women's paid work, but also a more specific sense of women's social advancement. Margaret's narrative is illustrative of Catriona MacDonald's conclusion that Scottish women's history 'has generally styled the lineage in a surprisingly Whiggish manner, presuming unrelenting progress and improvement in each generation'.[83] While Margaret's recollection reflects a liberal feminist influence by depicting a series of progressive generational advances in terms of social esteem and labour market roles, Jennifer McCarey's perspective qualifies narratives of women's relative advancement.

A key element in narratives of women's advancement was the improved status given to women's employment, especially married women's work. Marian Macleod grew up in Law, South Lanarkshire. Her grandfather had worked as a miner in the area when it lost its last local collieries in the early phase of nationalization. Marian commuted to work in administration roles for engineering firms, first for Honeywell at Newhouse, North Lanarkshire, and later at Motherwell Bridge. She cited both as providing 'jobs for life' for women workers such as herself and a friend and colleague, Wendy:

> They had Bellshill, they had a place. And at Honeywell they had about three different factory units they had that one that sits, do you know it on the main road? But the one up from there that ran along the way that was bloc sixteen, Wendy was in there.

---

[81] J. D. Stephenson and C. G. Brown, 'The view from the workplace: women's memories of work in Stirling, *c.* 1910–*c.* 1950', in *The World is Ill Divided: Women's Work in Scotland in the Nineteenth Century and Early Twentieth Century*, ed. E. Gordon and E. Breitenbach (Edinburgh, 1990), pp. 7–28, at p. 24.

[82] J. Benson, *Affluence and Authority: a Social History of Twentieth-Century Britain* (London, 2005), p. 31.

[83] C. M. MacDonald, 'Gender and nationhood in modern Scottish historiography', in *The Oxford Handbook of Modern Scottish History*, ed. T. M. Devine and J. Wormald (Oxford, 2012), pp. 602–19, at p. 603; Margaret Wegg, interview.

...

That was the other thing when we came out of school, you actually reckoned you had a job for life. If you got a decent job you could stay in it practically as long as you wanted to. I left Honeywell for other reasons and went to Motherwell Bridge just about the time we got married. But even there you could have been there for a long, long, while. But then they started having pay offs in early eighties. That was when they started having redundancies, and that was a really big thing in the area.[84]

It is evident from Marian's testimony that economic security was extended to some women as well as men between the 1940s and 1970s. Just as male respondents were shocked by the announcements of redundancies at their workplaces, Marian felt shaken when Motherwell Bridge began to lay workers off. Redundancies undermined the certainty that she had previously felt when working there. However, these developments took place within the context of the highly gendered dimensions of employment practices discussed above. A *Scotsman* article from 1967 covering employment in the expanding electronics industry illuminated the outlook of factory managers, detailing that 'dexterity and intelligence tests' were utilized by US multinationals in their recruitment of women workers. The feminine qualities of 'highly flexible' intricacy and 'careful control' were prized, while clean production processes and relatively high wages incentivized labour.[85] Before the late 1960s, comments from subsidiary management to government officials indicate there was a positive relationship between management and women workers. C. J. A. Whitehouse, of the Board of Trade, noted in 1960 that Sunbeam's 'female labour in East Kilbride could not be bettered'.[86] Whitehouse commented again in 1965 that despite experiencing difficulties with male workers, whose occupational traditions were threatened by mass production practices, women workers 'are good'. The quality of female labour justified the subsidiary negotiating male grievances and endeavouring to recruit skilled toolmakers.[87]

Honeywell publicly commended its women workforce. The director of their Newhouse plant was quoted in the *Herald* during 1960 as having said 'that their quickness in picking up detail of the work was about the best he

---

[84] Duncan and Marian Macleod, interview.

[85] NRS, SEP 4/2337 press cutting folder, 'Dexterity and intelligence tests for women workers', *Scotsman*, 21 Feb. 1967.

[86] NRS, SEP 3/567/18, C. J. A. Whitehouse, Board of Trade, to Mr Macbeth, Board of Trade, subject: Sunbeam Electric Ltd, 12 Dec. 1960.

[87] NRS, SEP 4/567/54, C. J. A. Whitehouse, Sunbeam Electric Limited, 3 May 1965.

had found in his career'.[88] In later years, when assembly plants experienced industrial relations turbulence, it was male workgroup rivalries which were at the centre of conflict. For instance, during 1978, an unofficial strike by the plant's thirteen toolmakers led to over 500 assembly line workers being laid off. The dispute was instigated by the toolmakers to maintain their pay differentials over male maintenance engineers who had recently been granted parity.[89] Male 'craft sensibilities' were pivotal to explaining industrial disputes within multinational subsidiaries, but also often limited the scope of trade union activism to sectionalist interests.[90]

These dynamics were not fully pervasive and tended to weaken as women workers became more assertive. By the late 1960s, women's work in engineering had developed norms and expectations which awarded it a comparable moral economy status to male industrial employment. This shift in attitude from 'gratitude' to employers towards workgroup 'possession' of workplaces and employment mirrored the process that Cowie uncovered among the RCA workforce in Bloomington, Indiana.[91] In Scotland, women also made up a large proportion of the workforce in electronics plants. Their experience of factory production shifted attitudes from a feeling of commitment to the company towards conceiving of the factory in terms of the firm's obligation towards the community which both sustained and depended upon the plant. Moral economy sentiments were emboldened by the understanding that assembly goods factories were brought to Scotland with the assistance of public money through regional policy. These feelings were especially marked in communities that had exchanged employment in heavy industries for assembly factories. Subsidiary plants became strongly embedded within the settlements that depended on them for employment. Women workers expected that they would be provided with secure and well-paid labour. Management's imposition of market logic was resisted by countermovements that asserted communitarian norms against the balance sheets of distant multinationals.

When employment stability and comparatively high wages were not offered, there were instances of industrial action. For instance, at Berg, manufacturers of air brakes in Cumbernauld, the failure to provide consistent employment triggered a strike which became a struggle for union

---

[88] NRS, SEP 4/1629, *Herald*, 4 Oct. 1960.

[89] 'Future of Sunbeam plant in jeopardy', *Herald*, 15 Sept. 1978, p. 3.

[90] W. Knox and A. McKinlay, 'American multinationals and British trade unions, c. 1945–1974', *Labor History*, li (2010), 211–29, at p. 225.

[91] J. Cowie, *Capital Moves: RCA's Seventy-Year Quest for Cheap Labour* (New York, 2001), p. 4.

recognition in June 1970. The dispute was sparked following the temporary layoff 'at short notice' of ten women due to delays in machinery deliveries. They were joined by other sections of the workforce. The 'internal works committee' that had traditionally overseen dialogue between workers and management in the non-union factory broke down. An ad hoc strike committee was established and evolved into a branch of the Amalgamated Engineering Union.[92] During 1969, over 1,000 workers at the largely female engineering workforce of Birmingham Sound Reproducers (BSR) plant in East Kilbride struck for union recognition. Their struggle for 'dignity and justice' through the exercise of collective union 'voice' reached militant proportions. As in major coal disputes, strikebreakers were the subject of crowd actions, and picket lines were virulent. Union recognition was achieved in November, after one and a half months of action. It only came about after government intervention, which was stimulated by BSR's status as a significant generator of export income.[93]

BSR's strategic importance bolstered the workforce's moral economy sensibilities: policymaker promises of a secure industrial future working to produce competitive goods contributed to expectations that the plant would be embedded through the exercise of collective trade union voice. Following the strike, the East Kilbride Development Corporation registered it as a major event but also sought to present the outcome as leading to a future of peace. In February 1970 it stated that, 'until recently industrial relations have not been a major problem', but pointed to the 'labour difficulties of last autumn' as having changed this. The eventual granting of union recognition led to a hope that 'harmonious' relations could be re-established.[94] Events at BSR and other plants had a political and cultural influence which is evident in John McGrath's 1980 play, *Blood Red Roses*. The play is set in East Kilbride. Its main character, Bessie McGuigan, is a communist shop steward in a multinational-owned engineering plant. She struggles against a combination of class and patriarchal oppression within her workplace but also in her efforts to raise a family as a single mother and against the forms of male chauvinism within the labour movement described by Jennifer McCarey.[95]

---

[92] NRS, SEP 4/4251/23, S. Palmer, File note: Berg Manufacturing (UK) Ltd Cumbernauld, 6 Nov. 1970.

[93] J. Boyle, B. Knox and A. McKinlay, '"A sort of fear and run place": unionising BSR, East Kilbride, 1969', *Scottish Labour History*, liv (2019), 103–25, at pp. 111–12.

[94] NRS, SEP 4/568/16, East Kilbride Development Corporation corporate dinner, 2 Feb. 1970, 30 Jan. 1970.

[95] J. McGrath, *Six-Pack: Plays for Scotland* (Edinburgh, 1996), pp. 201–76.

Promises associated with factories brought to Scotland via regional policy, which was popularly understood as an exchange for employment within heavy industry, influenced demands for better pay and conditions by encouraging the development of an 'ownership' consciousness.[96] John Slaven recollected that within Tannochside, the Caterpillar factory's status as a stable and long-term employer was widely acknowledged:

> There was a very strong union in it … The Caterpillar locally was a place that was known to be a good place to try and get employed in. There was a feeling that it was a good place tae work. And, I think that was very much replicated that even years later people will still talk about that factory.[97]

This encompassed a form of industrial citizenship which. As in the NCB, employment security was extended to incorporate career opportunities. In the case of Caterpillar women were included as well as men, with John's mother joining his father in the factory in 1970:

> My mum went in basically after she had me and my sister and just got a job because my dad worked in there. And my mum went in tae the pay bill. She had never worked in pay bill, who used computerized programmes. And she was trained to be a computer operator. We're talking about the late sixties, early seventies. So, I suppose what that gave it was a sense ae opportunity. So, because you could go in, they valued that. It was a sorta place you could kinda make a bit ae a life and I think you could go in even at entry level the wages were quite good. But there was progression, and there was opportunities there. So, I think that was one ae the things they valued.[98]

John described how his mother became a shop steward at the plant and played a role within the 1987 occupation of the factory against closure. Caterpillar epitomizes the role of ownership consciousness upon divestment, which was confirmed in the workers taking physical control of the factory between January and April 1987 before ultimately accepting closure upon better redundancy terms:

> At one level it's terrible, but my mum really will admit she had the time of her life. She absolutely loved the occupation. She was very involved, tremendous sense ae purpose. It was a very hands-on, labour-intensive, occupation. It had

---

[96] J. Phillips, 'The "retreat" to Scotland: the Tay Road Bridge and Dundee's post-1945 development', in *Jute No More: Transforming Dundee*, ed. J. Tomlinson and C. A. Whatley (Dundee, 2011), pp. 246–65, at pp. 252–3; E. Gibbs and J. Phillips, 'Who owns a factory? Caterpillar tractors in Uddingston, 1956–1987', *Historical Studies in Industrial Relations*, xxxviii (2018), 111–37, at p. 111.

[97] John Slaven, interview.

[98] John Slaven, interview.

to be. Big huge site, a lot ae logistics of things having to get done, money to get collected. So, it was like having a full-time job. My mum was up and down to London all the time, demos, delegations etc. So, at one level it actually seemed quite an exciting time. I have to be honest, there was absolute sense that it wasnae going to be successful. Fae the outside, I think there was maybe a bit of hope that something would be salvaged from it. But I suppose at one level, there was a sort ae sense ae, y'know. There was tremendous support I mean there really was tremendous support. The many ways it probably did have a kinda feeling of an end of the era. Caterpillar shut in '87, people knew the Craig [Ravenscraig steelworks] was going. It just went. It did have a kinda Custer's Last Stand feel about it to be brutally honest wi ye, but at the time. Y'know, it was, I think it was quite fun.[99]

At Caterpillar, the strength of community embeddedness John described the factory as having, and the value of the employment opportunities it provided, were communicated in the four-month occupation. Opposition to closure was bolstered by the fact the company had received extensive public support, including a high-profile £62 million grant to retool the factory the previous year which was prominently welcomed by the secretary of state for Scotland, Sir Malcolm Rifkind. This was emphasized during the occupation in a banner hung on the front of the factory which stated 'Caterpillar and Rifkind say "Yes" to £62 million now CLOSURE! WHY?'[100] Although the occupation was led by male stewards, the fact that it was John's mother who was active at Caterpillar, after his father had been made redundant earlier in the decade, is also indicative of changing patterns of wage-earning. The 'end of an era' feeling, and the sense that a vital community resource was being removed, had also been visible during the final closure of Burroughs' large computer plant in Cumbernauld the previous year. It further confirms the effects that intensified deindustrialization had upon women employed in assembly plants. Unlike at Caterpillar, the chief public spokesperson for the workforce was a woman. Veronica Cameron served as the engineering union convenor at the plant. She concluded that management had deliberately 'set up' the closure through denying the factory scheduled production of the A5 mainframe computer. Cameron asserted the workforce's moral economy claim to the plant and the employment it provided by stating that Cumbernauld had been 'built around the Burroughs factory'. She articulated a sense that Burroughs, as a beneficiary of regional policy, and

---

[99] John Slaven, interview.

[100] C. Woolfson and J. Foster, *Track Record: the Story of the Caterpillar Occupation* (London, 1988), p. 41.

through its use of Cumbernauld labour over three decades, had incurred social obligations towards the town.[101]

Women workers who had sustained households through industrial employment were displaced by deindustrialization and exercised agency to contest and physically oppose closures. In their opposition to closure, women trade unionists mobilized identities invested in their workplace and community. This was visible in the occupation of Lovable Lingerie in Cumbernauld in 1981, as well as at Plessey's factory in Bathgate, West Lothian, in 1982.[102] These factors were also apparent in two major disputes which won support across the Scottish and British labour movement; the 1981 occupation of Lee Jeans in Greenock and the 1993 Timex strike in Dundee in which largely female workforces opposed closure in the first instance, and the radical restructuring of remuneration before eventual divestment in the latter.[103]

Another example of women's activism in response to industrial closures was Margaret Wegg's involvement in the 1984–5 miners' strike. The existing literature on the strike has appraised women's involvement, emphasizing the leadership role women played at local level in community struggles to preserve employment. This was distinct from the more pronounced gendered division of labour in the 1926 lockout, when men dominated political leadership. During 1984–5, Women Against Pit Closures had a prominent role in building support for the strike as well as taking on more traditional tasks such as running soup kitchens.[104] A similar story is told in Maggie Wright's documentary film *Here We Go: Women Living the Strike*, which is largely narrated through the voices of Scottish women who were involved in supporting the strike. Many of them were miners' wives who had little prior political experience, but others were experienced trade unionists and political activists.[105] Margaret's story broadly fits within

---

[101] NRS, SEP 4/4070/14, *Herald*, 8 Jan. 1986; SEP 4/4070/25, *Glasgow Herald*, 14 Feb. 1986.

[102] P. Findlay, 'Resistance, restructuring and gender: the Plessey occupation', in *The Politics of Industrial Closure*, ed. T. Dickson and D. Judge (Basingstoke, 1986), pp. 70–1.

[103] A. Clark, 'Stealing our identity and taking it over to Ireland: deindustrialization, resistance and gender in Scotland', in *The Deindustrialized World: Confronting Ruination in Postindustrial Places*, ed. S. High, L. MacKinnnon and A. Perchard (Vancouver, 2017), pp. 331–47, at p. 335; B. Knox and A. McKinlay, 'The union makes us strong? Work and trade unionism in Timex, 1948–1983', in Tomlinson and Whatley, *Jute No More*, pp. 286–7.

[104] J. Spence and C. Stephenson, '"Side by side with our men?" Women's activism, community, and gender in the 1984–1985 British miners' strike', *International Labor and Working-Class History*, lxxv (2009), 68–84, at pp. 74–9.

[105] *Here We Go: Women Living the Strike*, TV2day, M. Wright, 2009.

this narrative of empowerment and activism. She explained that, 'Once they started up the kitchen then, as I say, got roped in to goin giving the wimmin's point of view of what was happening during the strike. That was, you know, I says, you began to take an interest in the political side'.[106] It was the first time Margaret had been actively involved in the labour movement. Her memories centred on unexpectedly addressing a crowd in Clydebank, to the west of Glasgow, during 1984:

> I got roped in to going to meetings and speakin at them and, you know, which I wasn't too thingmae at the time! You know, I'd never done anything like that. As I say that was in the middle o the strike when that happened, you know. That they asked, would I go? The very first one was, they said was only a gathering of about six or seven people you know. Turned out it was in the shoppin, the centre in Clydebank, in the actual centre. And as I say, that was very first time I'd ever spoke.[107]

Margaret recalled her activism stimulated a confrontation with patriarchal attitudes, as she had to justify her activities against her neighbour who had previously objected to her working. He now felt threatened by her newfound confidence and involvement in public life. But Margaret also defined her own activities against those of a younger generation of women who she felt had already taken on a more liberated role:

> The younger ones coming up like younger than me, they would you know, by then the attitude wasn't the same you know, it wasn't the same. Whereas some of them were really 'aw no no no she's in the hoose, she watches the wean' *laughs* you know that was their attitude you know. That one ae the boys I used to have, him and I used to go at it. Yeh, because we used to take, during the strike some of the miners were Bellshill and they had the Miners' Welfare. And they used to have nights, concert nights and different things. And one of the men that sort of run it, run the club, he used to give us so many tickets for the women to take you know, and had a minibus. The strike centre had a minibus so one of the men used to drive was it about eight or ten of us, the women, through to have a night out, you know. And he [John Shaw, her neighbour] used to say that, he says, 'every time you go out with my wife she comes in you know with more ideas you know just stop putting the ideas in her head'. I said 'she doesnae need to be' oh no I says 'there's none of this under the thumb sort of thing you know'. I says, 'she should have her own ideas and her own thinking, not what you want', you know. He used to say that 'I wish you'd stop putting ideas into her head she's getting too independent' you know what I mean. But as I say what can you, that's the way I looked at it. As I say

[106] Margaret Wegg, interview.
[107] Margaret Wegg, interview.

I'd never been politically minded or anything like that. Although I worked and that, the house was ma, you know, the core of my life at the time, you know.[108]

There is a clear distinction between Margaret's role and that of many of the women profiled in research on the miners' strike in that she had worked in the industry. Margaret was made redundant from the pit canteen upon the final closure of Cardowan colliery in October 1983. She was not alone in this position within the Scottish coalfields. Canteen workers played a leading role in organizing women's involvement in the miners' strike in Midlothian and Ayrshire. Liz Marshall, who worked at the canteen in Killoch colliery in Ayrshire, clearly articulated her own activities as motivated by a struggle to maintain her own employment and not simply as support for the men: 'as a canteen worker … I would get no big redundancy. So, I needed a job, and for me, it was always about a job'.[109] The 1984–5 miners' strike was a direct struggle for jobs for a small number of women, or, as in Margaret's case, their activism was influenced by a history of employment within the industry as well as by family and community connections.

Margaret's involvement in trade union activism continued after the strike when she attended conferences as a representative of miners' wives along with NUMSA officials. Her activism was not the case of women who 'broke away from domestic bondage' detailed in some 'heroic' accounts of the strike.[110] Previous experience of employment was important to shaping Margaret's perspective, but her strike activism was a major life-changing set of events for someone who had not been involved in labour movement activism before the strike. This separates Margaret from the experienced activists who instigated the Women Against Pit Closures movement across the UK and within Scotland.[111] Margaret's memories of being supported by progressive union members is also indicative of the politics that prevailed within the NUMSA. The Scottish Area proposed admitting miners' wives as union members following the end of the strike in 1985, but were defeated by opposition from other parts of Britain.[112]

---

[108] Margaret Wegg, interview.

[109] G. Hutton, *Coal Not Dole: Memories of the 1984/85 Miners' Strike* (Glasgow, 2005), pp. 6, 35, 54.

[110] V. Allen, 'The year-long miners' strike, March 1984–March 1985: a memoir', *Industrial Relations Journal*, xl (2009), 278–91, at p. 284.

[111] F. Sutcliffe-Braithwaite and N. Thomlinson, 'National Women Against Pit Closures: gender, trade unionism and community activism in the miners' strike, 1984–5', *Contemporary British History*, xxxii (2018), 78–100, at pp. 80–9.

[112] Phillips, *Scottish Coal Miners*, p. 257.

Although sharing the collective trauma of deindustrialization, as workers and community members, the gendered dimensions of women's narratives contain a clear distinction from those of their male counterparts. Margaret's memories of the miners' strike are prefaced by comments about how a younger generation of women had rejected a role which centred their life and sense of self on domestic roles. As discussed above, her daughter, who had taken on further educational studies and management training before her death in the mid 1980s, personified this perceived change. Among male respondents, the shift towards female employment was not universally viewed positively. Peter Downie's comments within the all-male Moodiesburn focus group, which consisted of a retired miners' group, indicated unease with both the decline of male employment and the rise of women's work:

> We're a nation of women workers now. We've got the big Morrisons, Aldis aw these big places, big factories, employing three and 400 women and the men is lying in the hoose puttin the women oot to work. That's what's wrong wi the industry today, and that's what's wrong wi the country today.[113]

These comments demonstrate the retention of patriarchal attitudes in relation to the appropriate division of labour within areas where the economy was formerly strongly characterized by male industrial labour. In some cases, younger male workers have continued aspiration to gain employment in manual jobs which were perceived as maintaining a higher income and more respectable status than service jobs. Scott McCallum reflected that his profession, joinery 'wisnae ma planned option obviously but it's a trade'. While it was not employment in coal mining and removed him from his family's tradition, Scott also gladly reflected that joinery had given him work in a local factory in Stepps for fifteen years before he began employment with Dundee Council.[114] George Greenshields is a Labour councillor for Clydesdale South who comes from a mining background and had previously worked in local opencast mines. His account of economic changes in the area emphasized the expansion of the service sector and the feminization of the workforce. George noted increasing employment in social care in particular: 'I think most ae the women became employed through all these kindae care in the community like Auchlochan [Retirement Village], y'know that kindae type thing. You see aw the girls goin aboot in their South Lanarkshire uniforms and so on goin round aboot in the care in the community'.[115]

---

[113] Moodiesburn focus group.
[114] Walkerdine and Jimenez, *Gender*, p. 118; Scott McCallum, interview.
[115] George Greenshields, interview with author, Coalburn Miners' Welfare, 11 Feb. 2014.

Yet alongside the positives associated with expanded employment opportunities, it is notable the jobs emphasized, particularly retail work and social care, are typical of feminized sectors in their insecurity and low wages. This tends to support Arthur McIvor's conclusion that as the west of Scotland continued its transition from an economy marked by significant industrial sectors towards one increasingly dependent on services, it tended to be characterized by poorly paid women's employment.[116] Jennifer McCarey, who works as an organizer for the public sector union, Unison, articulated her own concerns about women's contemporary position within a deindustrializing economic environment. She argued that labour movement decline had been a significant step backwards in disempowering a strong base of feminism:

> I cannot think of a time it has not been harder for women. As soon as I got involved to now. I think it's probably a bit worse now cause I think that there was a real grow in kind of feminism and women's identity in the movement in the eighties and I think in the nineties that kind of got lost. People need that, a kind of feminist agenda, and things. I think it's probably got a wee bit worse. Seem to be going to a lot of meetings and there's platforms of five guys and only one woman and thinking I remember a time when that wouldn't even have been tolerated, and now we seem to be back there.[117]

In Jennifer's testimony, this fed into contemporary concerns over trade union activism, including frustration that Unison had not taken a more confrontational stance on equal pay claims. She extended this to concerns over the organizing of women who work within social care in casualized conditions and the role of her union in mobilizing them and winning more secure employment conditions. The loss of secure industrial employment and weakened trade unions have had a strong but distinct effect on women. Women's employment opportunities have expanded in an economic environment marked by the contraction of traditional male employment in heavy industries and the expansion of predominantly female service jobs. A generation of women workers has faced the challenges of insecurity and the growing likelihood of being the sole or dominant earner in households as a result. This contrasts with their parents' experience of comparative labour market prosperity for three decades after the Second World War.

Gender relations in the Scottish coalfields were highly integrated with the industrial labour market. Over the nineteenth and first half of the twentieth

---

[116] A. McIvor, 'Gender apartheid? Women in Scottish society in the twentieth century', in *Scotland in the Twentieth Century*, ed. T. M. Devine and R. J. Finlay (Edinburgh, 1996), pp. 188–209, at p. 206.

[117] Jennifer McCarey, interview.

century, a demarcation of domestic and public spheres developed through women's exclusion from underground work. These distinctions were consistently transgressed, but nevertheless they were significant in sustaining a patriarchal culture. The long process of deindustrialization disrupted the economic basis of the gendered division of labour through colliery closures and industrial diversification, as well as the expansion of service employment. Nevertheless, the male attachment to masculine occupational identities and breadwinner status retained salience into the late twentieth century, even as its material reality was eroded. Working-class affluence was often obtained through women's engagement in paid work, which secured a moral economy status for employment in assembly goods plants. Women's industrial employment was embedded into the social life of coalfield areas, which was signified by the growing confidence of female trade union activists and the support they received from male counterparts. But women continued to negotiate patriarchal power structures and expectations in the workplace, at home and in public life. These persisted, even as men faced mass unemployment and the complete disintegration of the conditions in which their occupational cultures originated. The disembedding of the economy from the regulation of communitarian routines during the last two decades of the twentieth century was experienced as a profound challenge to both male and female workers. However, gender dynamics have strongly distinguished how it was remembered. Men's narratives emphasize social redundancy whereas women's testimonies were more likely to discuss new employment opportunities, highlighting both their liberatory potential and the losses incurred through the transition to increasingly precarious service sector work.

5. Generational perspectives

Successive generations have negotiated the cultural, political and economic changes associated with deindustrialization in the Scottish coalfields since the mid twentieth century. There are important differences between the meanings that they attributed to the closure of coal mines, steelworks and factories. This chapter analyses the perspectives of three distinct generations in the Scottish coalfields. The 'interwar veterans' were at the helm of the newly formed National Union of Mineworkers Scottish Area (NUMSA) when coal mining was nationalized in 1947, after experiencing class conflict and mass unemployment in the 1920s and 1930s. During the 1960s, the interwar veterans were succeeded in workplace and community leadership by the maturing 'industrial citizen' generation, whose formative experiences took place under social democratic economic management. Finally, a generation of 'flexible workers' faced intensified deindustrialization during the 1980s and 1990s and adapted to less secure labour markets. The examination of these generational distinctions proceeds through an analysis of the transitions between the interwar veterans and industrial citizens, and then between the industrial citizens and flexible workers. A final section assesses the complex interplay of inheritance and conflict between generations. Each generation learned from its predecessor, but with a critical view of their failures as well as their successes. Cultural changes engendered by shifts in political economy also differentiated ambitions and values. Nevertheless, a strong sense of connection to family, community and occupational historical experience as well as commitments to maintaining workplace and trade union traditions shaped coalfield generations, even as coal mining employment itself was marginalized.

As collieries closed, the history of their distinctive role in shaping identities within former coal mining settlements and in shaping moral economy approaches towards industrial employment remained formative to consciousness in the coalfields. Experiences of the transition to a diversified industrial structure, and thereafter towards a service sector-dominated economy through intensified deindustrialization, were received through the lens of earlier experiences. Existing Scottish coalfield scholarship demonstrates the value of generational analyses. Alan Campbell has developed a method of studying coalfield generations by linking them with forms of workplace organization, associated industrial relations,

'Generational perspectives', in E. Gibbs, *Coal Country: The Meaning and Memory of Deindustrialization in Postwar Scotland* (London, 2021), pp. 155–86. License: CC-BY-NC-ND 4.0.

and political outlooks.[1] His research on the Scottish miners concludes in 1939. Jim Phillips's recent contributions insightfully trace generations of miners' socio-political attitudes through industrial reorganization under nationalization.[2] These approaches are extended in this chapter by incorporating perspectives from workers in other sectors. It includes a focus on the assembly goods plants that to some extent replaced coal employment between the 1940s and 1970s as well as to the career trajectories of miners after the last major colliery closures during the 1980s and 1990s.

Generational consciousness is the product of age cohorts becoming aware of themselves as a distinct entity defined against others based on collective experience.[3] Karl Mannheim's foundational theory emphasizes the role of major life experiences that generally take place before groups mature to their thirties, and which formatively shape political and social outlooks. These events are experienced within 'interior time', defined as 'the time-interval separating generations'.[4] Distinct phases in economic restructuring were fundamental to generational consciousness within the Scottish coalfields through establishing perspectives shaped by the conditioning of interior time. The extension of state economic management during the 1940s, which incorporated both a commitment to full employment and the nationalization of coal, and the later abandonment of this regime in favour of the operation of liberalized markets from the late 1970s, were central to experiences and perceptions of deindustrialization. In both phases, differentiated age groups had distinct vantages on the same events. They were formative for a younger generation and interpreted through longer life-histories by an older cohort.

Episodes of generational succession are pivotal to understanding the evolving politics of Scottish mining trade unionism over the second half of the twentieth century, especially shifting attitudes towards the nationalized coal industry. Labour and political historians have often emphasized the

---

[1] A. Campbell, 'Traditions and generational change in Scots miners' unions, 1874–1929', in *Generations in Labour History: Papers Presented to the Sixth British–Dutch Conference on Labour History, Oxford 1988*, ed. A. Blok, D. Damsma, H. Diedriks and L. Heerma van Voss (Amsterdam, 1989), pp. 23–37, at pp. 27–34; A. Campbell, *The Scottish Miners, 1874–1939*, i: *Industry, Work and Community* (Aldershot, 2000) pp. 159–60.

[2] J. Phillips, 'Economic direction and generational change in twentieth century Britain: the case of the Scottish coalfield', *English Historical Review*, cxxxii (2017), 885–911; J. Phillips, *Scottish Coal Miners in the Twentieth Century* (Edinburgh, 2019).

[3] S. Scherger, 'Concepts of generation and their empirical application: from social formations to narratives – a critical appraisal and some suggestions', *CRESC Working Paper*, cxvii (2012), pp. 3, 9.

[4] K. Mannheim, *Essays on the Sociology of Knowledge* (London, 1952), p. 282.

dynamics of conflict which can be associated with generational differences. During the 'long Italian 1968' an 'old left' and 'new left' clashed over their regard for constitutional democracy.[5] Not dissimilarly, in early 1970s America, a young cohort of industrial workers rebelled against the system of highly bureaucratized trade unionism that an older cohort regarded as its major achievement.[6] These conflicts were fought over objections to alienation in the workplace as much as the more piecemeal objectives of wages and conditions usually associated with trade unionism. They were highly influenced by large-scale societal changes that acted to mould expectations of economic wellbeing and democratic engagement as well as perceptions of social justice.

There were parallels in the Scottish coalfields. The succession of Michael McGahey to the NUMSA's presidency in 1967 represented the ascension of a generation that had matured under nationalization. This transition ushered in a harder stance against colliery closures, and expectations of pay and conditions which were determined by comparing miners to other industrial workers in the context of rising 'affluence'.[7] This change did not take place without conflict, during which the older generation, like their Italian and American counterparts, defended the integrity of the social democratic structures which they had built. However, cohorts within the twentieth-century Scottish coalfield were also strongly moulded by the intergenerational transmission of historical consciousness. Family and workplace socialization bound the moral economy and facilitated major collective mobilizations in national coal strikes during the 1970s and 1980s.

Transition I: interwar veterans to industrial citizens

Table 5.1 denotes an ideal anatomy of the transformation in the Scottish coalfield's economic structure, workforce and the shifting political priorities under which economic changes were managed. These variables were central in shaping differing generational experiences and perspectives. The interwar generation held trade union leadership from the beginning of nationalization up to 1967. Its chief interior time experiences were the miners' lockouts of 1921 and 1926, and the economic and social dislocation associated with mass unemployment and the victimization of trade unionists during the

---

[5] S. J. Hilwig, '"Are you calling me a fascist?" A contribution to the oral history of the 1968 Italian student rebellions', *Journal of Contemporary History*, xxxi (2001), 581–97, at p. 590.

[6] J. Cowie, *Stayin' Alive: the 1970s and the Last Days of the Working Class* (New York, 2010), pp. 24–5, 60–5.

[7] Phillips, 'Economic direction', p. 902.

Table 5.1. Generation, temporality and employment structure in the Scottish coalfields

Generation	Interwar veterans	Industrial citizens	Flexible workers
Employment regime.	Paternalist private employers. Sectarian and anti-trade union.	Managed economy. Public ownership of coal and later steel. Increased trade union voice and government commitment to full employment.	Intensified deindustrialization. Abandonment of economic management, diversification commitments and full employment policies.
Industries/ labour market	Coal, steel.	Coal, steel, assembly plants.	Public and private sector services.
Formative experience.	Interwar unemployment and nationalization.	Rising expectations within both traditional industries and new workplaces. Labour movement mobilization during 1960s and 1970s.	Defeat in the 1984–5 miners' strike and dismantling of social democratic joint regulation and full employment commitment. Long-term unemployment.

1920s and 1930s. Interwar veterans saw the operation of joint regulation within the nationalized industry as a key achievement which granted employment stability. The ascendance of the industrial citizen generation was institutionally confirmed by Michael McGahey succeeding Alex Moffat as NUMSA president in 1967. Their consciousness was formed in the context of rising working-class affluence during the decades following the Second World War. Industrial citizens had a more ambivalent perspective on nationalization than interwar veterans and exhibited a less quiescent attitude towards the National Coal Board (NCB). This generation was characterized by both mounting opposition to closures and participation in industrial action during the late 1960s and early 1970s.

The distinction between these two generations' experiences is exemplified by a comparison of memories from the oral testimonies collected for this study. Willie Allison, who was born in 1933, elaborated positively upon the stability and social justice of nationalization in comparison to the struggles miners faced in the private industry, emphasizing the social control wielded by employers. Illuminating the distinction between the interior time experiences of interwar veterans and industrial citizens, Willie compared the victimization and unemployment his father, who was also a miner, faced in his working life. Willie noted that despite major pit closures, he was never unemployed after starting work with the NCB in 1948:

> I was brought up in a wee village called Lanerig in Stirlingshire. My first memory's ma daddy coming from work and telling ma Mammy we're to get oot the house because the coal owner, the pits, were privately owned. Because the coal owner sacked a guy cause he got his leg broken. And they went on strike and [were] to get oot o the hoose that night. And ma mammy went along tae the people that owned the hoose, and owned the mines, at the top o the hill as we called it and leathered the man and the woman and got jailed seven days for it. And ma Daddy after that got a job wi Mortons. Alan Morton played wi Rangers. Ma daddy said the Mortons were the best employers he had in his life. But he had to walk seven miles to work. Couldnae even afford a bike. We lived in a house, a dry toilet outside, a well outside the door. Just, set-in beds. Miners worked hard and aw they had was enough to pay the store book on a Friday and back tae work on a Monday. Nationalization came aboot, it was the greatest thing that ever happened tae miners. There were some decent coal owners, but the biggest majority were bastards for a better word. Ma father died at 61 wi a heart attack. He never lost a shift. Good livin. He was a good Catholic and he put his bunnet on the bus and all the rest ae it. After I got married, I moved to Plaines in Airdrie. Near Airdrie. But I worked. I was never unemployed. But I worked a lot ae units. But they aw shut. Loads ae mines shut and I went tae England came back went to McCaskey mine for a short time, then started in Boglea in Greengairs.[8]

Willie's emphasis on the social change associated with nationalization was shared in all testimonies. His comments highlight that the price of resistance to employers' power was felt by whole families during the interwar period. Women also directly experienced these aspects of class relations in employment settings. Jessie Clark recalled her experiences of domestic service aged fourteen as exploitative but also marked by differences accorded to age distinctions:

> I didn't like it at all. The first job I had was with two old ladies *laughs* who were the oldest woman was about the age I am now you know. But when I was fourteen, in these days old women were really old! There was a generation gap that you don't get now a days, y'know. And although I was just eight miles away from my home, I only got home once a month. I got a day off once a month and I was paid two pounds a month was ma wages.[9]

---

[8] Moodiesburn focus group, retired miners' group, The Pivot Community Centre, Moodiesburn, 25 March 2014.
[9] Jessie Clark, interview with author, residence, Broddock, 22 March 2014.

Jessie left domestic service to work in the Douglas Castle pit canteen because she 'just didn't want to be somebody's skivvy'. She reflected upon how the generation of interwar veteran women who preceded her had their opportunities limited through both class and gender oppression. Women were taught domestic science instead of complex maths at school, and few mining families could afford to send their children to grammar school if they passed entrance exams. The post-Second World War expansion of the welfare state provided Jessie with enhanced opportunities however, and she was able to find employment in social work.[10] Jessie's employment trajectory is indicative of the improved material and ideological conditions following the Second World War when manual workers and their families benefited from a 'redistribution of social esteem'.[11]

Abe Moffat's autobiography conforms to a narrative of progressive political and industrial reform during the 1940s. Moffat's outlook was pivotally shaped by his earlier formative experiences which followed his demobilization at the end of the First World War. He recollected maturing in the context of the major struggles in the Fife coalfield between 1919 and 1926.[12] These formative experiences were followed by division between miners' unions, mass unemployment and the blacklisting of communists, including both Abe and his brother, Alex. The Moffats' subsequent support for a unified miners' union and respect for its official structure under nationalization was pivotally shaped by these earlier episodes of personal and collective disruption. For Moffat and other interwar veterans, the 1940s experience of state intervention within coal mining, and its collaborative rather than hostile relation with organized labour, was coloured by the stark contrast with the previous two decades. Moffat viewed the conversion of the luxurious Gleneagles hotel in Perthshire to a miners' convalescence home as evidence that 'During the [Second World] War Scottish miners were for once treated as human beings'.[13]

Maintaining the benefits of relative employment stability within the nationalized industry's structures of joint administration was the chief concern of interwar veterans from 1947. The conservative dimensions of these trends were anticipated in Moffat's role as an advocate of productionism during the Second World War. From the summer of 1941, in the context of the fascist threat to the Soviet Union, Communist

[10] Jessie Clark, interview.

[11] R. McKibbin, *Classes and Cultures: England, 1918–1951* (Oxford, 1998), p. 161.

[12] M. Ives, *Reform, Revolution and Direct Action amongst British Miners: the Struggle for the Charter in 1919* (Chicago, Ill., 2016), pp. 83–111.

[13] A. Moffat, *My Life with the Miners* (London, 1965), p. 74.

Party of Great Britain (CPGB) members were integral to minimizing industrial conflict in the Scottish coalfield. During the autumn of 1943, Moffat arranged to meet with miners imprisoned in Barlinnie. Their arrest followed an unofficial stoppage at Cardowan colliery that had led to over thirty miners being detained and fined under Order 1305, which made striking illegal due to emergency wartime conditions. The refusal of around half of the arrested strikers to pay escalated the dispute and led to further walkouts across the Lanarkshire coalfield. In collaboration with another Fife miner, the later under-secretary of state for Scotland, Joseph Westwood, Moffat acted to defuse the situation and ensured an orderly return to work.[14] Moffat retained a pronounced hostility to unofficial strike action, which contributed to later generational conflicts within the NUMSA.

Expectations of dialogue and the maintenance of procedure stimulated as well as muted interwar veteran criticisms of the nationalized industry. This was exemplified by the NUMSA general secretary, Bill Pearson, when he condemned the NCB's 'mysterious ways of working' during the closure of Baton colliery in 1949. Pearson expressed disappointed that the Board's Labour Department had not involved the union in organizing transfers.[15] NUMSA representatives had similar misgivings ten years later when they criticized the NCB's refusal to meet representatives to discuss the announcement of twenty pit closures in Scotland, although more profitable opencast mining would expand.[16] Thomas Fraser, the MP for Hamilton, was among those who communicated grievances. Fraser won the seat with strong support from the Scottish miners' union after standing as a Labour candidate at a 1943 by-election.[17] He clearly articulated the interwar veteran understanding of nationalization as a measure of economic security, emphasizing that he expected guarantees provided to miners who had relocated to Scottish Special Housing Association homes on promises of forty years of secure employment would be maintained in work. Fraser's contentions were articulated in moral economy terms which juxtaposed the forces of the double movement, the pressure for

[14] Moffat, *My Life*, p. 78.

[15] National Records of Scotland, Edinburgh (NRS), Coal Board (CB) 222/14/1/21A, Notes of proceedings between the Scottish Divisional Coal Board and the National Union of Mineworkers Scottish Area (NUMSA) regarding the proposed closure of Baton colliery, held at no. 58 Palmerston Place, Edinburgh on Monday 8 May 1950.

[16] The National Archives (TNA), POWE 37/481, I. Horrobin, parliamentary secretary, Whitehall, to T. Fraser, Westminster, 5 March 1959.

[17] Moffat, *My Life*, p. 73.

financial gain versus integrity and social well-being: 'The National Coal Board's moral obligation to the mineworkers should be stronger than any opencast contract'.[18]

While Abe Moffat and Bill Pearson collaborated with state and industry authorities to end the unofficial strike during 1943, younger men who shared their political affiliations were standing on picket lines. They included Michael McGahey and other young communists. McGahey was sacked from Gateside colliery for his involvement in the industrial action. Generational markers were stronger than party political allegiances in explaining attitudes towards social partnership. Younger trade union activists across Labour, communist and Trotskyist divides struck because they did not carry a worldview formed by the 'initial adult experiences' of defeat and division that shaped interwar veterans.[19] The experience of wartime and post-war full employment, and rising living standards, raised material expectations among industrial citizens. Scottish miners of this cohort were strongly motivated to act on matters of health and safety, as well as miners' pay.[20] Mick McGahey, who was born in 1955, emphasized that his father had campaigned for health and safety improvements and oversaw cooperation between the NUM and the NCB during the 1960s and 1970s. This advanced beyond the earlier cooperation associated with the nationalized industry and came about due to both growing concern within the NUM and the development of the Board's expertise.[21] Nationalization's provision of improved material conditions and its mechanisms of joint regulation facilitated the demands for further improved conditions:

> After nationalization there was, I think, an understanding between the unions in the coal mines and those people that were running the nationalized coal industry. And that was that they had tae take it to another level. Take it to a different stage. There was mair concern about safety. My faither was the person who campaigned rigorously within the NUM, and within the coal industry, for self-rescuer masks. It was only up to the kinda like the mid-sixties you didnae have anything like that. So, the self-rescuer mask. You just slipped it on and you could breathe and get oot o the pit, and it was ma faither that

---

[18] TNA, POWE 37/481, Notes of a meeting with a deputation from the Scottish Area of the National Union of Mineworkers held in the parliamentary secretary's room, Thames South, Thursday 26 Feb. 1959, 5.15 pm.

[19] Phillips, 'Economic direction', p. 897.

[20] A. Portelli, *They Say in Harlan County: an Oral History* (Oxford, 2010), p. 308.

[21] A. McIvor and R. Johnston, *Miners' Lung: a History of Dust Disease in British Coal Mining* (Aldershot, 2006), pp. 205–8.

drove that. And by that point there was that relationship, which was a good working relationship between the union and the employer based on a common purpose.[22]

The demand for all miners to be equipped with self-rescuers rose to prominence following the fire at Michael colliery in 1967 which killed nine miners. The NUMSA successfully pressed for a public inquiry and the NCB made provisions for self-rescuers afterwards.[23] Willie Doolan, a former miner from Moodiesburn, reflected ruefully in 2019 that if self-rescuers had been available, they could have saved some of the forty-seven lives lost in the Auchengeich colliery fire which took place not far from his home in 1959. The achievement of major health and safety improvements under nationalization was in part the product of generational learning. During the late 1950s, major disasters replaced fatalities from smaller accidents as the dominant cause of deaths in Scottish mining. Maturing through these years stimulated this cohort's campaigning on health and safety issues.[24]

Industrial citizens' expectations were galvanized by the improved conditions and full employment delivered by the social embedding of the economy after 1945. In Polanyi's terms this incorporated the efforts to have labour standards 'determined outside the market' through the intervention of 'public bodies' including empowered trade unions, the NCB, and legislation.[25] Commitments to social partnerships were also strongly evident in managerial testimonies. Ian Hogarth was part of the industrial citizen cohort. He was born in 1928 and grew up in Springboig, to the east of Glasgow. Ian's father worked as an accountant for Bairds and Scottish Steel, which gave him some acquaintance with mining. He recalled the 'hungry thirties, I remember when I was at the primary school. I would say more than half my schoolmates' fathers were unemployed'.[26] Ian subsequently graduated with a metallurgy diploma from Strathclyde Technical College in 1951. He later joined the NCB and went on to become a management trainee at Lady Victoria colliery in Midlothian before transferring west to

---

[22] Mick McGahey, interview with author, Royal Edinburgh Hospital, Edinburgh, 31 March 2014.

[23] J. Phillips, 'The closure of Michael colliery in 1967 and the politics of deindustrialization in Scotland', *Twentieth Century British History*, xxvi (2015), 551–72, at p. 552.

[24] E. Gibbs and J. Phillips, 'Remembering Auchengeich: the largest fatal accident in Scottish coal in the nationalised era', *Scottish Labour History*, liv (2019), 47–57, at pp. 47–9.

[25] K. Polanyi, *The Great Transformation: the Political and Economic Origins of Our Time* (Boston, Mass., 2001), p. 258.

[26] Ian Hogarth, interview with author, National Mining Museum, Newtongrange, 28 Aug. 2014.

work at Cardowan in Lanarkshire. Ian became the first Area ventilation engineer for the Central West Area before transferring to the Lothians in 1959 and remained at the Board's Scottish headquarters at Green Park in Edinburgh until retiring in 1987. Although he had vague memories of the interwar depression, it was the new nationalized industry that shaped Ian's approach to management. Ian's outlook was moulded by the mutual respect he found between coal miners and managers. Jimmy Henderson, the Cardowan undermanager, arranged to have older miners employed in suitable work, including one man who suffered from pneumoconiosis and another who had lost a leg in the First World War. Ian remembered both these men as industrious and loyal workers.[27] This sense of social responsibility foregrounded later efforts to develop more comprehensive compensation and health and safety measures. Ian elaborated on the commitment he felt towards those who worked under him by relaying that he rebuked men working under him for describing themselves as 'only miners', imploring 'you're highly skilled men'.[28]

Other managers remembered a similar sense of societal obligation. Peter Mansell-Mullen began his managerial training in Nottinghamshire during the 1950s after graduating from politics, philosophy and economics at Oxford. Peter explained that he joined the Board because he 'had a fairly privileged existence and I wanted to put something back'. Peter combined these inclinations with an interest in nationalization, which there was 'a great feeling' of sympathy for at the time. He went on to become the NCB's director of manpower, but, like Ian, described working to bring down barriers between himself and other employees, including other managers who were 'largely people who'd come up inside the ranks' of the industry.[29]

The introduction of formalized managerial structures under nationalization offered far greater opportunities to entrants of the industrial citizen generation than those which had been available to interwar veterans. These developments contributed to a stronger sense of occupational identity among mine managers. Under public ownership, managers were no longer tied to the private industry's often relatively small-scale and paternalistic firms. Their new-found prestige encouraged managers to be 'protective of the nationalized industry'.[30] These inclinations were further cemented

---

[27] Ian Hogarth, interview.

[28] Ian Hogarth, interview.

[29] Peter Mansell-Mullen, interview with author, residence, Strathaven, 3 Oct. 2014.

[30] A. Perchard and K. Gildart, '"Run with the foxes and hunt with the hounds": managerial trade-unionism and the British Association of Colliery Management, 1947–1994', *Historical Studies in Industrial Relations*, xxxix (2018), 79–110, at p. 81.

by managers' connections to mining communities affected by closure programmes. Ian Hogarth noted his concern about the fate of miners who experienced mass unemployment: 'the pride these people had pride in their job, don't take it away from them'.[31]

Sam Purdie's recollection of being informed by his superiors during the mid 1960s that he would not be promoted to a colliery chief mining engineer in the Scottish coalfield due to the rate of closures further confirms the impact of deindustrialization across the coal industry workforce. Closures disrupted the expectations of labour market stability and career advancement that the industrial citizen generation had anticipated in the nationalized coal industry. Sam indicated that he took the moral economy sentiments, as well as industrial hard-headedness, from the coal industry into the Marathon oil rig construction yard at Clydebank where he later worked. Although he explained efforts at simplifying craft pay scales left him with 'scars', Sam also recalled having a strong working relationship with Jimmy Reid who was the engineering union representative at the yard. This partnership extended to lobbying Tony Benn as secretary state for energy during the late 1970s.[32]

As Mick McGahey's comments indicate above, there was a trend towards convergence between workers and management through a generational commitment to work towards improved conditions. Ian Hogarth felt health and safety was taken 'very much more seriously because nationalization stopped this problem of the coal owners ignoring things that were involved with safety'. He gave the example of methane monitoring, which was given recognition through the Board appointing him among its first cadre of Area ventilation officers. Hogarth further emphasized that under nationalization, legislation on methane levels was strictly followed whereas it had been flouted due to profitmaking imperatives under private ownership, and through the intimidation of underground officials by coal owners.[33] Ian Terris began work at Cardowan colliery in 1952. Like other industrial citizens, he was attracted by the offer of secure employment and advancement in the industry. Terris attained this through training as a shotfirer and later transferred to Rothes colliery in Fife during 1959. His autobiography confirms that the industrial citizen generation placed demands on the nationalized industry, and underlined expectations associated with health and safety. These were directed at management as well as fellow workers. At Rothes, Terris became a 'marked rebel' for demanding oilskin overalls to work in damp conditions,

---

[31] Ian Hogarth, interview.
[32] Sam Purdie, interview with author, UWS Hamilton campus, 3 May 2018.
[33] Sam Purdie, interview.

but he also recalled a physical altercation with a tunneller after Ian found him smoking underground.[34] Terris's journey of social mobility saw him ultimately leave the mining industry. He used his technical training as a basis to study engineering at Stow College in Glasgow before starting a new career as a maintenance service officer.[35]

Transition II: industrial citizens to flexible workers

The intensification of deindustrialization from the late 1970s drew sharp dividing lines between industrial citizens and flexible worker generations, which can be demonstrated by comparing trajectories. Willie Hamilton, who was born in Shotts during 1936, had a career that was characteristic of the industrial citizen miner's capacity for advancement in the coal industry and outside it. Willie followed his father, uncles and brother into the industry, starting at Kingshill 3 colliery in Shotts after leaving school aged fifteen during 1951. He recalled the NCB was perceived as providing a 'job for life'. Willie became an NCB official at Kingshill 3 and then Polkemmet in West Lothian, progressing from a shotfirer to a deputy and later an overman.[36] However, there were significant qualifiers to this advancement in terms of self-presentation. Willie was keen to emphasize that he retained strong connections with the men who worked under him, which was underpinned by a strong sense of class solidarity and shared investment in the moral economy. He underlined this through reference to the 1972 miners' strike. Along with other officials, he engaged in 'giein the pickets a lift doon the road' to Kingshill 3 from Shotts.[37]

Willie recalled that unlike in other pits, there was little animosity between miners and colliery officials during the strike. Officials such as himself attended for work but then did not cross picket lines, allowing more senior staff to act as safety cover. Like Ian Terris, Wille exemplified the industrial citizen generation's capacity for occupational and social mobility when he left the shrinking coal industry for utilities management. However, he retained a coal mining identity which was perhaps shaped by both the needs of telling a composed life story and the effects of historical hindsight. Willie remembered arguing with a Ravenscraig steelworker in a Shotts pub over the impending coal strike during the early 1980s. He warned the Ravenscraig man: '"You cannae see beyond the end of yer nose"

[34] I. Terris, *Twenty Years Down the Mines* (Ochiltree, 2001), pp. 60–74.
[35] Terris, *Twenty Years*, pp. 127–8.
[36] Marion and Willie Hamilton, interview with author, residence, Shotts, 19 March 2014.
[37] Marion and Willie Hamilton, interview.

[The man replied] "how?" I says, "If the pits shut how much steel does the pits use in a year? Million ton?" "aw they use a lot" "there's gonna be one more steelworks no needed if there's a million ton no used it'll be the Craig that shuts", and so it was'.[38]

Intensified deindustrialization during the last two decades of the twentieth century eroded opportunities for the ascendant generation. Flexible workers entered a labour market characterized by falling rates of industrial employment. Willie Hamilton remembered that among his peers, many men had opted to work in assembly plants including the Cummins engineering factory in Shotts where Willie himself worked for a short spell in 1959. Other large plants within travelling distance included the Organon and Honeywell at the Newhouse industrial estate as well as Caterpillar in Tannochside. Willie noted that the Cummins and Caterpillar plants had now closed, while both Organon and Honeywell have both severely reduced their employment levels.[39] The contraction of coal mining further contributed to the demise of opportunities for industrial careers. Alan Blades, who was born in 1962, a generation apart from Willie, described how at Bedlay, the NCB had provided opportunities to obtain skilled jobs through apprenticeships. His father was able to gain promotion by undertaking night school courses in shotfiring. Alan's advancement in the industry was thwarted by closures and privatization. Between 1979 and 1997, he worked at Bedlay, Polkemmet and then at the Longannet complex. In contrast to his older brother, Alan was unable to attain the rank of faceworker.[40] Scott McCallum had a more pronounced experience of exclusion from coal mining opportunities. He was ten years old when Cardowan colliery closed during 1983. The pit was within walking distance of the family home and employed his father and brother at the time. Subsequently, the 1984–5 miners' strike became a formative experience for Scott, as well as explanation for his own inability to follow his grandfather, father and brother into coal mining:

> It was part of growing up. You had tae fight for what was going. We took part in it. Goin to Edinburgh. Doing marches at the time. [We] had our stickers, oor badges, on. You had to get involved in it. Primary seven. I was just gettin ready to go to high school … You were still young. I was what? Twelve. But it was like, Margaret Thatcher was the enemy was drained in your head. Thatcher was closing the pits. I was brought up to think she was a hate lady. What she's done is ruin your living conditions … We visited the soup kitchens very rarely, but I remember standing in queues waiting for butter and there was stuff that

---

[38] Marion and Willie Hamilton, interview.
[39] Marion and Willie Hamilton, interview.
[40] Alan Blades, interview with author, resident, Airdrie, 26 Feb. 2014.

came from Russia and Germany, France. Selection boxes from France. It was tough. Really hard.[41]

The miners' strike was a formative experience within other flexible worker testimonies. Brendan Moohan was a nineteen-year-old underground worker at Monktonhall colliery in Midlothian when the strike began. Brendan was dismissed for gross misconduct in February 1985 following his arrest while picketing the previous June. His narrative underlines the strike as a transformational experience: 'as a young guy, I didn't have any views on it [pit closures]. Until 1984 I didn't have strong opinions about really anything'.[42] Brendan's testimony was characteristic of 'activist' accounts of the strike.[43] He emphasized that the broad coalition of strike supporters had a significant change in his own consciousness as a young man who had previously been comparatively apathetic and lived in a secluded and socially conservative community.

Revisionist scholarship questions the extent of changes in social relations within the coalfield associated with miners' strike activism,[44] and relates it to longer-term alterations in economic structure.[45] Emphasis on long-run and slow-moving developments does not negate the subjective significance that men who were young miners at the time have ascribed to experiences of the strike over three decades later in terms of its impact on their perspective and in moulding a sustained worldview.[46] Connections made through these experiences have often played an important role in shaping their subsequent life course in other respects. Brendan first met his future wife, Angela, through her engagement in solidarity action with the strike and both have retained a sustained involvement with the labour movement since.[47] Resistance to workplace closure had a similar effect on other workers. Bob Burrows recalled the difficult decisions he faced when sitting on the hardship committee during the occupation against the closure of the Caterpillar tractor factory in Tannochside, North Lanarkshire, during 1987:

[41] Scott McCallum, interview with author, The Counting House, Dundee, 22 Feb. 2014.

[42] Brendan Moohan, interview with author, residence, Livingston, 5 Feb. 2015.

[43] D. Kelliher, 'Constructing a culture of solidarity: London and the British coalfields in the long 1970s', *Antipode*, xlix (2017), 106–24, at pp. 106–9.

[44] F. Sutcliffe-Braithwaite and N. Thomlinson, 'National Women Against Pit Closures: gender, trade unionism and community activism in the miners' strike, 1984–5', *Contemporary British History*, xxxii (2018), 78–100, at pp. 79–80.

[45] J. Phillips, *Collieries, Communities and the Miners' Strike in Scotland, 1984–5* (Manchester, 2012), pp. 32–3.

[46] D. Featherstone and D. Kelliher, *'There was just this Enormous Sense of Solidarity': London and the 1984–5 Miners' Strike* (London, 2018).

[47] Angela Moohan, interview with author, residence, Livingston, 5 Feb. 2015.

> There were three or four of us would sit down and listen to everybody's individual case. And this cousin of mine come in and put his case forward. I knew he was working. He didnae disclose that. I then said, 'you're working', and he says 'aye, yep, okay'. He didn't get any money from us, and to this day, thirty years later, he's never spoken to me. That's hard isn't it? That's very hard. That's the stance ah took. That's the level of commitment. That's the fairness. And that has stood me right through the thirty years.[48]

Bob explained that the skills he learned on the hardship committee served as an apprenticeship for his later career as a local authority personal debt counsellor. He underlined the shock that the transition entailed. Bob went from a well-paid job at Caterpillar to starting work at a local authority homeless shelter. In this role he was exposed to the impact of intensified deindustrialization through the growing poverty which affected service users, but also in the form of his diminished pay and conditions. Bob found his wages almost halved despite working longer hours, and he was reliant on Family Income Support to make ends meet.[49]

While he detailed major changes in circumstances, Bob insisted on the retention of a moral economy perspective which was reinforced by the experience of opposing accelerated deindustrialization in the factory occupation. A sense of major transition across the periods between the interwar private industrial economy, the postwar managed economy and the liberal market economy of the late twentieth and early twenty first centuries was apparent in several other testimonies. Margaret Keena and Barbara Goldie were both born in 1931 and grew up in mining families in Cambuslang, South Lanarkshire. They vividly contrasted interwar conditions with those which followed the relative economic affluence and stability in the decades after the Second World War that they benefited from as members of the industrial citizen generation. Margaret recalled her cousin returning home from working in the pits in the Cambuslang area in his dirty work clothes, and underlined the relatively backward conditions that prevailed in the workplace and the home:

> Wullie finished at two o'clock in the day and he come in. And all you say was those two big eyes, and he'd be filthy. And Bella would be saying to us 'right c'mon, Wullie' And it was a big tin back in the room. 'Get oot, get oot Wullie's to get his bath', and he'd this lamp on his hat and what have you.[50]

---

[48] Bob Burrows, interview with author, Tannochside Miners' Welfare, Tannochside, 20 Jan. 2017.

[49] Bob Burrows, interview.

[50] Barbara Goldie and Margaret Keena, interview with author, Cambuslang Bowling Club, Cambuslang, 8 Dec. 2014.

These comments were made within a wider discussion profiling how interwar experiences entered interior time and framed generational understandings. Barbara remembered her mother suffered from tuberculosis before she declared that 'miners are heroes' for experiencing harsh conditions underground. The dialogue then moved to considering the injustices and dangers within the mines. Several local deaths were mentioned, while Margaret recalled an uncle who was blinded in an accident underground during the 1930s and received minimal compensation.[51] Margaret's sister and Barbara's brothers and sister worked at the local Hoover factory after the Second World War. The experience of Cambuslang's transition from a coal and steel town to a diversified industrial structure, and then to a deindustrialized economy, was succinctly summarized in a discussion of the contribution the factory made to the town between the 1940s and 2005, when it closed:

> Barbara Goldie: That kept Cambuslang going.
>
> Margaret Keena: A lot of local people got work there. Conditions totally different from the pits and the mines I gather. And the wages were, no that I ever knew. They got very well paid.
>
> Barbara Goldie: Well, I had a sister and two brothers that worked in the Hoover.
>
> Margaret Keena: Well, I had a sister that worked in the Hoover.
>
> Barbara Goldie: And I mean at the end up, I mean it was the only factory that was going in Cambuslang. Then they closed it.
>
> Margaret Keena: Aye they went away again. That was it, down again, everybody looking for work.[52]

Hoover's factory was associated with both safer and cleaner work and higher wages than had been provided in the private coal mining industry. It was embedded in the community, providing employment for families, with jobs for women as well as men. The exit of 'the Hoover', after a prolonged rundown, ended significant industrial production in Cambuslang and contributed to rising unemployment in the area. John Slaven, who grew up in Birkenshaw in North Lanarkshire, had similar memories, but from the perspective of a representative of the flexible worker generation that entered the labour market during the intensified deindustrialization of the 1980s. John was born in 1965, after his parents relocated from Glasgow so that his father could take up work in the adjacent Caterpillar factory at

---

[51] Barbara Goldie and Margaret Keena, interview.
[52] Barbara Goldie and Margaret Keena, interview.

Tannochside. His mother later also took up administrative employment at the plant. John summed up his feelings on the factory as representative of social improvements and economic modernization in comparison to the coal mining employment which had previously dominated the area:

> You have to understand the physical environment of Caterpillar. It was a big shiny new factory. It looked modern. The front ae it looked modern. It looked American. And they made these brilliant tractors that were sold all over the world. Why would they go? You werenae down an old decrepit pit that's seam was running out. You were in this big modern factory that had computer design pay systems. That were making big yella tractors that everybody knew aboot. You know, they could pay high wages.[53]

Lanarkshire's evolution under the diversified industrial economy was accompanied by social advancement. John felt that Birkenshaw and the surrounding area of North Lanarkshire was 'a sorta prosperous working-class place,' with recently constructed 'good quality houses'. Tannochside was at the heart of investment brought by regional policy with both Ranco, another American engineering firm, and the nationalized British Steel Corporation steelworks at Ravenscraig, in close proximity. A strong sense of belonging was built around residing in the area and working within these factories, which indicates that the social democratic order of industrial citizenship extended beyond the publicly owned mining and steel industries. John referred to the Caterpillar factory as having built a 'sense of identity' characterized by a strong trade union and a 'paternalistic' management that supported social clubs, family days out involving workers' children, and sporting activities.[54] As with Hoover's rundown and final closure in Cambuslang, John remembered the closure of Caterpillar and the broader and sudden decline of industrial employment during the 1980s as a dislocating process that established his generational status by creating a significant distance from his parents' experience. Employment at Tannochside peaked during the 1970s, before Caterpillar was badly affected by the global economic downturn of the early 1980s. The company sustained substantial losses between 1982 and 1984.[55] John's testimony indicates the trauma associated with the series of local closures and job losses which ensued over the 1980s. The rundown of industrial employment in Lanarkshire altered his own life course through the absence of opportunities for someone who had left secondary school

---

[53] John Slaven, interview with author, STUC Building Woodlands, Glasgow, 5 June 2014.

[54] John Slaven, interview.

[55] M. C. McDermott, *Multinationals: Foreign Divestment and Disclosure* (Maidenhead, 1989), pp. 101–2.

without many qualifications. Following long-term unemployment, John chose to leave for a job on the railways in London during 1985:

> I remember that was quite a profound shock. And it was very fast, fae being the optimism went all the way through the sixties, seventies. And I'm no just saying that retrospectively. I can remember it. This kindae feeling, ach things are gonna be okay, and it was very short. I mean a good example you know the first redundancies were in 1979, 1980, by '85 when I left school I had tae leave. I had to go to London because there was absolutely nothing there.[56]

John's sentiment that 'there was absolutely nothing there' for young workers, and a sense that he 'had' to leave for London provides a dramatic comparison with his earlier description of the relative affluence and stable industrial employment that had made Tannochside attractive to incomers. These reflections echo with Alessandro Portelli's explorations of the 'the short and violent life of the Industrial Revolution'. In Italy's steelmaking centres, the industrial era was contained within a single lifetime, and even in American 'heartlands' easily traceable through a single family's memory.[57] Scotland's coalfields experienced major industrial development during the nineteenth century. However, the period between its zenith in production and employment terms during the 1910s and early 1920s and the cessation of deep coal mining in 2002 correlates with life-expectancy. The diversified industrial economy endured even more briefly, with its political and economic foundations fatally undermined by the late 1970s. Deindustrialization reveals the volatile patterns of resource extraction, investment and divestment that prevail within capitalist economies. These relations expose the 'aura of permanence' associated with the physicality of industrial activities' presence in the built environment, as well as their centrality to communal experiences.[58] Large-scale industrial employment and the social routines and cultural identities constructed around it shaped the expectations of successive cohorts. Scotland's status as an 'industrial nation' remained 'common sense' among policymakers as well as a dominant public perception into the late twentieth century.[59] This anchored moral

---

[56] John Slaven, interview.

[57] A. Portelli, '"This mill won't run no more": oral history and deindustrialization', in *New Working-Class Studies*, ed. J. Russo and S. L. Linkon (Ithaca, N.Y., 2005), pp. 54–9, at pp. 54–5.

[58] J. Cowie and J. Heathcott, 'Introduction: the meanings of deindustrialization', in *Beyond the Ruins: the Meanings of Deindustrialization*, ed. J. Cowie and J. Heathcott (Ithaca, N.Y., 2003), pp. 1–15, at p. 6.

[59] J. Tomlinson and E. Gibbs, 'Planning the new industrial nation: Scotland, 1931 to 1979', *Contemporary British History*, xxx (2016), 584–606, at p. 598.

economy outlooks within the coalfields. In the early 1980s, communal expectation remained that industrial employment would continue to sustain communities.

Inheritance and conflict

Generational successions are marked by cultural inheritances that are tempered by changes in the socioeconomic environment. Successive cohorts' interior time experiences stoke modifications to or the rejection of parental expectations, but elements of tradition passed through coalfield generations. Paternal family bonds were an important part of coal mining identities. Scott McCallum recollected that within his family in Cardowan village there was an understanding that aged sixteen '[you] go there [to the pit], that was what ma father did, what ma grandfather did. It was just what you did'.[60] Antony Rooney recalled that within the Bellshill area this was secured through custom: 'at that time the miners had an agreement, y'know, the oldest son in a family had to be more or less guaranteed a job'.[61] Brendan Moohan described the NCB as having a policy of providing miners' sons with employment during the early 1980s. However, he entered the mining industry to escape labour market adversity rather than through active choice:

> I didn't want to go down the pit, but ultimately ma dad got a form because it was either the pit or the forces and I wisnae doin the forces. So, I decided, okay, we'll go for it. Go and work in Monktonhall. Back then, if you had a member o the family like a father or an uncle or whatever, then you were deemed as coming from a mining family. Therefore, you'd be more likely to be recruited. There'd have to be something fundamentally wrong for them not to recruit you![62]

Yet miners often wished for their sons to enter alternative industries. Anthony Rooney's father rejected miners' customary generational succession: 'ma father wouldnae let any of us go anywhere near the pits'. Both Anthony and his brother found jobs in assembly engineering, with Anthony going on to become a shop steward at Caterpillar. Anthony displayed an ambivalence to coal mining. He summed up his attitude to underground work by stating that mining 'was a dangerous dirty job', and went on to reflect on the social injustices suffered by his father who died of lung disease: 'All ma father got out of it was I carried him down the stair in a

---

[60] Scott McCallum, interview.
[61] Anthony Rooney, interview with author, Morrisons café, Bellshill, 24 Apr. 2014.
[62] Brendan Moohan, interview.

box, sixty years old'. Just four years earlier, Anthony's father had been forced to retire early after the closure of Bardykes colliery in Cambuslang. But Anthony also proudly presented a picture of his father in his work clothes and indicated he felt a strong family connection to the industry: 'I've never forgot my upbringing, y'know. I've definitely got an affinity with the miners and that. As I say, my whole family was on both sides, my mother's side coal mining and my father's'.[63] Anthony inherited a connection to trade unionism and Labour party politics, but in the context of the enhanced opportunities the diversified industrial economy provided to the industrial citizen generation. Scott McCallum's memories were also coloured by his father suffering death from an occupational illness. A sense of injustice was further consolidated by the denial of compensation because Scott's father was a smoker, despite clear evidence that mining had contributed to his condition:

> He was a smoker and he decided to stop smoking after Christmas, New Year one year. He started bringing up blood. He forgot to flush the toilet one time. It wis black and it smelled like a charcoal barbeque. And he had to come oot, put his hands up and admit that this was coal dust. But he was a smoker and you couldnae well prove anythin at the time. It wis coal, stains o black coal and the smell. He suffered through that tul he died.[64]

In Scott's case, his father's experience consolidated a sense of dislocation associated with the absence of the opportunity to enter the mining industry after intensified closures. However, other respondents portrayed an attitude that more resembled Anthony Rooney's father's ambitions. Willie Hamilton recalled:

> When I left school, I wanted to be a motor mechanic but there was nothing going. But the only thing was the steel industry. And ma Dad and ma brother, was in the pit. So, ma Dad got me a job in the pit. I vowed none of my family was going underground. It was a dirty filthy job. I got used to it, and the men who worked, most men, I'd say ninety-five per cent, ninety-six per cent, got used tae working in it, you know? Where you said to yourself 'I wouldn't let', it was tradition to follow your dad into yer employment. If he worked in the steel industry you went. If he worked as a joiner, you went, you know. So, most of my family went to work in the mines.[65]

---

[63] Antony Rooney, interview.
[64] Scott McCallum, interview.
[65] Marion and Willie Hamilton, interview.

Willie's case was characteristic of the expanded expectations held by the industrial citizen cohort. He passed on his ambitions for alternative work opportunities to his sons. None of them worked in coal mining or heavy industry, but one quite closely followed Willie's own ambitions by becoming an electronic engineer.[66] Pat Egan recalled that his father, who like Willie was a miner of the industrial citizen generation, expressed disappointment at his decision to join the NCB, going as far as trying to actively prevent him from working at Bedlay colliery:

> I was an apprentice butcher fir a year. I couldnae stick it and went in the following year fir the pit and found oot my faither woudane put my name doon for it. He hated it. Years later, I got a degree fi university at twenty-eight or thirty. My dad wis just like that 'you could ae had that years ago'.[67]

Pat remembered that many of his friends from the Lanarkshire village of Twechar had opted to work for the NCB upon leaving school: 'a lot o them did go, it wis fun'.[68] Pat's refusal of educational opportunities as a young man in favour of taking up colliery employment, and his father's misgivings, conform to something of an established coal mining trope. The most influential Scottish example is in William McIlvanney's novel, *Docherty*. McIlvanney's book centres on the struggle of an Ayrshire mining family to maintain dignity in the face of abject poverty and the dangers of coal mining under the privately owned industry of the early twentieth century. Themes of generational succession and conflict loom large. Tam Docherty agonizes over having to allow the youngest of his three sons, Conn, to join him in the local pit. Conn himself is keen to leave a school where he feels shunned for his class status and is eager to attain the mantle of manhood and respectability that come with being a miner and household provider. His ultimate entry to coal mining is presented as a significant moral and cultural defeat for Tam. Conn joining him underground is both a concession to material impoverishment but also the crushing reproduction of class society.[69] Dick Gaughan's song, 'Why old men cry', released in 1998, similarly reflects on his father's encouragement to look beyond the coal mines where he was employed, but is highly qualified in light of the pit closures that affected Midlothian during the 1980s and 1990s:

---

[66] Marion and Willie Hamilton, interview.
[67] Pat Egan, interview with author, Fife College, Glenrothes, 5 Feb. 2014.
[68] Pat Egan, interview.
[69] W. McIlvanney, *Docherty* (London, 1975).

> I walked from Leith to Newtongrange
> At the turning of the year
> Through desolate communities
> And faces gaunt with fear
> Past bleak, abandoned pitheads
> Where rich seams of coal still lie
> And at last I understood
> Why old men cry
>
> My father helped to win the coal
> That lay neath Lothian's soil
> A life of bitter hardship
> The reward for years of toil
> But he tried to teach his children
> There was more to life than this
> Working all your life
> To make some fat cat rich[70]

Gaughan's lyrics are indicative of the improved status that coal mining employment had obtained by the late twentieth century. When Pat defied his father's wishes and began working at Bedlay in 1977, at the age of seventeen, two years after *Docherty* was published, mining was no longer a poverty wage payer. However, parallel dynamics are apparent: a father urging his son to look up towards formal education rather than down colliery shafts, and young men drawn towards the local social status and collectivism of coal mining. McIlvanney also introduces the major theme of a larger, politicized, generational conflict. These revolve around the aspirations of the generation succeeding Tam. Both of his older sons demonstrate their dissatisfaction with poverty. Angus betrays Tam's trade union values by becoming a contractor. Meanwhile Mick, having been injured on the western front, becomes a communist and an advocate of relentless class warfare which chafes with Tam's socialist humanism.[71] McIlvanney presented *Docherty* as 'a book that would create a kind of literary genealogy for the people I came from'.[72] At the time he was writing it, renewed generational tensions were shaping the coalfields. The forms of conflict over lifestyle, aspiration and morality that McIlvanney located in Edwardian Ayrshire manifested themselves within the NUMSA in an organized form and through more amorphous cultural attitudes towards employment.

---

[70] D. Gaughan, 'Why Old Men Cry', on *Redwood Cathedral* (GreenTax, 1998).

[71] McIlvanney, *Docherty*.

[72] W. McIlvanney, *Surviving the Shipwreck* (Edinburgh, 1991), pp. 217–18.

The industrial citizen generation's discontents, which were expressed in a collective form during the wage strikes of 1972 and 1974, were analysed by an NCB researcher, J. C. H. Mellanby Lee, in a paper published during 1973. Lee was considerably influenced by contemporary industrial sociology.[73] He accepted the core foundations of Alan Fox's pluralism by acknowledging the distinct legitimate interests within workplaces, and understood 'constructive' or 'organized' conflict as a necessity in coal mining.[74] Lee's chief concerns lay in the potential damage caused by 'unorganized' conflict, especially by miners' use of absenteeism as a 'weapon' with which to express dissatisfaction. These were highly connected to demographic changes in the workforce which had shifted expectations of coal mining's economic and social rewards. Reprising the language of the *Affluent Worker* study of Luton car workers published five years earlier, the Board's research noted the increasingly 'instrumental' orientation of miners to work in mechanized and bureaucratized cosmopolitan collieries.[75] These attitudes were juxtaposed to those of miners 'a generation ago', in 'small mining villages' where social status had counted for more than financial compensation.[76]

Comparisons between coal and the automotive manufacturing sector were also made by miners and their representatives. At a meeting between ministers and trade union leaders in 1961, the NUM general secretary, Will Paynter, juxtaposed the exodus of workers from coal with car workers' commitment to their sector. Despite a cyclical sectoral downturn, vehicle builders 'had faith in the industry's future and wished to stay in it'.[77] These sentiments developed over the decade but were most strongly communicated by workers younger than Paynter. A focus group of retired miners in Moodiesburn, North Lanarkshire, held during 2014, made explicit reference to miners' pay and conditions relative to car workers in their memories of the 1972 and 1974 wages strikes. Those present were members of the industrial citizen generation: men born during the 1930s and 1940s who

---

[73] TNA, Coal 101/488, J. C. H. Mellanby Lee, Operational Research Executive (Scotland), 'A paper for consideration by the Scottish Area Monday and Friday Absence committee', 8 Aug. 1973.

[74] A. Fox, *Industrial Sociology and Industrial Relations* (London, 1966), pp. 4–5.

[75] J. Goldthorpe et al., *The Affluent Worker: Industrial Attitudes and Behaviour* (London, 1968), p. 39.

[76] TNA, Coal 101/488, J. C. H. Mellanby Lee, Operational Research Executive (Scotland), 'A paper for consideration by the Scottish Area Monday and Friday Absence committee', 8 Aug. 1973.

[77] TNA, POWE 14/857, Notes of a meeting on fuel policy at Admiralty House at 4.30 pm on 2 Feb. 1961.

entered the nationalized industry with expectations of economic rewards and employment stability.[78]

Peter Downie contributed to the discussion by recalling that his wages nearly doubled as a result of the award that followed the 1974 strike, which took place while he was hospitalized following an industrial accident: 'In 1974 ma wages at that time was 35 pound, and I went in tae the hospital and when I came oot the hospital and started walking again, my wages were 65 pound'.[79] His memories of hospitalization reinforce the NUM's 'special case' for extensive wage rises in the early 1970s, which underlined the dangers and physical effort miners' endured to produce Britain's chief indigenous fuel. The NUM's arguments rested on explicit reference to miners' deteriorating wage position relative to other male industrial workers.[80] Michael McGahey summed up the industrial citizen generation's assertive position at the 1972 NUMSA annual conference which was held shortly after the strike ended. With an element of implicit criticism directed at leaders of the interwar veteran generation, he asserted that 'for too long in the past our members were conditioned to believe that the contraction of the mining industry was an inevitability and that pit closures and the stockpiling of coal had weakened the bargaining power of the miners'. Furthermore, McGahey emphasized the NUMSA's objective of achieving further improvements in pensions, decreases in working hours and a lower retirement age.[81]

The roots of those sentiments lay in the experience of closures during the late 1950s and early 1960s. These episodes contributed to the frustrations of the industrial citizen generation and to conflict within the NUMSA which was presided over by interwar veterans until 1967. A key flashpoint took place in 1959 when the Board announced the closure of Devon colliery in Clackmannanshire. Over fifty men endured a stay-down strike between Tuesday 23 June and Thursday 25 June.[82] They were supported by over 15,000 miners from across the Scottish coalfield, including all of Alloa's sixteen collieries, as well as pits in Lanarkshire, West Lothian and Fife.[83]

---

[78] Moodiesburn focus group.

[79] Moodiesburn focus group.

[80] J. Arnold, '"The death of sympathy": coal mining, workplace hazards, and the politics of risk in Britain, *c.* 1970–1990', *Historical Social Research*, xli (2016), 91–110, at pp. 95–6.

[81] National Mining Museum Scotland archives, Newtongrange, Midlothian (NMMS), NUMSA, Minutes of Executive Committee and Special Conferences from 28 June 1971 to 14/16 June 1972, pp. 598–9.

[82] TNA, POWE 37/481/90, 'Back to work at Scottish pits to-day', *Herald*, 26 June 1959.

[83] TNA, POWE 37/481, Percy, Ministry of Power, to J. S. McClay, 25 June 1959; 37/481/91, 'Talks to-day on pit closure', *Scotsman*, 29 June 1959.

Press coverage profiled the strikers' youthfulness. The *Scotsman* described the youngest participant, seventeen-year-old James McGuigan, as 'quite a determined young man'. Other participants noted in the press included James Craig junior, the twenty-five-year-old son of the Devon NUM branch secretary.[84] Phillip Stein reported another instance of intergenerational cooperation in the CPGB-affiliated *Daily Worker*. Fifty-two-year-old Guy Bolton brought food and encouragement to his twenty-four-year-old son George, a CPGB member and the NUMSA's future president. Guy affirmed the determination of the young men underground: 'the mood of the lads is such they will have to be dragged on to the cast'.[85] Stein detailed that the strikers had deployed 'flying squads' that brought out other collieries. Over 1,000 miners had marched on the NCB's Alloa Area headquarters in Whins, demonstrating in both a carnivalesque and subversive manner: 'Leader of the demonstration, William McDougall, turned round to the men and shouted "Well lads, we are democratic – which way are we going?" There was an ear-splitting "In there" and the policemen were swept aside'.[86]

The unofficial action of 1959 anticipated the grievances articulated over the 1960s and the 1970s as the industrial citizen generation ascended within coalfield politics. Strikers rejected the industry's managed contraction and challenged the authority of the NCB, as well as union leaders of the interwar veteran generation who broadly accepted both. Abe Moffat recalled six years later that he was in London when he heard about the events at Devon and began discussions with officials from the NCB's Scottish division. Moffat confirmed that as at Barlinnie sixteen years before, his role was as an arbitrator who sought to defuse tension through ensuring dialogue between the union and employers. This was the only method by which he could 'persuade the men to come up the pit'.[87] The limitations of Moffat's achievement, and of his generation's conception of a moral economy centred on procedure, were confirmed by the NCB's disquiet over his claim that 'the Scottish Divisional Coal Board was prepared to negotiate with the Union on the closure'. Moffat had to rescind this claim at the NCB's insistence. An exchange of letters between Board officials confirmed 'all that the Scottish Coal Board have agreed' was 'to listen to what the men have to say after work has been resumed'.[88] Reports of the meeting, which

---

[84] TNA, POWE 37/481/83, '60 miners stage stay-down strike', *Scotsman*, 24 June 1959.

[85] TNA, POWE 37/481/82, P. Stein, '"Stay-Down" to Save Pit', *Daily Worker*, 24 June 1959.

[86] TNA, POWE 37/481/82, Stein, '"Stay-Down" to Save Pit', *Daily Worker*, 24 June 1959.

[87] Moffat, *Life*, p. 185.

[88] TNA, POWE 37/481, Letter to Mr Jarrat, 26 June 1959.

took place upon the return to work, detail that an NUM deputation led by Moffat specified disquiet over the movement of families by men who had left Lanarkshire to work in Alloa.[89]

The Devon strike communicated a general pattern of discontent that was especially focused among younger men who had been promised secure employment at modernized collieries in central Scotland. On the morning of 13 November 1961, nine Scottish collieries were affected by strike action against the announcement of further pit closures.[90] Unlike in 1959, during December 1961 and January 1962, Scottish delegate conferences voted for a day of official strike action against the industry's contraction.[91] The strike was to be timed alongside a lobby of parliament in March, but was eventually rescinded by the NUMSA's executive committee in favour of building inter-industry unity against wage restraint. However, this exposed tensions within the NUMSA: 3,000 miners in the Clackmannanshire area joined an unofficial walkout. A group of Fishcross strikers reportedly met NUMSA leaders with chants of 'throw them in the pond'.[92] Events between March and June 1967 subsequently confirmed the differentiation in outlook between the interwar veteran generation of miners represented by the Moffats and the younger cohort who coalesced around McGahey. In March 1967, an NUMSA delegate conference voted overwhelmingly, by sixty-two votes to twelve, in favour of a resolution from the NUM Economic Sub-Committee, which stated:

> The only solution is for the industry to concentrate as quickly as possible upon the most efficient and profitable pits and to adjust total capacity to potential demand levels. To fight for the survival of grossly uneconomic pits and for high levels of development expenditure on these pits is to place upon the industry a burden that ultimately will make necessary an even greater degree of contraction.[93]

Delegates saw closures as an unfortunate necessity. E. Tannahill from Kingshill 3 commented that in the last two decades Lanarkshire had lost forty collieries and now only eight remained, 'despite all the campaigns, all

---

[89] TNA, POWE 37/481/88, M. R. Foster, 'Miners to resume', *Scotsman*, 26 June 1959.

[90] J. Phillips, 'Deindustrialization and the moral economy of the Scottish coalfields, 1947 to 1991', *International Labor and Working-Class History*, lxxxiv (2013), 99–115, at p. 102.

[91] NMMS, NUMSA, Minutes of Executive and Special Conferences, 12 June 1961 to 6/8 June 1962, pp. 338, 353.

[92] NMMS, NUMSA, Minutes of Executive and Special Conferences, 12 June 1961 to 6/8 June 1962, pp. 502, 509.

[93] NMMS, NUMSA, Minutes of Executive Committee and Special Conferences from 27 June 1966 to 14/16 June 1967, pp. 279–80, 323.

the mass meetings and all the demonstrations'. Alex Moffat bluntly stated that 'whether we liked it or not, the mining industry was going to contract', and securing production on a financially profitable basis was the only option for survival.[94] Just three months later, in June 1967, the NUMSA annual conference voted unanimously to overturn this position. In the absence of the ill Alex Moffat, Michael McGahey delivered the customary Area conference presidential address, anticipating his election to the post upon Moffat's retirement, which followed soon after. The Ministry of Power had intimated that objectives of capital investment towards a 200-million-ton annual capacity target for 1970 had been reduced to 140 million. By 1980 this would be reduced further to only 80 million across all holdings in England, Wales and Scotland.[95]

McGahey stated that opposition entailed 'refusing to cooperate in the total rundown of the industry'. His stance was based on moral economy arguments which counterpoised financial measurements to social needs: 'I reject the present approach taken in many quarters which would make the cost of coal the sole criterion for determining the future size of the mining industry'. In terms redolent of the double movement, McGahey contrasted financial measurements with the 'disruption of mining communities' that would be caused by closures and redundancies. McGahey portrayed the government's attitudes towards coal as wanton neglect of Britain's prime domestic source of fuel. The NUMSA's objective was to secure mining employment through an energy policy that would serve to 'guarantee coal its proper share in the energy market, and to protect the long-term interests of the people we represent'.[96] Despite the ultimately unanimous vote, some delegates raised objections relating to the failures of previous attempts to oppose closures. However, the overwhelming sentiment was that the latest direction of closure and policies was a threat to the industry that had to be resisted. R. Baird from Cardowan claimed the industry was haemorrhaging workers because there was 'no security' in coal mining employment. T. Cullen, the delegate for Gartshore 9/11, reaffirmed this, stating that the 'executive should oppose all closures including partial closures'.[97] Tannahill, the delegate from Kingshill 3 who had spoken in support of Alex Moffat

[94] NMMS, NUMSA, Minutes of Executive Committee and Special Conferences from 27 June 1966 to 14/16 June 1967, pp. 319–20.

[95] NMMS, NUMSA, Minutes of Executive Committee and Special Conferences from 27 June 1966 to 14/16 June 1967, pp. 396, 411.

[96] NMMS, NUMSA, Minutes of Executive Committee and Special Conferences from 27 June 1966 to 14/16 June 1967, p. 411.

[97] NMMS, NUMSA, Minutes of Executive Committee and Special Conferences from 27 June 1966 to 14/16 June 1967, p. 396.

four months previously, argued that the revised target was tantamount to the government reneging on promises of a stable future for the industry: a 200-million-tonne output would have limited the impact of closure through the expansion of capacity elsewhere. As a result, he voted in favour of the change of position.[98]

During the 1980s, industrial citizen generation leaders still sought more generous redundancy and retirement terms to help rebalance the workforce in the context of closure threats and high rates of youth unemployment.[99] These suggestions were popular with activists who argued for statutory rights to retirement as an alternative to the government's selective usage of redundancy payments and early retirement offers. John Mitchell of Frances colliery in Fife moved a composite motion at the 1983 NUMSA conference that argued for retirement at fifty-five for underground workers and sixty for surface workers at a rate that allowed former miners to 'live in dignity'.[100] These efforts accord with Gildart's view that over the second half of the twentieth century, for some miners 'the mine became a place they longed to escape from'.[101] The alienation that was visible in the industrial sabotage and absenteeism which so concerned Board officials in the early 1970s was communicated more constructively in rising demands that the nationalized industry would provide improved and safer conditions as well as a longer retirement and shorter working life. However, these demands also indicate that the industrial citizen generation retained a moral economy conception of lifetime employment in coal mining. Brendan Moohan's narrative, perhaps influenced by his father's experience of working with the NCB before retiring due to ill health, emphasized a generational distinction in outlook and socioeconomic circumstances:

> To the generation before mine it was cradle to grave. It was, you left school, you went to the pit, until you retired, and a couple of years later you died. That was the path, that was the deal. I was not really the type of teenager to have looked to view ma life that way. I was a bit more adventurous. But I mean you know, if the industry hadnae of been closed down the chances are I probably still would be there.[102]

---

[98] NMMS, NUMSA, Minutes of Executive Committee and Special Conferences from 27 June 1966 to 14/16 June 1967, pp. 413–14.

[99] TNA, Coal 31/138, Meeting held at 11am on Wednesday 25 Feb. 1981, Thames House South, London.

[100] NMMS, NUMSA, Executive Committee Minutes, July 1982 to June 1983, p. 654.

[101] K. Gildart, 'Mining memories: reading coalfield autobiographies', *Labor History*, i (2009), 139–61, at pp. 145–6.

[102] Brendan Moohan, interview.

Brendan underlined elements of generational conflict as well as reciprocal relationships in reflecting on his experience of social life in mining communities and the form of social conservatism that prevailed within them. This centred on constructions of respectability and masculinity. Brendan recollected that he found the atmosphere of the Lothians coalfields stifling due to his style of dress and the music he listened to. To a young man who questioned authority, the highly formalized rules-based order of the mining industry and its associational life, including Miners' Welfares, were an imposition to be resisted:

> When I was nineteen, I thought they were old farts. But you know, when I look back. I mean these guys with blazers that sat in smoked filled rooms were actually protecting a form of socialism. You know, in their own unique way they were the guardians of the community, and actually what they represented had to be fought for to be achieved over centuries. So, there was, looking at them wi that set ae eyes as opposed to the set ae eyes I had at nineteen when they looked down on the way that ah dressed. And I thought that I'd rather die than dress the same as they would.[103]

Involvement in the 1984–5 miners' strike brought some reciprocation between Brendan and the enforcers of community customs. He recalled joining other union members in arranging to expel a strikebreaking miner from the Royal Musselburgh Golf Club, which was owned by a miners' charity:

> I remember during the strike, a guy in Preston Pans who was a member of the club and an active golfer had actually scabbed. And a meeting was called to expel him from the club, and we all turned up, cause as far as we were concerned it was a Miners' Welfare Institute and therefore his actions affected the welfare of the mining community! So you know, this guy was pleading that he was being picked on because he went back to work. And we were like 'yeh, you are, absolutely!'[104]

However, political and cultural generational tensions remained among activists. Later in his testimony, Brendan specified that the 'old farts' he had in mind were often Communist Party members. His narrative indicated the importance of the link between generational experiences and political standpoints, confirming the hostility between older 'Stalinist' NUM officials and younger 'Trots' like himself who were predominantly organized in the Militant Tendency within the Labour party. Brendan recalled Chris Herriot, the NUM Youth Delegate at Monktonhall, and Joe Owens, who

---

[103] Brendan Moohan, interview.
[104] Brendan Moohan, interview.

was the pit delegate at Polkemmet during the 1984–5 strike as standout examples of younger Militant members who gave a greater heterogeneity to internal politics within the NUMSA during the 1980s. This included a marked presence at Monktonhall colliery where some young miners would 'leave copies of the *Militant* [newspaper] lying about' areas that men had their breaks underground.[105]

Jessie Clark's husband, Alex, perhaps epitomized the type of leader from the generations of industrial citizens that Brendan clashed with in his youth but came to respect in later life. Alex Clark worked in collieries in South Lanarkshire from the late 1930s until 1953 when he became a full-time CPGB organizer. He played a prominent role as a local young trade unionist and communist activist during the early period of nationalization. Jessie was keen to stress Alex's intellectual and artistic involvement based on a working-class culture of self-education that defied the official expectations:

> So, the whole attitude was that the authorities and maybe even some of the teachers not them all just didn't have any expectations for these kids you know. The boys are aw gonna be miners, doesnae matter where do you get miners that read poetry, y'know? Which is untrue! Because in actual fact I married one who taught me about poetry. To really appreciate it you know, but that was the attitude.[106]

Alex was also a member of the Lesmahagow male voice choir. This laid the foundations of his career development as he became involved in Scottish performing arts and culture as a full-time official for the theatre union, Equity, and through the Scottish Trades Union Congress Arts and Entertainment Committee.[107] Brendan qualified his description of conflict by describing older and younger activists drinking in the same pubs and noting that he used to borrow books from more senior union members who encouraged younger men to read novels as well as political literature.[108] Bill McCabe described a similar culture at Caterpillar, recalling an older steward at the plant lending him his copy of the *Ragged Trousered Philanthropist*. Bill portrayed becoming a shop steward before he was twenty as 'the natural progression', following in the trajectory of his brother and father at the plant.[109] Brendan has in some senses also followed a similar course to an

---

[105] Brendan Moohan, interview.

[106] Jessie Clark, interview.

[107] A. Clark, 'Personal experience from a lifetime in the communist and labour movements (part 2)', *Scottish Labour History Review*, xi (1997–8), 14–16, at p. 15.

[108] Brendan Moohan, interview.

[109] Bill McCabe, interview with author, Tannochside Miners' Welfare, 20 Jan. 2017.

earlier generation of political and cultural leaders, having become a poet. He has published an anthology of poems about his experience of the 1984–5 strike. The final poem in that collection, 'Vision', opens by acknowledging the weight of past generations stating that 'Others have gone before, and will follow me'. Brendan presents the strike is as a coming-of-age event, prefiguring 'a flammable future'.[110] Contrastingly, the Ayrshire poet, Jim Monaghan, narrated the strike and its aftermath with an emphasis on loss. Sharing Brendan's later-found appreciation of the culture that sustained the men of the industrial citizen generation, as well as younger flexible workers such as his brother, who was an active picketer during the 1984–5 strike, he narrates the 'United Colours of Cumnock':

> My town was once a red town,
> another miner dead town,
> a men who fought and bled town,
> wae brave and stalwart wives.
> That's red that came from meeting rooms,
> from folk that worked the pumps and looms,
> when borough bands played different tunes,
> and marched – for better lives.
>
> But now my town's a grey town,
> a fifty mils a day town,
> a watch life slip away town,
> a tunnel wae nae light.
> That's grey that weeps from dying eyes,
> bewildered parents, children's cries,
> wae skinny erms and stick like thighs,
> and nae strength left – tae fight![111]

Like Brendan's reappraisal, Monaghan's perspective looks 'beyond the ruins' of industrial society. The emotional space provided by temporal distance from the 1984–5 strike and the major closures which followed it, have allowed for a form of generational reckoning that necessarily entails a broader appraisal of the industrial era and coalfield culture. A form of critical nostalgia is visible in both men's accounts. Coal mining is remembered for its costs in human lives and industrial diseases, as well as the instilling of a socially conservative culture. Yet a strong sense of bereavement for lost cohesion and collective mobilization are also emphasized. Monaghan's

---

[110] B. Moohan, *The Enemy Within* (Livingston, 2012).
[111] J. Monaghan, 'United colours of Cumnock', *Concept*, viii (2017) <http://concept.lib.ed.ac.uk/issue/view/216> [accessed 20 Dec. 2019].

critique of the present is amplified through Cumnock's dramatic departure from its coal mining past. This confirms that accelerated workplace closure and industrial job losses were interior time experiences for the flexible worker generation. Coal mining and its associational culture continue to inform perspectives within Scottish coalfield areas today. The socioeconomic structure marked by increased rates of economic inequality that has developed in the wake of deindustrialization is understood and criticized through outlooks with roots in the collective memory of colliery employment. The norms and values that powerfully shaped the moral economy therefore also have outlasted the industrial structure in which they developed.[112]

Deindustrialization's effects are still being transmitted across generations within the Scottish coalfields. Collective memories from earlier periods are passed on within families and communities. Customs and expectations with origins in older eras shape appraisals of the contemporary labour market. Over the second half of the twentieth century, three distinct generational cohorts coalesced around distinct formative interior time experiences that were central to determining their distinct political outlooks and cultural attitudes. The moral economy of the nationalized period was formatively shaped by the interwar veteran generation who stewarded the NUMSA during the early period of nationalization and placed extensive value on custom and institutional stability. Industrial citizens who were socialized during the early years of colliery contraction were less reverential towards the nationalized industry. Their expectations included both employment security and that rewards for the hard and dangerous physical labour of coal mining would be at least equal to the economic rewards offered in assembly goods factories. A final generation of flexible workers matured during the growth of mass unemployment and the bitter final struggles against major industrial closures which characterized the 1980s. Representatives of this generation were resistant to the conservatism and rules-based order of coalfield communities. However, their reflections on the effects of intensified deindustrialization also demonstrate their shared investment in the moral economy that they mobilized to defended against the atomizing effects of liberalized market forces.

---

[112] S. Condratto and E. Gibbs, 'After industrial citizenship: adapting to precarious employment in the Lanarkshire coalfield, Scotland and Sudbury hardrock mining, Canada', *Labour/Le Travail*, lxxxi (2018), 213–39, at p. 239.

6. Coalfield politics and nationhood

Coalfield experiences were central to the politics of class and nation in twentieth-century Scotland. Both trade union membership and Labour party parliamentary representation were disproportionately concentrated in the coalfields. Miners often exercised political leadership, especially in villages and towns whose economies and social structure were overwhelmingly reliant on coal mining. This was often based on highly parochial associations. The struggle to unify miners across Scotland and Great Britain was long fought and always subject to geographical tensions. Nationalization and coalfield restructuring considerably altered coal's political economy. Scottish nationhood figured significantly in responses to the administration of the industry by the National Coal Board (NCB), which was headquartered at Hobart House in London. The context of a single employer across the UK created both dangers and opportunities for the National Union of Mineworkers Scottish Area (NUMSA). It facilitated the consolidation of a stronger Scottish miners' union, and also encouraged the adoption of a national countermovement to the NCB's pursuit of financial and technical priorities that threatened the future of Scottish coal mining. This chapter begins by analysing the confluence of locale, class and nation in the NUMSA's political culture. Under nationalization, a stronger and more centralized union structure was constructed, and the Area was given a left-wing or militant identity by its predominantly Communist-affiliated leadership. Section two examines how Scottish national consciousness framed industrial politics under nationalization, underlining the NUMSA's shift towards more pronounced support for home rule during the 1960s as the NCB centralized. However, this nationalist orientation was always qualified by a class politics grounded in 'Unionist' sensibilities regarding the integrity of the nationalized industry and UK energy policy, trade union unity and affiliation to the Labour party. The final section discusses the Scottish miners' gala, which was a major innovative feature of postwar Scottish coalfield culture. By hosting a large-scale annual event which reprised local traditions on a national plain, the NUMSA was able to mould occupational and class consciousness onto Scottish nationhood. But it did so within the context of the social democratic infrastructure provided by the nationalized coal industry and British labour movement connections.

'Coalfield politics and nationhood', in E. Gibbs, *Coal Country: The Meaning and Memory of Deindustrialization in Postwar Scotland* (London, 2021), pp. 187–223. License: CC-BY-NC-ND 4.0.

Collective memories rooted in small-scale locales were central to conceptions of community within the Scottish coalfields, but the bonds established within coal mining villages and towns also served as a basis for the mobilization of a less geographically restricted industrial identity. Under nationalization, experiences located in localized community and workplace settings were rearticulated through the institutions and consciousness of nationhood.[1] The protective countermovement which challenged the NCB's pursuit of closure programmes that threatened the integrity of the Scottish coalfields was substantially framed in terms of national identity. Between the 1940s and the 1960s, moral economy sentiments came to rest on expectations of sympathetic action from devolved Scottish institutions which were assumed to be less socially and politically distant.

The NUMSA's political culture built on community and workplace experiences to emphasize the broader class and international dimensions of events in the coalfields. These efforts entailed a struggle to overcome the long history of divisions between jealously guarded autonomous regional associations. An increasingly centralized Scottish miners' union emerged under nationalization, centred on a strengthened executive apparatus. This was shaped by deliberation on interwar experiences. Defeat in the 1926 lockout exposed the Miners' Federation of Great Britain's weakness as a relatively loose amalgamation of county unions. Another lesson was drawn from the failure of the communist-led separatist 'red union', the United Mineworkers of Scotland (UMS).[2] Abe Moffat, who had been general secretary of the UMS, was elected president of the National Union of Scottish Mineworkers in 1942 after the UMS was wound up. Moffat's election began a sustained Communist Party of Great Britain (CPGB) presence at the top of the union. His presidency was marked by a commitment to maintaining a unified organization for Scottish miners as part of the UK-wide NUM, incorporating continued support for the Labour party.[3]

Abe and his brother Alex, who succeeded him as president between 1961 and 1967, combined a social democratic approach on industry matters with a communist perspective on international affairs. In Alan Campbell and John McIlroy's term, the Moffats' 'Stalinist–Labourism' consolidated a commitment to 'restraint and moderation' in the context of partnership

[1] J. Phillips, 'The meanings of coal community in Britain since 1947', *Contemporary British History*, xxxii (2018), 39–59, at pp. 42–3.

[2] S. Macintyre, *Little Moscows: Communism and Working-Class Militancy in Inter-War Britain* (London, 1980), pp. 61–72.

[3] A. Campbell and J. McIlroy, 'Miner heroes: three communist trade union leaders', in *Party People, Community Lives: Explorations in Biography*, ed. J. McIlroy, K. Morgan and A. Campbell (London, 2001), pp. 143–68, at 143, 158.

with the NCB', including the acceptance of major closures and coalfield restructuring.[4] Building a disciplined commitment to the NUMSA across the nation's coalfields required a sustained effort from activists. The union was relatively successful in achieving this, in part through appealing to memories of the early twentieth century, which shaped understandings of subsequent struggles.

In common with other UK coalfields, the events of 1926 strongly entered historical consciousness and became especially important to framing understandings of the 1984–5 strike in Scotland.[5] Mick McGahey was a third-generation communist miner. Like his father and grandfather before him, Mick faced repression in the form of arrest and job loss for his involvement in picketing at Bilston Glen colliery in Midlothian where he worked as a surface worker.[6] His testimony included a strong recollection of familial connection to miners' struggles:

> My grandfather was involved in the 1926 general strike. He got sent to jail. He did six months in the jail. My grandmother got evicted. Family oot the pit owner's hoose. And they ended up in Kent. And they moved aboot the coalfields in England. And eventually came back to Scotland and settled in Cambuslang.[7]

However, this historical consciousness developed in advance of 1984. Mick's father, the NUMSA president, Michael McGahey, welcomed G. Solovyev from the Central Committee of the Soviet Miners' Union to NUMSA's annual conference in 1972. McGahey stated that Scottish mineworkers had been 'reminded time and time again by their fathers and grandfathers' of the assistance Soviet miners gave to their British comrades during 1926.[8] Other conference speeches also referred to these earlier events.

---

[4] J. McIlroy and A. Campbell, 'Coalfield leaders, trade unionism and communist politics: exploring Arthur Horner and Abe Moffat', in *Towards a Comparative History of Coalfield Societies*, ed. S. Berger, A. Croll and N. La Porte (Aldershot, 2005), pp. 267–83, at p. 276.

[5] D. Nettleingham, 'Canonical generations and the British left: narrative construction of the British miners' strike, 1984–85', *Sociology*, li (2017), 851.

[6] A. Campbell, 'Scotland', in *Industrial Politics and the 1926 Mining Lockout: the Struggle for Dignity*, ed. J. McIlroy, A. Campbell and K. Gildart (Cardiff, 2009), pp. 173–98, at pp. pp. 184–5.

[7] Mick McGahey, interview with author, Royal Edinburgh Hospital, Edinburgh, 31 March 2014.

[8] National Mining Museum Scotland archives, Newtongrange, Midlothian (NMMS), National Union of Mineworkers Scottish Area (NUMSA), Minutes of Executive Committee and Special Conferences from 28 June 1971 to 14/16 June 1972, p. 621.

For instance, Alex Day greeted the same conference as a representative of the Scottish Trades Union Congress (STUC) general council with a reference to the labour movement having had 'a forty-six-year-old debt to repay' before the solidarity action which enabled NUM to win significant wage improvements in the industrial dispute earlier that year.[9]

The NUMSA was profiled by the Moffats' allies as a significant innovation with political and cultural as well as industrial implications. In his 1955 history of the Scottish miners, Page Arnot lauded the Moffats' achievements in building a unified Scottish miners' union. He emphasized the political direction that their leadership gave to the NUMSA. Arnot also highlighted the role the union played in 'encouraging the development of music, art and the theatre' to address the 'cultural needs' of the mining community, including through the Scottish miners' gala. The Scottish miners also obtained a wider resonance through their union's support for the Edinburgh People's Festival and the communist singer-songwriter Ewan MacColl's theatre workshop, as well as Unity Theatre's activities in Fife. In addition, the NUMSA organized a concert given by the American communist singer Paul Robeson at the Usher Hall in Edinburgh in May 1949. A large part of the audience was made up of miners from across the Scottish coalfields.[10] Arnot's emphasis on international connections and political campaigning was shared in the NUMSA's youth activities. These incorporated the establishment of residential political schools aimed at young miners. Delegates enrolled on 'an educational course not only on technical mining matters but on general social and political questions'. From 1954 this message was also spread in the pages of the *Scottish Miner* which was published monthly and represented the achievement of the long-held aim of a newspaper for mining trade unionists in Scotland.[11]

Mick McGahey recalled the important role these activities played in shaping trade union activism:

> The Communist Party ae Britain played a massive role in training and development and education. Whenever you became active in the National Union ae Mineworkers in Scotland, the first thing you did whether you were the youth delegate, whether you were on the committee, didn't matter what role you had. The first thing they did was send you on a training course. You went to the Salutation Hotel in Perth for a weekend school. And it was aboot Marx, it

---

[9] NMMS, NUMSA, Minutes of Executive Committee and Special Conferences from 28 June 1971 to 14/16 June 1972, p. 643.

[10] R. P. Arnot, *A History of the Scottish Miners: from the Earliest Times* (London, 1955), p. 413.

[11] Arnot, *Scottish Miners*, p. 413.

was aboot Engels, it was aboot Lenin. It was aboot the ownership ae the means ae production. It was aboot the politics behind why does the government behave like that, why do we behave like that. It was a complete package ae political education that I don't think exists nowadays in any organization. Any trade union organization. That's what made it strong.[12]

These activities developed within Stalinist–Labourism's parameters. Under the Moffats, the NUMSA displayed a protective or even suspicious approach to political pluralism. The Area rejected an invitation to courses from the Workers Educational Association at its 1950 conference where it was explained that the NUMSA ran its own residential political school in Dunoon that provided a 'week's intensive education not only in the policy of the organization but in the economy of the country generally'.[13] Even NUMSA activists who declined to join the CPGB recognized the importance and distinctiveness of the outlook the Scottish Area's leadership provided. For instance, Nicky Wilson, who became active in the Scottish Colliery, Enginemen, Boilermen and Tradesmen's Association (SCEBTA) as a young electrician during the 1970s and is a longstanding member of the Labour party, referred to 'the leadership we had in the union in the past in Scotland who believed in bringing on young people and that didnae happen in other areas'. Nicky posited that this formed a critical distinction between trade unionism in the Scottish coalfields and the NUM's 'moderate' areas. He emphasized that the NUMSA's commitment to political education of young members differentiated it from other Areas, including Yorkshire, which he described as historically having been a 'right-wing area' despite the later influence of Arthur Scargill. Nicky underlined that the Yorkshire Area, unlike the NUMSA, had 'no political education hardly at all'.[14]

There are elements of composure in these recollections. In the aftermath of defeat in 1985, CPGB members and supporters within the Scottish and South Welsh Areas of the NUM articulated criticisms of 'Scargillism', which was seen as politically naïve and not attuned to the changed climate of the 1980s.[15] This was given greatest prominence after George Bolton succeeded Michael McGahey as president in 1987. In 1985, shortly after the miners' return to work, Bolton argued that Scargill, along with his

---

[12] Mick McGahey, interview.

[13] NMMS, NUMSA, Minutes of Executive Committee and Special Conferences, 20 June 1949 to 2 June 1950, p. 631.

[14] Nicky Wilson, interview with author, John Macintyre Building, University of Glasgow, 10 Feb. 2014.

[15] P. Ackers, 'Gramsci at the miners' strike: remembering the 1984–1985 Eurocommunist alternative industrial strategy', *Labor History*, lv (2014), 151–72, at pp. 153–4.

prominent English supporters, including the Labour MPs Tony Benn and Dennis Skinner, were exponents of 'a range of thinking in the movement that's not caught up with the reality of Thatcher and Thatcherism, and the state of British politics'.[16] The rift in the NUM's internal politics perhaps consolidated the NUMSA's support for devolution. These developments were concurrent with the formation of a cross-class civil society coalition in support of a devolved Scottish parliament. The strike's outcome appeared to sever traditional class-based strategies, and the polarization of the 1987 general election furthered national contentions regarding the UK's fragmenting politics.[17] Despite these influences, Nicky's recollections concur with the description of the NUM's organized left put forward by Vic Allen, a vocal supporter of Arthur Scargill. Criticisms of the NUM's right wing or 'moderate' leadership during the 1950s and 1960s initially came from Scotland, South Wales and Kent but the left was able to gain crucial footholds in Yorkshire, partly through the efforts of Scottish miners who had transferred southwards.[18]

Nicky Wilson was not isolated in his recollections. Tommy Canavan was another longstanding Labour party member and trade union activist at Cardowan who exemplified the reach that CPGB perspectives and history had within NUMSA. He was keen to emphasize the importance of the *Scottish Miner*. When interviewed, Tommy presented a copy of the paper from 1967 which had a historical article discussing Willie Gallacher's role as a Communist MP for the coalfield constituency of West Fife, as well as a column making the case for the establishment of a Scottish parliament. In Tommy's view, the publication played an important role in publicizing the union's perspective.[19]

The NUMSA's policies and support for international solidarity campaigns was highly influenced by the CPGB's outlook. This included a high prominence given to opposing nuclear weapons under the presidencies of both Moffat brothers as well as McGahey. Under Communist leadership, the NUMSA also opposed British military involvement in Greece and the Korean War and supported anti-colonial struggles in Kenya and Malaya.[20]

---

[16] D. Priscott, 'The miners' strike assessed', *Marxism Today*, Apr. 1985, pp. 21–7, at p. 24.

[17] E. Gibbs, 'Civic Scotland vs communities on Clydeside: poll tax non-payment *c.* 1987–1990', *Scottish Labour History*, xlix (2014), 86–106.

[18] V. Allen, *The Militancy of British Miners* (Shipley, 1982), p. 139.

[19] Tommy Canavan, interview with author, residence, Kilsyth, 19 Feb. 2014.

[20] NMMS, NUMSA, Minutes of Executive Committee and Special Conferences, 23 June 1947 to 8 June 1948, pp. 462–3, 497; Minutes of Executive Committee and Special Conferences, 20 June 1949 to 2 June 1950, p. 785; Minutes of Executive Committee and Special Conferences, 18 June 1951 to 20 June 1952, pp. 664–5.

There was also a clear occupational dynamic to this internationalism. Scottish and Soviet miners renewed their longstanding connections through the exchange of delegations in 1967.[21] A Soviet delegation returned in 1972, while the NUMSA arranged visits to the East German coalfields the same year.[22] The Scottish miners also lent support to struggles in other coalfields, most prominently to the major communist-led French miners' strike of 1947.[23]

The NUMSA's internationalism developed concurrently with the STUC's politics, which were formatively shaped by miners and CPGB activists. In 1968, the STUC welcomed a delegation from Hungary that included Antal Simon, general secretary of the Hungarian Mineworkers' Union. They visited East Kilbride among other locations across Scotland.[24] These fraternal links were also evident in STUC delegates making numerous trips to Comecon, including attendance at Budapest May Day in 1972 as well as visits to East Germany and Czechoslovakia during the early 1970s.[25] The Scottish miners further extended their connections with their Hungarian counterparts through an invitation for Scottish miners and their children to holiday in the country. Support was also granted to anti-colonial and national democratic struggles. The STUC's sustained opposition to Apartheid South Africa included a relationship with Jon Gaetsewe of the South African Congress of Trade Unions in western Europe. Gaetswere recurrently appears in STUC annual reports as a point of contact who assisted in providing Scottish financial support to South African trade unions.[26] The NUMSA was at the forefront of arguing for expanding the STUC's international activities. At the 1973 conference, Michael McGahey moved a resolution in support of the Spanish Workers' Commission, a Communist-affiliated trade union federation, after the imprisonment of ten trade unionists by the Franco regime. McGahey's intervention came after one of the Commission's leaders, Carlos Elvira, addressed the conference.[27]

---

[21] Minutes of Executive Committee and Special Conferences from 27 June 1966 to 14/16 June 1967, pp. 429–30.

[22] Minutes of Executive Committee and Special Conferences from 28 June 1971 to 14/16 June 1972, p. 597.

[23] NMMS, NUMSA, Minutes of Executive Committee and Special Conferences, 23 June 1947 to 8 June 1948, pp. 255–6.

[24] STUC, *Annual Report 1967–1968*, lxxi (1968), 191–2.

[25] STUC, *Annual Report 1972–1973*, lxxvi (1973), 233.

[26] STUC, *Annual Report 1972–1973*, lxxvi (1973), 193; *1978–1979*, lxxxii (1980), 548.

[27] STUC, *Annual Report 1972–1973*, lxxvi (1973), 456.

In some cases, NUMSA interventions were key in shaping STUC policy. At the 1953 conference, Abe Moffat spoke in favour of a motion that concluded 'congress wishes the people of Kenya and Tanganyika every success in their fight for land, liberty and happiness'.[28] Moffat argued this was within the best traditions of the British labour movement: 'He always had understood it was the basic policy of the labour and trade union movement in Britain to fight against all colonial wars and exploitation and he hoped that was going to continue to be its policy'.[29] The STUC general council opposed the resolution. G. Hamilton of the Transport and General Workers' Union articulated the mainstream position which echoed both the pressures of Cold War alignment and the legacy of sympathy for imperialism within the labour movement, as well as British belligerence in the conflict.[30] Hamilton stated that, 'Before Kenya and Tanganyika could have self-government the British had much more to contribute towards the education of the African peoples'.[31] The motion passed, which is perhaps surprising given the context of a live struggle in which British soldiers and settlers were engaged in armed conflict with the Kenya Land and Freedom Army.

Despite these successes, the records of both the NUMSA and STUC also contain elements of disquiet and opposition to the NUMSA leadership. At the 1957 NUMSA conference, anti-communist delegates moved resolutions in opposition to the NUMSA having a relationship with organizations proscribed by the Labour party, implicitly including the CPGB. Another motion was moved in opposition to the communist-led World Federation of Trade Unions (WFTU).[32] P. McCann, a delegate for the Lanarkshire colliery, Gartshore 3/12, moved both resolutions. He alleged that the WFTU was 'communist-dominated' and that further, 'the real intention' of the cultural organizations and ties the NUMSA held with Comecon nations, was to aid regimes such as Czechoslovakia, Poland and East Germany which had all recently been responsible for violently suppressing democratic opposition.[33] Although both resolutions fell, their presence

---

[28] STUC, *Annual Report 1952–1953*, lvi (1953), 260.

[29] STUC, *Annual Report 1952–1953*, lvi (1953), 260.

[30] John McIlroy, '"Every factory our fortress": Communist Party workplace branches in a time of militancy, 1956–79, part I: history, politics, topography', *Historical Studies in Industrial Relations*, x (2000), 99–139, at p. 113.

[31] STUC, *Annual Report 1952–1953*, lvi (1953), 260–2.

[32] STUC, *Annual Report 1952–1953*, lvi (1953), 769.

[33] NMMS, NUMSA, Minutes of Executive Committee and Special Conferences from 18 June 1956 to 5 to 7 June 1957, pp. 766–9.

indicates that the CPGB's hegemony within NUMSA was never complete. Oppositional forces won delegate elections in Scottish collieries. More pertinent anti-communist trends had been apparent ten years earlier when a leadership-supported conference resolution opposing British involvement in Greece and the Labour government's growing ties with the United States fell.[34] This result perhaps reflected the conflict on this occasion between the CPGB and Labour party positions, undermining the broad unity between Communists and the Labour left that usually prevailed within the NUMSA. These developments took place in the context of a general build-up of Cold War tensions within the British labour movement. Labour party supporters of nuclear armaments and affiliation with the American-led western powers who went on to establish the North Atlantic Treaty Organization (NATO) were opposed to CPGB supporters of the Soviet-led Eastern Bloc, which later formed the Warsaw Pact, and who backed unilateral disarmament.[35]

Another setback was the NUMSA's failure to overturn the STUC's ban on the Scottish USSR Society in 1951.[36] There were other episodes of discord during the 1940s and 1950s, including an affirmation of the STUC's opposition to the WFTU in 1958.[37] However, later exchanges of delegations with the Soviet Union indicate that the CPGB and its allies were able to overturn hostility to communist-led regimes as well as garner support for international causes. Significant successes included the STUC committing to opposition to American involvement in Vietnam.[38] The STUC was therefore an important avenue through which the NUMSA were able to overturn social democratic orthodoxies in relation to international alignment as well as industrial strategy and constitutional politics. These findings question definitions of labourism contingent on a broad support for British foreign policy.[39] Major sections of the post-1945 Scottish labour movement were in fact open to outlooks shaped by opposition to pro-NATO perspectives. The activities of communists within unions, most prominently and continuously at the helm of the NUMSA, encouraged the development of increasingly critical policy stances. They also passed at

[34] NMMS, NUMSA, Minutes of Executive Committee and Special conferences, 8 July 1946 to 11 June 1947, pp. 38–9.

[35] R. Samuel, *The Lost World of British Communism* (London, 2006), pp. 205–6; McIlroy, '"Every factory our fortress" (part 1)', p. 113.

[36] STUC, *Annual Report 1950–1951*, liv (1951), 22–3

[37] STUC, *Annual Report 1957–1958*, lxi (1958), 297–305.

[38] STUC, *Annual Report 1967–1968*, lxxi (1968), 135, 192.

[39] J. Saville, 'Labourism and the Labour government', *Socialist Register*, iv (1967), 43–71; R. Miliband, *Parlimanetary Socialism: a Study in the Politics of Labour* (London, 1973).

conferences due to the growing discontent with British social democracy's failure to deliver social advances and economic security in the context of deindustrialization. This was most marked during the Wilson and Callaghan governments of the 1960s and 1970s when the STUC developed a devolutionary perspective and became increasingly opposed to the Labour party's pro-NATO Cold War alignment.

Left-wing opposition to the international aspects of Stalinist–Labourism was also apparent. Lawrence Daly was a prominent critic of CPGB policy following his resignation from the party in 1956. Before he became the NUMSA general secretary in 1965, and the UK-wide NUM general secretary in 1968, Daly established the Fife Socialist League with support from other members of the 'new left' who had also rejected the CPGB's slavish adherence to Soviet policy and the suppression of internal debate. He won over 5,000 votes in West Fife during the 1959 general election and came third, defeating the CPGB candidate in Willie Gallacher's former constituency.[40] Daly's political positions caused tensions within the NUMSA. On 15 April 1957, the executive committee noted that the *Scottish Miner* had refused to publish Daly's article commenting on the Soviet invasion of Hungary on the grounds that it 'dealt with internal differences of Communists'.[41] At the Area's annual conference two months later, Daly inquired as to why his article had not been published in the *Scottish Miner*, arguing that 'the essential point at issue was the question of the democratic rights of members'. Daly was not a lone voice. Supportive delegates included a representative from Priory colliery in Blantyre. Abe Moffat responded rather perversely given the politicized character of the *Miner*'s coverage. The NUMSA president claimed that 'the paper had not been established for the purpose of discussing the policy of the Communist or any other political party'. Moffat conflated two separate issues in order to present his stance as one for unity by arguing 'the miners' paper could not be used for the purpose of attacking anyone's religious or political belief'.[42]

Despite the appearance of rigidity in Abe Moffat's response to Daly, the invasion of Hungary shook old certainties. Abe's brother, Alex Moffat, briefly left the Communist Party, but he rejoined before assuming presidency of the NUMSA in 1961.[43] In the years after 1956, a 'broad left' strategy

[40] M. Kenny, *The First New Left: British Intellectuals after Stalin* (London, 1995), pp. 40–1.

[41] NMMS, NUMSA, Minutes of Executive Committee and Special Conferences from 18 June 1956 to 5 to 7 June 1957, p. 607.

[42] NMMS, NUMSA, Minutes of Executive Committee and Special Conferences from 18 June 1956 to 5 to 7 June 1957, p. 714.

[43] N. C. Rafeek, *Communist Women in Scotland: Red Clydeside from the Russian Revolution to the End of the Soviet Union* (London, 2008), p. 141.

displaced Stalinist–Labourism within the NUMSA's political culture. These developments were part of a wider trend in the CPGB's orientation, which was especially marked within trade unions. Communists showed a greater willingness to undertake activities in coalition with Labour supporters and other left-wing allies on industrial and political matters through holding a less sectarian or restrictive attitude over Cold War alignments and commitments to official labour movement structures.[44] The CPGB's criticism of the Soviet invasion of Czechoslovakia in 1968 indicated a very different attitude to that which it had maintained over Hungary in 1956. While Abe Moffat had barracked Lawrence Daly for his opposition to the latter, Michael McGahey, who had succeeded Alex Moffat as NUMSA president in 1967, was among the members of the CPGB's political committee that voted to oppose the Soviet suppression of the Prague Spring. Jimmy Reid, who also voted for the resolution, recollected the episode as a major rupture which led to McGahey's father, Jimmy, a founding member of the CPGB, shutting his house door in his son's face after stating, 'So you and Jimmy Reid condemned the Soviet Union'.[45]

Michael McGahey's less restrictive attitude was also visible in support for the liberation struggles in Vietnam and South Africa through which the CPGB aligned itself with anti-Stalinist political forces in broader solidarity campaigns. Willie Doolan felt that these positions entrenched the NUMSA's support for devolution, which he remembered evolving through discussions held during meetings between Scottish CPGB trade union activists in Edinburgh during the 1970s. McGahey, 'was very critical of Czechoslovakia. And he was critical of the Soviet invasion of Afghanistan too. Mick believed that a country should have its own say. Mick was an ardent campaigner for a Scottish Parliament'.[46] These shared commitments encouraged a closer relationship between McGahey and Daly, which partially overcame their differences about the CPGB. Daly was at the forefront of solidarity efforts with Vietnam. When Daly moved a resolution at the 1967 NUMSA conference calling on the British government to condemn America's involvement in the war he referred to his recent visit to the North Vietnamese coalfield and seeing villages 'which had literally been wiped out by systematic bombing attacks'. The resolution passed

---

[44] R. V. Seifert and T. Sibley, *Revolutionary Communist at Work: a Political Biography of Bert Ramelson* (London, 2012), pp. 110–14.

[45] J. Reid, 'Mick McGahey', *Herald*, 2 Feb. 1999 <http://www.heraldscotland.com/sport/spl/aberdeen/mick-mcgahey-1.307647> [accessed 21 Dec. 2019].

[46] Willie Doolan, interview with author, The Pivot Community Centre, Moodiesburn, 12 March 2014.

unanimously.[47] Under McGahey and Daly's leadership, a shift within political culture took place with respect to both education efforts and public platforms the NUM provided at events such as the Scottish miners' gala. Prominent representatives from across the British labour movement and political parties were welcomed by the Scottish miners.

This pluralistic development was indicative of the broad left turn within the CPGB's industrial strategy during this period, but also had a basis in the 1930s Popular Front experience of broad political alliances. Through this change of strategy, the Communist International had abandoned an ultra-left 'third period' position which denigrated social democrats as 'social fascists' to seeking unity against fascism.[48] In Scotland, the CPGB, allied with the Independent Labour party and Scottish National Party (SNP), advanced a Popular Front at the 1935 election.[49] In West Fife, Willie Gallacher united left-wing political forces to defeat a Labour candidate who was on the right of the party and unseated a Conservative MP. Gallacher's success was built on the dismay felt by local Labour supporters and Communists alike after the Tories won the seat with around a third of the vote in 1931, which was a smaller vote share than the combined total of their parties.[50] The Popular Front's programme emphasized broad-based political support for home rule, a strategy that was reprised within the NUMSA under McGahey, in coordination with the STUC. John Kay, who was the CPGB's Scottish industrial organizer between the mid 1960s and early 1990s, recollected that the Moffats had been 'imposing hard nuts'. He counterpoised this approach with the one developed under McGahey, which saw NUMSA pursue a strategy based on building alliances across political divisions within the left and labour movement. Under McGahey, 'the miners led the way', by devising increasingly diverse platforms at both NUMSA education events and the annual Scottish miners' gala. Speakers included left-wing trade unionists from across the UK and Liberal and SNP representatives who shared the NUMSA's support for devolution and opposition to nuclear weapons and power stations.[51]

---

[47] NMMS, NUMSA, Minutes of Executive Committee and Special Conferences from 27 June 1966 to 14/16 June 1967, pp. 440–1.

[48] H. Graham and P. Preston, 'The Popular Front and the struggle against fascism', in *The Popular Front in Europe*, ed. H. Graham and P. Preston (Basingstoke, 1987), pp. 1–19, at pp. 3–4, 13–14.

[49] J. Foster, 'The twentieth century', in *The New Penguin History of Scotland: from the Earliest Times to Present Day*, ed. R. A. Housing and W. W. Knox (London, 2001), pp. 417–96, at p. 445.

[50] Foster, 'The twentieth century', p. 445.

[51] John Kay, interview with author, residence, Bishopbriggs, 8 Aug. 2014.

Deindustrialization and national consciousness

Local and regional distinctions proved a powerful and enduring feature of coalfield politics and trade unionism. The Marxist and Scottish nationalist labour historian, James D. Young, celebrated the miners in a 1976 special edition of *New Edinburgh Review* dedicated to mining culture. Young commended miners as 'the most militant and class-conscious' workers in Britain. He regarded their 'parochialism' as a key source of strength, which had given solace to secluded communities that possessed 'a record of consistent resistance to capitalism and capitalist social values'.[52] These sentiments accord with narratives about the militancy of single-industry mining settlements. Lumphinnans in West Fife, where the Moffats grew up, is among the examples of 'Little Moscows' detailed in Knotter's account of 'small place Communisms'.[53] Jessie Clark similarly described the South Lanarkshire mining village of Douglas Water as 'a little Moscow'. This choice of terminology underlined the role of CPGB activists as trade union representatives and their central involvement in the community's highly collectivized social and political life.[54] Despite the celebratory tone associated with memories of pit village radicalism, the strength of local attachments threatened the development of a coherent trade union that could sustain the strains of political and industrial conflict. A more profound national unity among miners was made practicable by the experience of coalfield restructuring and the building of larger cosmopolitan collieries, but these developments also potentially increased antagonisms and differences within the workforce.

The moral economy's embedding of customary rights within local employment complicated the process of transfers and threatened the acceptance of incomers. Within receiving collieries, transferring miners represented threats to jobs and traditional arrangements in industrial relations. Coalfield restructuring therefore created tensions between locales in the rationing of industrial employment. Pat Egan recollected that divisions between sections of the workforce from different areas of Scotland were highly visible when he transferred to the Longannet complex following the closure of Bedlay in 1982:

> When you came through to Longannet, it was a sorta divide on where people came fae. People fae the Hillfoots [Alloa] as they were called, wouldnae vote

[52] J. D. Young, 'The British miners, 1867–1976', *New Edinburgh Review*, xxxii (1976), 5–9, at p. 5.

[53] A. S. Knotter, '"Little Moscows" in western Europe: the ecology of small-place Communisms', *International Review of Social History*, lxi (2011), 475–510, at p. 478.

[54] Jessie Clark, interview with author, residence, Broddock, 22 March 2014.

for Fifers. And Fifers wouldnae vote for Hillfooters regardless ae ability. That was the union elections and stuff. Aye, it was crazy. Then of course when we [Lanarkshire miners] came through we were known as the Jimmys *laughs* That's what we were called! We did in time eventually move away fae that cause we were fae aw o'er, cause you had then people fae the Lothian coalfields.

Pat's comments indicate the importance of a sense of belonging attached to pits within the settlements surrounding them. For Pat, who grew up in Twechar, a relatively isolated village in North Lanarkshire, Bedlay was a colliery invested with community and familial significance. Upon transfer to Longannet he was displaced into 'the middle ae nowhere'.[55] Similar dynamics relating to coalfield restructuring and fear about the increasing scarcity of employment had also applied in North Lanarkshire during the 1960s. After the closure of Gartshore 9/11 colliery during the late 1960s, Peter Downie transferred to Bedlay. At Bedlay, he was met with pronounced antagonism from a local miner who saw the incomers as a threat to his economic security:

As a matter of fact, when we went tae Bedlay up fae Grayshill tae Bedlay. Jimmy Cleland, a man that I stayed up beside. I was brought up beside his mother and faither all of ma days. And he says, 'you's are effin up here to steal oor jobs get away back tae wherever you come and get a job ae your oan'. We'd no other pit we could tae bar the gas hole [Cardowan].[56]

The NUMSA was fragmented by occupational as well as geographical distinctions. SCEBTA retained organizational independence as an affiliate to the union. As late as 1972, the SCEBTA general secretary Frank Gormill extended fraternal greetings to Michael McGahey and the NUMSA at the Area's conference, underlining his union's commitment to 'cooperate to the fullest extent' in the context of industrial action and strike mobilizations. However, Gormill reiterated his union's opposition to amalgamation despite its membership shrinking as collieries closed over the 1960s.[57] Nicky Wilson worked as an electrician at Cardowan before he transferred to Longannet. Nicky recalled that solidarity between workgroups, and closer organizational links, developed incrementally through experiences of industrial action and by necessity as the industry contracted:

---

[55] Pat Egan, interview with author, Fife College, Glenrothes, 5 Feb. 2014.

[56] Moodiesburn focus group, retired miners' group, The Pivot Community Centre, Moodiesburn, 25 March 2014.

[57] NMMS, NUMSA, Minutes of Executive Committee and Special Conferences from 28 June 1971 to 14/16 June 1972, p. 635.

One Friday backshift, and we were where the big workshops wis. You'd the engineers' workshop and the electricians' workshop at the end. So, we used tae get oot oor tools and walk through the engineers' workshop to get tae the pithead and go underground. Well, you'd the two guys ah worked wi, and some o the engineers were aw standin, because some aw them weren't gettin overtime over the weekend they said. Well, 'that's it we're no goin doon'. So they were goin home and so ah says, well 'ahm goin home an aw'. They says, 'no, you're an electrician'. But ah says, 'we're the same union, ahm goin home'. So ah did it, and after that they aw wanted me to be the delegate for some reason.

So, ah did get involved. Ah didn't have ambitions to be involved at that stage although ah wis on the committee, the union committee at the pit. Although the delegate we hid had been there for about thirty years at the pit so it wis probably about time for him to move on anyway … And then, 1989, what happened was that although the two unions were still there, they formed an administration body tae deal with national business of the NUM Scotland Area which looked after the tradesmen's and the miners' union in Scotland.[58]

Nicky and Pat's testimonies emphasized gradual progress towards unity through a restructuring process that brought Scottish miners into greater contact across regional and demarcation boundaries. Union activism asserted a common interest. The NUMSA's sustained left-wing leadership and support for industrial action from SCEBTA tends towards supporting their conclusions. However, there are lacunae, most notably perhaps the faltering attempts to win backing for strikes against pit closures during the early 1980s. This was most marked in late 1982 when the NCB announced the closure of Kinneil colliery in West Lothian and twelve miners commenced a 'stay-down' strike at the pit. The strikers surfaced on Christmas Day following failure to gain support at other collieries. At a delegate conference three days later, Michael McGahey commended the 'heroic' action of the strikers and bemoaned the 'lack of response from the Scottish coalfield'. David Hamilton, a delegate from Monktonhall pit in Midlothian, was more bluntly pessimistic, stating that 'never at any time had he envisaged a situation developing where a large number of men at Monktonhall would pass a picket line'.[59] These divisions foreshadowed later tensions over closure threats, which surfaced across the UK fifteen months or so later when the miners' strike began in March 1984.[60] Before this, the 'highly precarious'

---

[58] Nicky Wilson, interview.

[59] NMMS, NUMSA, Executive Committee Minutes, July 1982 to June 1983, pp. 319–27, 333, 337.

[60] J. Emery, 'Belonging, memory and history in the North Nottinghamshire coalfield', *Journal of Historical Geography*, lviii (2018), 77–89, at pp. 81–2.

unity between NUM areas had already broken over attempts to garner support for industrial action.[61]

These distinctions were matched by internal differences within the NUMSA. John McCormack was the NUM delegate for Polmaise colliery in Stirlingshire, which was threatened by the intensification of closures. He criticized the executive committee for having 'dragged their feet' on initiating solidarity action.[62] McCormack repeated these criticisms in a pamphlet published in 1989 where he claimed the NUMSA leadership was 'working hand-in-hand with the Coal Board regarding the closure' of Kinneil.[63] McCormack's perspective reflected Polmaise's relatively exceptional status within the Scottish coalfield. There was no CPGB connection at Polmaise and it was the last remaining traditional village pit in Scotland, with the workforce largely concentrated in nearby Fallin. During December 1982, the NUMSA's leadership was undermined at pit level in larger cosmopolitan collieries such as Monktonhall. The refusal of miners across Scotland to support the stand at Kinneil precluded building up to a UK-wide strike. But the NUMSA continued its opposition to closures, including at Cardowan in 1983, where the workforce ultimately rejected strike action after months of targeted demoralizing campaigning by the NCB.[64] The comparatively solid response of the NUMSA to the 1984–5 strike, when only Bilston Glen colliery in Midlothian saw significant strikebreaking before the autumn of 1984, reaffirmed Scotland's status as a 'left' or 'militant' coalfield.[65]

Scottish miners' collective self-image was shaped by a strong occupational identity and national allegiances. These sustained the NUMSA's claim that the union held a leading position within the broader labour movement. In this respect, the NUMSA paralleled the NUM's South Wales Area. The South Wales Area president, Emlyn Williams, attempted to build industrial action against pits closures during the 1980s through appeals to a history of militancy that extended back to the early twentieth century. He also drew attention to the threat intensified deindustrialization posed to the

---

[61] H. Francis and G. Rees, '"No surrender in the valleys": the 1984–85 miners' strike in South Wales', *Llafur*, v (1989), pp. 41–71, at p. 43; V. Allen, 'The year-long miners' strike, March 1984–March 1985: a memoir', *Industrial Relations Journal*, xl (2009), 278–91, at p. 283.

[62] NMMS, NUMSA, Executive Committee Minutes, July 1982 to June 1983, p. 336.

[63] J. McCormack (with S. Pirani), *Polmaise: the Fight For a Pit* (WordPress version, 2015) <https://polmaisebook.wordpress.com/> [accessed 21 Dec. 2019], p. 10.

[64] TNA, Coal 89/103, Bill Magee, press release: Cardowan ballot results, NCB press office, Edinburgh, 26 Aug. 1983.

[65] J. Phillips, 'Material and moral resources: the 1984–5 miners' strike in Scotland', *Economic History Review* lxv (2012), 256–76, at p. 264.

Welsh economy.[66] Michael McGahey similarly implored miners to provide leadership to the Scottish labour movement through rhetoric couched in class and national terms. At the 1983 NUMSA annual conference, he argued that 'peripheral areas' were being targeted to the detriment of 'an abundance of coal reserves that will last the Scottish nation for one, two or three centuries'.[67] Furthermore, McGahey argued that 'the defence of jobs in the mining industry is directly linked with other industries in Scotland', emphasizing the need for 'uniting our struggle' with steel and railway workers.[68] The miners' capacity for leadership was also observed by other trade unionists. Tom Dougan, a regional officer of the Amalgamated Union of Engineering Workers, addressed the conference from the general council of the STUC where he also expressed confidence that 'the NUM would be in the forefront of the fightback' against job losses. Dougan professed that the miners were at the core of 'a true socialist belief within Scotland', which he contrasted to the policies being pursued by the Thatcher government.[69] This framing was significant in shaping the historical memory of the 1980s, embedding national as well as industrial or class dynamics. Willie Doolan articulated this sense of the NUMSA's vanguard role thirty-one years later, in 2014:

> While maybe not the best educated, we understood the meaning ae struggle, we understood poverty, we understood the meaning ae hardship, and we were always involved in the struggle tae try and better that for the miners and their families. We weren't just doing this in a parochial trend. We expanded it tae other forms of workers. The likes ae in Britain, the steelworkers, the nurses. Where we seen workers in genuine struggle, we were always willin and proud to lend our hand in assistance to them. If we couldn't do it on the picket line or in political struggle with them we could we would try and help financially ... I'm not sayin that the Scottish miners are the be all and end all of the miners' union, but we played a big, big part in it in many years gone by ... I'm very proud of the role that the British miners have played in countless struggles that workers have been involved in both internationally and nationally.[70]

The conception of Scottish miners as standard-bearers of class struggle was actively mobilized by the NUMSA's leadership across the life of the union as a federated Area of the NUM. These efforts included an active appeal to

---

[66] B. Curtis, 'A tradition of radicalism: the politics of the South Wales miners, 1964–1985', *Labour History Review*, lxxvi (2011), 34–50, at pp. 35–7.

[67] NMMS, NUMSA, Executive Committee Minutes, July 1982 to June 1983, p. 635.

[68] NMMS, NUMSA, Executive Committee Minutes, July 1982 to June 1983, p. 633.

[69] NMMS, NUMSA, Executive Committee Minutes, July 1982 to June 1983, p. 678.

[70] Willie Doolan, interview 2014.

historical inspirations. For instance, the 1978 NUMSA annual conference included an obituary for Peter Kerrigan, a long-standing member of the CPGB and a Scottish volunteer who fought with the International Brigades during the Spanish Civil War.[71] The union also enjoyed a relationship with the Scottish Labour History Society who held exhibitions at the Scottish miners' gala in 1967, which incorporated items relating to the coalfields.[72] Miners' history was similarly visible in the CPGB's journal, *Scottish Marxist*. An edition from 1974 included a special section on Scottish miners that stretched back to an account of early attempts at labour organization in Ayrshire collieries by Christopher Whatley as well as extracts from Ian MacDougall's oral history of struggles in the Fife coalfields during the twentieth century.[73] This anticipated MacDougall's later publication of *Militant Miners* in 1981.[74] In the same edition, the NUMSA's general secretary, Bill McLean, provided a report on the 1974 strike which profiled the union as a leading force in building the strike among miners across Britain. McLean explicitly connected these events with the earlier history recorded in the previous two articles.[75]

Coalfield politics were a regular feature of *Scottish Marxist*. Two years earlier, Michael McGahey had written an article for the first edition of the journal. He critiqued the prevailing assumptions of energy policymakers with regards to the economics of coal following the NUM's successful industrial action for miners' wage increases. McGahey distilled a general lesson from these events, arguing that the arrest of thirteen pickets at Longannet power station, and their subsequent trial was 'a classic example of an endeavour on the part of the ruling class to take retribution for the success of a workers' struggle'.[76] Scottish communists also used *Scottish Marxist* to explore the connection between class politics and nationhood. Another article in the same edition cited the Upper Clyde Shipbuilders work-in and recent miners' strike as platforms to build the case for home

---

[71] NMMS, NUMSA, Executive Committee Minutes, 27 June 1977 to 14–16 June 1978, p. 601.

[72] NMMS, NUMSA, Minutes of Executive Committee and Special Conferences from 27 June 1966 to 14/16 June 1967, p. 104.

[73] C. Whatley, 'Some historical pointers: miners in 18th century Ayrshire', *Scottish Marxist*, vi (1974), 5–12; I. MacDougall, 'Reminiscences of John MacArthur, Fife militant', *Scottish Marxist*, vi (1974), 13–23.

[74] *Militant Miners: Recollections of John McArthur, Buckhaven; and Letters, 1924–6, of David Proudfoot, Methil, to G. Allen Hunt*, ed. I. MacDougall (Edinburgh, 1981).

[75] B. Mclean, 'The 1974 struggle', *Scottish Marxist*, vi (1974), 25–35.

[76] M. McGahey, 'The coal industry and the miners', *Scottish Marxist*, i (1972), 16–19, at p. 18.

rule.[77] George Montgomery, the NUM's mechanical and electrical safety inspector, presented a similar sense of a 'forward march' in the introduction to the *New Edinburgh Review* coal special edition four years later when he described miners as 'a once down trodden now organized section of the British working class'.[78]

However, deindustrialization remained a fault line within the British labour movement. Responses to incremental pit closure programmes served to underline national distinctiveness. One significant expression was the NUMSA's call for a public inquiry into the management of the Scottish coalfield during 1961. The STUC and the Scottish executive of the Labour party supported the demand as well as a significant number of local trades councils and Cooperative guilds.[79] The union attempted to claim ownership of a national as well as workforce interest by asserting that 'a Public Inquiry would be a defence of Nationalization and [the NUMSA] believes that the Scottish people who have given such wide support for a Public Inquiry will continue to do so as part of the fight to defend Scotland's Economy'.[80] A merger of concerns over industrial employment with local and national economic welfare was also apparent in the letters that the NUMSA's supporters sent to the NCB's Chairman, Alf Robens. For instance, Bothwell Constituency Labour Party in South Lanarkshire contacted Robens, stating that the party:

> Protests most strongly at the action taken by the National Coal Board, with regard to pit closures in Scotland. We support the NUM Scottish Area, in their demand for an inquiry into the administration of the coal mining industry in Scotland. We urge those responsible to consider the economic and social hardships that this policy of pit closures must relate throughout the mining areas of Scotland, and also the great damage to our Scottish economy and industrial development.[81]

Similarly, J. S. Campbell, the district clerk of Forth District, which is also in South Lanarkshire, explained to Robens that the experience of closures in the district, as well as its retained dependency on coal mining employment, informed their opposition to further contraction. Campbell implicated

---

[77] A. Murray and F. Hart, 'The Scottish economy', *Scottish Marxist*, i (1972), 23–35.

[78] G. Montgomery, 'Introduction', *New Edinburgh Review*, xxxii (1976), 1–3, at p. 3.

[79] NMMS, NUMSA, Minutes of Executive and Special conferences, 12 June 1961 to 6/8 June 1962, p. 267.

[80] NMMS, NUMSA, Minutes of Executive and Special conferences, 12 June 1961 to 6/8 June 1962, p. 238.

[81] TNA, Coal 31/96, W. Drennan, Bellshill, to Alf Robens, NCB, London, 31 Jan. 1962.

local concern within a political commitment to the industry in a coalfield area. While it had lost most of its pits, Forth retained a cultural affiliation with coal:

> Larkhall, one of the populace areas of this district was a large mining area at one time and therefore many of the people are still interested in the working of mines and many of the adult workers have to travel long distances from their work. There is certainly a transformation taking place in the working conditions of the district but there is still a large public opinion concerned with the production of coal.[82]

These developments indicate that the dislocation that was felt in contracting coalfields was given an increasingly national expression during the 1960s. Closures and coal industry reorganization directed from London were understood as a threat to the Scottish industrial economy, not just localized communities. This related to the continuing cultural influence of mining in the coalfields, even after the industry had departed. Opposition to closures took the form of a protective countermovement. Not only were job losses or closures protested, the NUMSA also claimed the mantle of nationhood to assert social order over the dislocation caused by the imposition of market logic from outside of Scotland. Intensifying closures also contributed to the growth of fissures between the Scottish Area and the UK-wide union. In part, this related to the attitudes maintained by its communist leaders and activists, which clashed with the worldview of 'moderate' or mainstream social democratic officials. Friction between the Scottish Area and UK leadership was apparent at a delegate conference called in response to the announcement of a pit closure programme in November 1961. The minute taker noted that delegates removed two suspected plain-clothes police officers from the meeting, before the executive committee called for a campaign of demonstrations in a 'fight to save Scotland's economy'. Delegates bemoaned the UK leadership as presiding over 'a fifth-rate union' unprepared to take industrial action over wages or closures or to support the campaign for a public inquiry.[83]

Responses to an NUMSA demonstration against pit closures in London during 1962 further exposed these differences. The NUM president, Sidney Ford, was reported to have described the procession as 'a circus'.[84] Minutes

---

[82] TNA, Coal 31/96, J. S. Campbell, Forth District Council, Larkhall, to Alf Robens, NCB, London, 29 Dec. 1961.

[83] NMMS, NUMSA, Minutes of Executive and Special Conferences, 12 June 1961 to 6/8 June 1962, pp. 215–29.

[84] NMMS, NUMSA, Minutes of Executive and Special Conferences, 12 June 1961 to 6/8 June 1962, p. 523.

Figure 6.1. Bob Starrett, 'Phase 3', *Scottish Marxist*, vi (1974) 24. Glasgow Caledonian University Archive Centre: Records of the Communist Party of Great Britain Scottish Committee/ © Bob Starrett

of the NUMSA executive committee meeting held on 26 March note a pointed critique of Ford. References to Ford's status as a career trade union official underline the class basis of the moral economy, and perhaps implicitly its national orientation too. Ford's response to the demonstration was 'Indicative of the fact that the President of the National Union had no working-class background and had never had to fight for wages or a job. In view of these circumstances, the Executive Committee felt that Mr Ford was more in need of their sympathy than their criticism'.[85] Ford

[85] NMMS, NUMSA, Minutes of Executive and Special Conferences, 12 June 1961 to 6/8 June 1962, p. 523.

was widely disliked on the NUM left due to his moderate stance. Rivalries between sections of the NUM amplified these political differences: Ford's background lay in the Colliery Officials and Staff Association (COSA), the NUM's white-collar affiliate. He defeated Alex Moffat in the 1960 union presidential election by a relatively narrow margin of under 10,000 votes. This made Ford's status as an 'office boy' even more grating to left-wing miners and officials with underground experience.[86]

The NUMSA's self-image as a harbinger of militancy increased through the role it played in stimulating national industrial action during the early 1970s, including through the prominence of Michael McGahey as a public face of industrial action. In 1974, *Scottish Marxist* published a cartoon that featured McGahey leading miners to dislodge the Heath government's incomes policy, which is shown in the figure above.

Continuities in the sentiments of NUMSA activists were evident at the Area's annual conference in 1978. Alex Timpany of Barony colliery moved a motion in opposition to the supposed quiescence of the NUM's national executive committee on pensions and lump sums for retired miners. Timpany condemned the executive for its moderation, which he related to its distance from the industry by referring to 'some of the fainthearted members of the NEC who purported to represent the interest of the members who toiled in the bowels of the earth'. Officials from the executive's 'right-wing majority' were 'so far removed from the rank and file' that they were incapable of understanding the injustices experienced by miners who suffered occupational illnesses and financial deprivation despite their centrality to British industry.[87]

NUMSA officials and activists also more explicitly cited the national as well as class or cultural distances raised by the conduct of industrial relations and the management of colliery closures on other occasions. During 1962, Alex Moffat made a presidential address to a delegate conference convened to discuss a possible one-day stoppage against closures in the Scottish coalfield. Moffat successfully argued in favour of focusing on a wages campaign with other industrial sectors in Scotland, partly through anticipating future dilemmas posed by the isolation experienced during fights against pit closures. However, Moffat was also keen to demonstrate his frustration with the UK leadership. In arguing with moderate members of the UK national executive over Scottish miners' right to organize an official strike under the banner of the NUMSA Moffat had drawn attention

---

[86] A. Taylor, *The NUM and British Politics*, ii: *1969–1995* (Aldershot, 2006), p. 25.

[87] NMMS, NUMSA, Executive Committee Minutes, 27 June 1977 to 14–16 June 1978, pp. 643–5.

to the significant disruption being experienced in the Scottish miners which far surpassed the costs of any industrial action: 'The president had reminded the [UK] national executive, however, that the one day protest stoppage would not cost the miners in Scotland what it had been costing the men transferred from one pit to another over the years since 1958'.[88] A COSA representative summarized growing discontent at a consultative meeting between management and trade unions preceding the closure of Gartshore 9/11 in North Lanarkshire during 1968. He argued that the imposition of closures, via programmes drawn up in London, represented a threat to democratic procedure and local employment: 'It used to be that the colliery manager had to plan out his own pit, then Area officials took control of this and now we find that the planning for the Pit is done 500 miles away. Handouts were all right, if unavoidable, but men wanted to work'.[89] These developments severely stretched the moral economy's basis in the shared culture of mining communities, and the social as well as financial value placed on employment.

Significantly, these comments came only a year after the NCB's major centralizing restructuring and indicate how the cumulative logic of deindustrialization encouraged a growing assertion of national as well as local and class-based discontent. Lawrence Daly had already begun arguing for strengthened industrial and political democracy in Scotland earlier in the decade. In a 1962 contribution to the *New Left Review*, Daly underlined the growing development of nationalist sentiment, which he accorded to 'planning' on a Britain-wide basis.[90] Anticipating analyses of 'internal colonialism' that gained influence over the following two decades, Daly agitated for an elected Scottish assembly to reverse the scenario whereby 'Scotland, with a record of many centuries as an independent nation, becomes an economic and political backwater'.[91] He singled out the coalfields as significant to a political solution, underlining that miners felt 'betrayed', not only the by NCB, but also through the NUM's toleration of closure and encouragement of cross-border transfers. Foretelling the hardening of opposition to closures that would take place under his and McGahey's stewardship of the NUMSA, Daly emphasized that 'what is

[88] NMMS, NUMSA, Minutes of Executive and Special Conferences, 12 June 1961 to 6/8 June 1962, pp. 447–8.

[89] National Records of Scotland, Edinburgh (NRS), Coal Board (CB) 300/14/1, Minutes of Special Consultative Committee Meeting of Gartshore 9/11 Colliery Consultative Committee (CCC) held in Grayshill Office on Thursday 18 Jan. 1968.

[90] L. Daly, 'Scotland on the dole', *New Left Review*, xvii (1962), 17–23, at p. 17.

[91] Daly, 'Scotland on the dole', pp. 22–23; M. Hechter, *Internal Colonialism: the Celtic Fringe in British National Development, 1536–1966* (London, 1975).

really surprising and impressive is the way in which the vast majority of the people refuse to be uprooted'.[92] This rejection of market logic laid the ground for a class-conscious case for home rule and an agenda centred on averting deindustrialization by embedding the economy in communitarian expectations. The SNP also perceived the link between colliery closures and the growth of Scottish nationalism as analysed by Daly. The party's leader, William Wolfe, wrote to Robens early the next year exclaiming discontent over the Board's failure to commit to the restoration of Barony colliery in Ayrshire during 1963. Wolfe argued failure to reopen Barony would not only damage the local economy but also fly against Scottish public sentiment:

> To the people of Ayrshire, particularly those unemployed and their families, whose hopes are centred on Barony, such a possibility is a grim prospect; and to the people of Scotland not only in the coal industry but everyone, the abandonment of Barony colliery, one of the largest and most attractive long-term development pits in Scotland, will be another sign of the ruthless spirit which members of the Scottish National Party at least recognise only too well.[93]

Wolfe's party prospered over the following decade through advancing a perspective which chimed with frustration at Scotland's perceived economic and political marginalization. These motifs accorded with miners' experience of centralization and colliery closures. The SNP's key electoral breakthrough took place in the Hamilton by-election of 1967. Winifred Ewing secured the Nationalist's first parliamentary representation since the end of the Second World War with a victory in the historic heart of the Lanarkshire coalfield.[94] While economic planning policies had up to this point quite effectively maintained economic security, they had done so at the cost of local control. This stimulated mounting concern over Scotland's industrial future, especially as the international economy faltered during the 1970s.

Following Ewing's election, the SNP became a significant political force in Scottish coalfield areas. Ayrshire was also at the centre of the short-lived Scottish Labour party established through Jim Sillars' break from the Labour party over the Wilson–Callaghan government's failure to deliver devolution.[95] Sillars had earlier faced the growing fear of Nationalist advance

---

[92] Daly, 'Scotland on the dole', pp. 19–22; R. S. Halliday, *The Disappearing Scottish Colliery: a Personal View of some Aspects of Scotland's Coal Industry since Nationalisation* (Edinburgh, 1990), p. 7.

[93] The National Archives (TNA), Coal 31/96, W. C. Wolfe, SNP to A. Robens, NCB, London, 31 Jan. 1963.

[94] J. Mitchell, *Hamilton 1967: the Byelection that Transformed Scotland* (Gosport, 2017).

[95] A. Perchard, '"Broken men" and "Thatcher's children": memory and legacy in Scotland's

in Labour circles during the 1970 South Ayrshire by-election when he was opposed by Sam Purdie. Purdie was a recent SNP convert. He had formerly been the SCEBTA delegate at Cairnhill mine as well as chair of the local trades council and had also been a union delegate to the South Ayrshire Constituency Labour party. In the latter capacity, Purdie had previously acted as the Labour party's election agent in South Ayrshire.[96] His SNP election campaign included highlighting alleged dangers to the future of Killoch colliery and the 2,300 jobs it sustained. Purdie wrote to Tony Benn, the minister of technology, highlighting the loss of 7,000 mining jobs in Ayrshire over the previous decade. Sillars portrayed his opponent as 'scaremongering' and assisting 'the anti-coal lobby'. Furthermore, Purdie's change of allegiances, which Sillars himself would ultimately follow eleven years later, was presented as the act of a renegade:

> He [Purdie] has claimed to speak for thousands of miners. Someone who has welshed on the Labour movement as he did will never have that honour. I have no doubt that he will be repudiated in his claim by the miners' own spokesmen. If the Ayrshire miners now warn him off the coalfield I for one could not blame them.[97]

McGahey supported Sillars by informing the *Scotsman* that there was 'nothing to cause concern about the future of the pit'.[98] Although Sillars won the by-election and comfortably defeated Purdie again at the general election later in the year, he was left 'haunted' by the former mining engineer's concession speech. Purdie challenged Sillars to justify Scotland being governed by a Conservative administration it had not voted for.[99] Sillars's invocation of class loyalty against national allegiance reprised longstanding Scottish labour movement motifs. John Taylor, Labour MP for West Lothian, addressed the 1950 NUMSA conference on behalf of the Labour party's Scottish Council. His contribution was highly critical of the recent Scottish Covenant campaign for a devolved Scottish parliament.[100] Despite the Covenant having been backed by the CPGB, the NUMSA and

---

coalfields', *International Labor and Working-Class History*, lxxxiv (2013), 78–98, at pp. 86–7.

[96] Sam Purdie, interview with author, UWS Hamilton campus, 3 May 2018.

[97] TNA, Coal 52/402/63, 'SNP candidate accused of "pit-closure rumours"', *Scotsman*, 27 Jan. 1970.

[98] TNA, Coal 52/402/63, 'SNP Candidate Accused of "Pit-Closure Rumours"', *Scotsman*, 27 Jan. 1970.

[99] J. Sillars, *Scotland: the Case for Optimism* (Edinburgh, 1986), p. 33.

[100] M. Keating and D. Beliman, *Labour and Scottish Nationalism* (Edinburgh, 1979), pp. 144–5.

the STUC, Taylor mockingly noted the support of 'such proletarians as the Duke of Montrose'. Furthermore, constitutional reform 'was a diversion and a trap'. Taylor was dismissive of territorial politics, stating that to socialists it 'did not matter much where we were governed from but by whom we were governed'.[101] In contrast, the NUMSA's longstanding commitment to home rule was grounded in an understanding of class that was sensitive to national dynamics and antagonistic to the effects of centralization.

Under McGahey's presidency, the NUMSA more overtly embraced Scottish nationalism, but nevertheless stopped short of the demand for separation. During the 1968 STUC annual conference, at which McGahey moved a motion in favour of a Scottish parliament, he distinguished 'healthy nationalism' from chauvinistic 'perverted nationalism'. The former was defined as 'love of one's own country, love of one's own people and pride in their traditional militancy and progressiveness'. McGahey's motion was remitted, which marked a pronounced shift in STUC policy towards devolution and paved the way for its later campaign in support of a Scottish assembly during the 1970s.[102] McGahey's STUC contribution readily acknowledged that his communist politics shaped his outlook. A strong continuity with the Popular Front is apparent in his pursuit of a broad alliance that appealed to 'questions of democracy, nationalism, patriotism and history'.[103] But during the late 1960s, appeals to 'subaltern' culture were also stimulated by the influence of the Scottish folk revival and the influence of analysis which championed plebeian traditions as a valuable resource for the left.[104]

Jimmy Reid redeployed McGahey's motifs during a 1974 CPGB election address in Clydebank which similarly distinguished between 'healthy nationalism' and 'jingoism'. Reid placed industrial job losses in a distinct national historical context by referring to twentieth-century 'Lowland Clearances'. Echoing the eighteenth- and nineteenth-century experiences of forced removal in the Highlands, Reid stated these developments could be approaching 'genocidal' if a Scottish assembly was not established to implement 'new policies in Scotland to reverse the trend and to preserve

---

[101] NMMS, NUMSA, Minutes of Executive Committee and Special Conferences, 20 June 1949 to 2 June 1950, p. 642.

[102] J. Phillips, *The Industrial Politics of Devolution: Scotland in the 1960s and 1970s* (Manchester, 2008), p. 38.

[103] J. Fyrth, 'Introduction: in the Thirties', in *Britain, Fascism and the Popular Front*, ed. J. Fyrth (London, 1985), pp. 9–29, at p. 14.

[104] N. Davidson, 'Gramsci's reception in Scotland', *Scottish Labour History*, xxxviii (2010), 37–58, at p. 38.

our culture and our nationhood'.[105] While McGahey's rhetoric avoided this hyperbolic language, he similarly clearly conferred the sentiment that nationhood ought to be carried forward under a labour movement banner. During 1978, a referendum on a Scottish assembly was legislated for under the Scotland Act. This heavily divided the labour movement. In response, McGahey asserted the need for a class-conscious Scottish national identity at the NUMSA's annual conference: 'We must never allow the Nationalists to appear to be the banner of the Scottish nation. That honour truly belongs to the Labour and trade union movement'.[106] John Taylor's logic from twenty-eight years previously remained influential within the Labour party. George Cunningham, a Scot and a Labour MP who sat for a London constituency, was an opponent of devolution who successfully moved a parliamentary amendment which established that a threshold of 40 per cent of eligible voters would have to vote in favour of the assembly for the referendum to mandate its establishment. The referendum, which took place in March 1979, ended in the debacle of a 'Yes' majority that failed to meet the threshold.[107]

Occupational and class consciousness spurred the development of demands for autonomy, but also limited them. These dynamics were encouraged by the emphasis on unity within the nationalized industry, and the harsh lessons that miners took from interwar experiences of fragmentation. McGahey's support for devolution at the STUC in 1968 was qualified with the reassurances that he 'rejected outright the theory of separating Scotland from the United Kingdom'. Although the NUMSA was committed to 'the decentralisation of power', McGahey also maintained that 'the miners' union was a [British] national union', that 'operated in a nationalized industry, which miners would never allow to be destroyed'.[108] These sentiments continued to predominate as the likelihood of devolution increased. The NUMSA's representatives at tripartite meetings that took place during Benn's term at the Department of Energy continued to argue in favour of significant autonomy and planning in Scotland. However, this was within an overall commitment to the nationalized industry at a UK-wide level and sharp opposition to the 'disintegration' of unitary frameworks.[109]

---

[105] J. Reid, *Reflections of a Clyde-Built Man* (London, 1976), pp. 121–2.

[106] NMMS, NUMSA, Executive Committee Minutes, 27 June 1977 to 14–16 June 1978, pp. 609–10.

[107] J. Mitchell, *Devolution in the United Kingdom* (Manchester, 2009), p. 125.

[108] STUC, *Annual Report 1967–1968*, lxxi (1968), 400.

[109] TNA, Coal 31/166, Department of Energy paper 14: tripartite energy consultations, 20 Feb. 1976, London, 1976.

Scottish miners' conceptions of class solidarities were shaped by the contradictory imperatives of resistance to centralization and support for the principle of a unified nationalized industry and trade union. NUMSA leaders and activists asserted a distinctive Scottish national identity while maintaining the Area as a bulwark of the left within the NUM and wider UK labour movement, and through its international connections. This was highly visible in the NUMSA's major events, including its annual conference, but especially the Scottish miners' gala which showcased coalfield culture through both Scottish and British national lenses.

The Scottish miners' gala

The importance John Kay ascribed to the Scottish miners' gala above chimes with the NUMSA's records and other testimonies. Elements of an 'invented tradition' are evident from the gala's inauguration in 1947. It aimed to consolidate conceptions of a unified Scottish mining community, the NUMSA's centrality to the labour movement, and popularize its leadership's political perspective. Eric Hobsbawm referred to invented traditions as practices and events 'which appear to claim to be old but are often quite recent in origin and sometimes invented'. His definition emphasized inventing traditions as necessarily tied to spreading conceptions of history and the contemporary politics through seeking a basis in the past:

> A set of practices normally governed by overtly or tacitly accepted rules and of a ritual or symbolic nature, which seek to inculcate certain values and norms of behaviour by repetition, which automatically implies continuity with the past. In fact, where possible, they normally attempt to establish continuity with a suitable historic past.[110]

Hobsbawm's examples included the FA Cup final and royal radio and television broadcasts. Broadly, he referred to phenomena where elites attempt to preserve authority in periods of major economic and social upheaval through appeals to a 'powerful ritual complex', which attempted to establish continuity with a mythologized national past.[111] However, Hobsbawm did consider that subaltern forces could similarly deploy such sentiments. For example, he argued James Connolly's *Labour in Irish History* was an example of an attempt to popularize 'a people's past' that contested dominant interpretations and would inspire a socialist approach to national

---

[110] E. Hobsbawm, 'Introduction: inventing traditions', in *The Invention of Tradition*, ed. E. Hobsbawm and T. Ranger (Cambridge, 1983), pp. 1–14, at pp. 1–2.

[111] Hobsbawm, 'Introduction', p. 6.

Figure 6.2. Miners' Gala Day 1969. Historic Environment
Scotland/© The Scotsman Publications Ltd.

liberation.[112] The Scottish miners' gala was a fuller deployment of 'invented practices' which linked the ideology of the NUMSA's leaders with their aim of establishing a distinct and united Scottish mining community in congruence with the practices of localized coalfield communities. At the national gala, attendees were invited to spectate and participate in the same activities that took place at local gala days. These included running races, football and boxing, pipe and brass band competitions as well as a 'Coal Queen' competition, entrance to which was confined to miners' daughters and partners.[113]

---

[112] Hobsbawm, 'Introduction', p. 10.

[113] NMMS, NUMSA, Minutes of Executive Committee and Special Conferences, 8 July 1946 to 11 June 1947, p. 8.

The Scottish gala involved the replication of village gala days on a larger basis. However, this reprisal of activities traditionally confined to a restricted *locale* communicated a different, *national*, purpose. Rather than taking place in a coalfield, the gala was always held in Edinburgh, shifting between Holyrood Park and Leith Links following a major procession through the city centre as demonstrated in figure 6.2. As well as holding the symbolism of Scotland's capital, Edinburgh was also where both NUMSA and the NCB Scottish division were headquartered, which further conferred the gala as an event that affirmed a distinct Scottish mining identity. Through the combination of local, national and international networks coalescing around NUMSA, the Scottish miners' gala paralleled the Durham miners' gala, which Hew Beynon and Terry Austin characterized as a 'political project'.[114] The Durham NUM Area general secretary, Sam Watson, used the gala to bring together a plurality of representatives from across the British coalfields, and the wider labour movement, including international speakers. In Watson's case this was to illuminate 'all elements in the new Labourist society' constructed after the Second World War, which emphasized his moderate social democratic outlook and support for liberal freedoms.[115] Watson was an internal political opponent of the NUMSA's leadership and a stalwart of the NUM's Labour mainstream wing. As a contracting coalfield, Durham was also potentially a rival to the needs and concerns of the Scottish Area. During 1961, Watson led the NUM national executive in voting down the NUMSA's appeal for the union to adopt its demand for an inquiry into Scottish colliery closures.[116]

In the case of the Scottish gala, similar mechanisms were used to project a distinctly different form of communist-influenced politics, but these also shared the Durham miners' gala's emphasis on occupational and territorial identities. The gala served as both a symbol and a key institution of an emergent Scottish national coalfield community. The NUMSA placed itself at the centre of the event, and as in Durham, the gala brought together politically diverse elements under the banner of the Scottish miners. This is demonstrated in figure 6.3, which shows Joan Lester, a junior minister in the Wilson government with responsibility for nursery education, speaking on a platform alongside representatives from the North Vietnamese trade

---

[114] H. Beynon and T. Austrin, 'The performance of power: Sam Watson, a miners' leader on many stages', *Journal of Historical Sociology*, xxviii (2015), 458–90, at p. 480.

[115] Beynon and Austrin, 'The performance of power', pp. 480–2.

[116] NMMS, NUMSA, Minutes of Executive and Special Conferences, 12 June to 6/8 June 1962, p. 237.

Figure 6.3. Gala Day 1969, featuring Joan Lester (Labour MP), Lawrence Daly, and Vietnamese Federation of Union representatives. Historic Environment Scotland/© The Scotsman Publications Ltd.

union movement at the 1969 gala.[117] Brendan Moohan recalled the gala as a major annual event that showcased the size of the industry and the common purpose of mining communities across Scotland who marched alongside their union banners and bands. In the context of a shrinking coal mining workforce, the gala affirmed the continuing industrial and cultural importance of coal mining:

---

[117] NMMS, NUMSA Minutes of Executive Committee and Special Conferences from 24 June 1968 to 18/20 June 1969, p. 92.

> As a child, everybody went to the gala day. And it was huge. And it would be at Holyrood. And it was enormous. And there was races. And there was the boxing ring and the various boxing clubs would be involved. I remember the men very often wore suits on that gala day. I can also remember the banners. Thousand, thousands of people representing their pits and their villages with their banners. And there would be brass bands.[118]

International delegations also made an impression on young attendees. Jackie Kay's poem, 'Last room in operations', which was written in tribute to Michael McGahey, profiles her memories of going to the gala with her adopted father John, whose memories of the gala are discussed above. Alongside links with the African National Congress (ANC)'s struggle against Apartheid, Kay remembers the NUMSA president introducing 'Jock and Tam fray Vietnam'.[119]

An 'imagined' Scottish mining community developed through the gala, extending beyond the localized communities built around individual pits and villages.[120] With parallels to Anderson's imagined communities of nationhood, the gala consolidated an imagined Scottish national coalfield community through an annual event in which collective participation and representation established connections between people above the expanse of a localized community. These occasions consolidated a sense of shared interest and emotional investment in the coal mining industry between individuals who by their very number and geographical proximity could not all be familiar to one another. A collective identity was maintained by the gala even as coalfield employment fell. The gala helped Antony Rooney to retain his familial connections to coal mining and linked it with his labour movement activities as a shop steward at the Caterpillar tractor factory in Tannochside, and as a Labour party activist in Bellshill:

> Used to have bus runs, y'know, buses, miners' trips. One ae the times I got a bus tae Edinburgh y'know. When I was older, I took my own kids, y'know, to the miners' Gala Day in Edinburgh every year. I was marching beside Tony Benn at one ae them, and Mick McGahey, walking through up in tae the Salisbury Crags. A lot of trade unionists there, aye. The miners' own Gala Day. Brought everybody together. Aw the mining communities. As I say there were always some important person, y'know. As I say, Tony Benn was there one year,

---

[118] Brendan Moohan, interview with author, residence, Livingston, 5 Feb. 2015.

[119] J. Kay, 'Last room in operations' (2014), unpublished poem used with permission of the author.

[120] B. Anderson, *Imagined Communities: Reflections on the Origin and Spread of Nationalism* (London, 2009), p. 6.

marching along wi him and Mick McGahey. And all the other trade unions, the miners, trade unionists fae the town.[121]

Antony Rooney and John Kay's memories indicate the involvement of key labour movement personalities in the gala, which affirmed its political importance. Tony Benn recalled in his diary that the 1977 event had a large attendance despite the weather, with Edinburgh 'freezing and raining' on the day.[122] The gala articulated both a Scottish mining identity and the unity of the British labour movement. This combination of distinctive Scottishness within a wider British context can be understood as an example of 'Unionist–Nationalism'. In Morton's history of nineteenth-century urban governance, local civic institutions provided the basis for the articulation of a Scottish national identity within Unionist sensibilities.[123] At the gala, in contrast, it was the social democratic infrastructure of the nationalized coal industry which facilitated the expression of a Scottish cultural and political consciousness within a British framework. These relationships are indicative of the mid twentieth-century British context where 'both national and transnational' civil societies functioned; the gala exemplified this via the invocation of cross-Britain solidarity in a Scottish political setting, including the speeches of labour movement figures from England and Wales, especially British national leaders of the NUM.[124]

However, the NUMSA's support for home rule reveals another dimension to the gala. This was very much justified on the basis of asserting democratic control over economic decision making in Scotland, that is to say making 'utilitarian' rather than 'existential' appeals to national feeling.[125] But cultural aspects of national identity were nonetheless clearly mobilized at the gala. Elements of the tension between a Scottish and British identity framed within the context of the devolved structures of the nationalized industry were evident. Pipe bands and Highland dancing competitions coexisted with the characteristic British mining iconography of banners and brass bands. In terms used by Tom Nairn, latent 'cultural "raw material" for nationalism' served to embed a political agenda which was largely a response to the uneven economic development of British capitalism after the Second World War, rather than an articulation of romantic nationalist aspirations.[126]

---

[121] Anthony Rooney, interview with author, Morrisons café, Bellshill, 24 Apr. 2014.

[122] T. Benn, *Conflicts of Interest: Diaries, 1977–80* (London, 1991), p. 177.

[123] G. Morton, *Unionist–Nationalism: Governing Urban Scotland, 1830–1860* (East Linton, 1999) pp. 7, 190–1.

[124] P. Ward, *Unionism in the United Kingdom, 1918–1974* (Basingstoke, 2005), p. 6.

[125] D. Torrance, *The Battle for Britain: Scotland and the Independence Referendum* (London, 2013), p. 184.

[126] T. Nairn, *The Break-Up of Britain: Crisis and Neo-Nationalism* (London, 1977), p. 144.

Figure 6.4. 1988 Gala poster. Historic Environment Scotland/© James Hogg.

To some extent, the annual galas could be interpreted as an indication of the vitality of Nairn's hope that the Scottish labour movement would 'build up its own nationalism' to oppose the SNP's socio-economic moderation. However, Scottish nationalism's containment within a Unionist structure also confirmed his disappointment with the role of Communist as well as Labour party politics.[127] The Scottish gala, like its counterpart in the South Wales coalfield, can be understood as an event in which the sensibilities of 'left labourism' predominated, but with an extended commitment to constitutional reform and an emphasis on internationalism.[128]

[127] T. Nairn, 'Three dreams of Scottish nationalism', *New Left Review*, xlix (1968), 3–18, at p. 17.
[128] Curtis, 'Tradition', p. 48.

The gala also highlighted the NUMSA's transnational connections, especially through networks provided by the CPGB. In 1956, a delegation of miners from Poland was present at the rally.[129] Delegates from the (North) Vietnamese Federation of Trade Unions addressed the rally in 1969 as part of NUMSA's international solidarity activities analysed above. Their presence was only achieved following a sustained political effort, which included drafting the services of Alex Eadie, the Labour MP for Midlothian, to secure visas for the representatives.[130] It also appears to have cemented links between NUMSA and Vietnam. A letter of thanks from the Federation was noted by NUMSA's executive committee on September 8 1969:[131]

> A letter was submitted from Vietnam Federation of Trade Unions thanking the Scottish miners for the warm hospitality extended to their delegation during their visit to Scotland. In addition, on behalf of the workers and Trade Unions of the Democratic Republic of Vietnam, they expressed their sincere appreciation for the sympathy and support being given to the Vietnamese people in their struggle against American aggression.[132]

A further exchange of communications took place when NUMSA sent a note of condolences upon the death of North Vietnam's president, Ho Chi Minh, later the same year. The Vietnamese Federation of Trade Unions warmly accepted NUMSA's message replying that:

> They had been deeply moved by the expression of thorough understanding and friendly solidarity of workers and Trade Union in Britain to the Vietnamese workers, Trade Union and people at this moment.
>
> With an ironlike determination to materialise at all costs the testament of their esteemed President with the ever stronger support and assistance of the British workers and people, and the world workers and people, the Vietnamese workers and people were convinced that their just struggle against US aggression, for national salvation, would win total victory.[133]

---

[129] NMMS, NUMSA, Minutes of Executive Committee and Special Conferences from 18 June 1956 to 5 to 7 June 1957, p. 607.

[130] NMMS, National Union of Mineworkers (Scottish Area) Minutes of Executive Committee and Special Conference from 24th June 1968 to 18/20th June 1969, Entry for executive committee meeting, 12th May 1969.

[131] NMMS, NUMSA, Minutes of Executive Committee and Special conferences from 27 June to 14/16 June 1967, pp. 465–9.

[132] NMMS, NUMSA, Minutes of Executive Committee and Special conferences, June 1969 to 15/16 June 1970, p. 55.

[133] NMMS, NUMSA, Minutes of Executive Committee and Special conferences, June 1969 to 15/16 June 1970, p. 166.

These exchanges were remembered over a decade later by the Vietnamese ambassador to Britain. Dang Nghiem Bai told a report from the *Labour Herald* that: 'I will always remember that the first group of western workers to support us were the Scottish miners'.[134] The advertisement contained in figure 6.4, publicizing the 1988 gala, indicates the longevity of the event which had begun forty-one years previously. It also indicates a continuity of traditional mining community leisure activities such as pipe bands and sporting events alongside political speeches as well as sustained commitments to causes of the left associated with the CPGB. Both nuclear disarmament and Scottish devolution were heavily profiled through the inclusion of CND and STUC representatives. However, it was support for the struggle against Apartheid which was the lead item. A speaker from the ANC led the billing on the South African-themed poster.

The Scottish miners' gala was a key institution in popularising and sustaining a distinct Scottish national coalfield community. It recast the emphasis of traditional gala days from a work and residence-centred locale to one based on a broader occupational attachment framed within the nationalized industry and the NUM's British and Scottish facets. Speakers recruited through networks around the NUMSA and CPGB combined industry-level, national and international political dimensions. The gala straddled the politics of occupation, class and nationhood, which were formative to a distinctive Scottish national mining community that developed through the nationalized industry and coalfield restructuring. Both the international connections and links to the UK-wide labour movement and coal industry that the gala furnished were representative of the factors that shaped and qualified the NUMSA's expressions of Scottish nationalism.

The Scottish miners' gala ultimately came to an end during the 1990s, following the intensification of colliery closures and the draining of the NUM's coffers as its membership fell.[135] A marginalized mining workforce struggled to claim the mantle of Scottish nationhood which it had articulated over the previous five decades. The first gala in 1947 exemplified the innovations in the NUMSA's political culture which developed under public ownership. Nationalization facilitated the construction of a stronger, more centralized, Scottish miners' union. Coalfield reconstruction undermined traditional parochial barriers. As the effects of centralization adversely impacted employment levels in the Scottish coalfield, the

---

[134] 'The struggle to build Vietnam', *Labour Herald*, 29 Apr. 1983, pp. 6–7.

[135] 'Last reminders of a dying breed', *Herald*, 31 May 1994 <https://www.heraldscotland.com/news/12694321.last-reminder-of-a-vanishing-breed/> [accessed 21 Dec. 2019].

NUMSA became increasingly pronounced in its support for home rule, but its Scottish nationalism was always contextualized by its defence of the nationalized industry and the NUM's unitary structures. Occupation, class and nationhood were fused in Scottish miners' consciousness, including within the internal politics of the NUM where the Scottish Area usually adopted a left-wing or militant stance. This was sustained by reference to historical traditions. The NUMSA's appeals to the past gave national coherence to localized experiences. These sensibilities were also apparent at the Scottish miners' gala which reprised the routines of local village gala days on to a larger, national, stage in Edinburgh. Speakers were drawn from across the UK and from international networks that the NUMSA was linked to through the CPGB. Through its contributions to the STUC, the NUMSA achieved successes in advancing support for foreign policy positions and home rule which considerably went against the mainstream social democratic grain. The consistency with which the leadership of a major union articulated support for nuclear disarmament, an oppositional Cold War alignment and support for home rule indicates that labourist political culture was programmatically malleable. Deindustrialization was formative to coalfield politics over the second half of the twentieth century. However, responses to it were dictated by longer historical memories, which were mediated through economic and political forces that encouraged a fusion of class and national consciousness. These influences continue to exercise potent effects in contemporary Scotland.

7. Synthesis. 'The full burden of national conscience': class, nation and deindustrialization

Deindustrialization in postwar Scotland developed over several decades from the 1940s. Extensive colliery closures were initially experienced on a localized basis as older, less productive pits were closed in eastern Lanarkshire under the nationalized industry's rationalization. The experience of labour market restructuring had a sustained impact in redrawing connections between residency and employment and disrupting communal bonds that had developed over a century of industrial production. Deindustrialization's lasting impact has formatively altered life courses, political consciousness and senses of place within the coalfields. These developments were the product of long-term policy decisions related to energy generation. During 1967, Ministry of Power officials gathered to discuss coal production plans over the next four years in the context of increased fuel competition and cost pressures on the industry. They wished to maximize opencast mining, which was the most profitable and labour-saving method. But they were constrained by opposition to coal job losses. The minute-taker bitterly noted that 'the industry cannot hope to become competitive if it is forced to carry the full burden of national conscience'.[1] National conscience can be understood in Polanyi's terms as the protective countermovement to financial and productionist logic: the nationalized industry owed social obligations towards its workforce and the communities which sustained Britain's economy.

While it may have appeared straightforward to diagnose national conscience from London, colliery closures altered territorial politics in Scotland. National and class consciousness were conjoined through the experience of administration under the National Coal Board's (NCB's) centralized structures and disappointment with UK energy policy. As localized experiences of closure accumulated, and the future of coal mining was called into question, pit closures increasingly became a matter for *national* discussion. This chapter presents a synthesis of this book's major themes. It begins by discussing the importance of national and generational contingencies in shaping memories

---

[1] The National Archives (TNA), Coal 31/154, Steering Committee on Fuel Policy, opencast coal production: 1967/71, 1967.

'"The full burden of national conscience": class, nation and deindustrialization', in E. Gibbs, *Coal Country: The Meaning and Memory of Deindustrialization in Postwar Scotland* (London, 2021), pp. 225–49. License: CC-BY-NC-ND 4.0.

of deindustrialization. The second section analyses the fusion of working-class and Scottish national consciousness over the latter half of the twentieth century. Both the spectre of southwards migration to the more profitable coalfield in the English Midlands and memories of the interwar depression loomed large in objections to accelerating deindustrialization and rising support for political autonomy. However, protest was never an entirely straightforward process in class or constitutional terms. Although moral economy claims on colliery employment were voiced most consistently by trade unionists, they were episodically supported by Scottish Unionist or Conservative representatives when they affected their constituents. In the final section, narratives of deindustrialization are analysed by contrasting politicized collective memories with life-story accounts from the oral testimonies which demonstrate the need for a nuanced appraisal. Critical nostalgia provides a valuable means to assess memories of deindustrialization: the transition to a services-dominated economy offered some former industrial workers and their children opportunities for social mobility, but the diminution of community cohesion, workplace stability and trade union power were nevertheless experienced as major and lasting losses.

Deindustrialization is an unsettling process. The closure of collieries, steel mills and factories reshaped the complex connections between work and residency that powerfully moulded social life within the Scottish coalfields. It also disrupted the imagined communities sustained through class consciousness and nationhood that strongly conditioned political outlooks. In their formative contribution to the historical study of deindustrialization, the American scholars, Jefferson Cowie and Joseph Heathcott, posited that it is 'more disorienting than overtly political'. Unlike episodes of industrialization, the experience of incremental job losses and disrupted social routines are 'more elusive than tangible', and do not afford the same collective responses as early experiences of proletarianization.[2] Disorientation was engendered by the erosion of collectivism's historic sources of strength in workplace solidarity and connections to communities, which in the Scottish context often had high public housing densities.[3] More recent accounts of deindustrialization underline 'rust belt' rebellions in spurring far-right or 'populist' politics within both North America and western Europe.[4]

[2] J. Cowie and J. Heathcott, 'Introduction: the meanings of deindustrialization', in *Beyond the Ruins: the Meanings of Deindustrialization*, ed. J. Cowie and J. Heathcott (Ithaca, N.Y., 2003), pp. 1–15, at p. 4.

[3] S. Condratto and E. Gibbs, 'After industrial citizenship: adapting to precarious employment in the Lanarkshire coalfield, Scotland and Sudbury hardrock mining, Canada', *Labour/Le Travail*, lxxxi (2018), 213–39.

[4] N. Isenberg, *White Trash: the 400-year Untold History of Class in America* (New York,

However, deindustrialization's political consequences should not be read as an automatic conveyor belt stemming inevitably from workplace closures. Huw Beynon et al's more optimistic account of enduring trade union membership rates in Wales emphasizes the legacy of industrial 'working-class visions of society', and the lasting memory of colliery employment.[5] Differences between national and regional experiences are illustrative of how heterogeneous social and political forces contest experiences of deindustrialization. They further indicate the need to understand long-term processes of economic change through detailed analyses of industrial relations, institutions, policymaking and ideology.[6] Wholesale economic reorientation profoundly changes social relations and culture. Deindustrialization incorporates a significant renegotiation of workplace relations as well as citizenship and conceptions of national belonging. These *longue durée* dimensions are revealed by the Scottish coalfields' experience of industrial contraction.

The temporalities of deindustrialization were strongly moulded and punctuated by changes in government policy towards managing labour markets, regional policy which attempted to attract inward investment and energy policies that oversaw the transition from a coal to multi-fuel energy economy. In his revisionist account of Britain's twentieth century, David Edgerton suggested that British nationhood reached its zenith during the mid twentieth-century experience of economic modernization under government direction. Edgerton emphasized the pursuit of technological advances, highlighting nuclear power stations and the NCB's 'super pits', as well as the development of mass production industries. These were the product of an activist or 'militant' project for national development in which the state was the prime mover.[7] The pursuit of financial liberalization during the late twentieth century undid a more pronounced sense of economic citizenship and shared objectives. This included government toleration, or even encouragement, of manufacturing job losses and regional inequalities,

---

2016), p. xxi; G. Thereborne, 'Twilight of Swedish social democracy', *New Left Review*, cxiii (2018), 5–26, at p. 10; J. Tomlinson, *Managing the Economy, Managing the People: Narratives of Economic Life in Britain from Beveridge to Brexit* (Oxford, 2017), p. 107.

[5] H. Beynon, R. Davies and S. Davies, 'Sources of variation in trade union membership across the UK: the case of Wales', *Industrial Relations Journal*, xliii (2012), 200–21, at pp. 200–1.

[6] G. Hospers, 'Restructuring Europe's rustbelt: the case of the German Ruhrgebiet', *Intereconomics*, xxxix (2004), 147–56, at p. 147.

[7] D. Edgerton, *The Rise and Fall of the British Nation: a Twentieth Century History* (London, 2018), pp. 309–38.

which continued after Thatcherism under New Labour.[8] In a more polemical style, David Marquand argues that a 'solidaristic' moral economy of Labourist social democrats and third way Conservatives structured British political economy before the triumph of 'market fundamentalism' ushered in an age of 'unreflective utilitarianism'.[9]

These perspectives on British history and economic change chime with the archival research and oral testimonies collected for this volume. Policymakers pursued a commitment to modernization and productivity increases through a comparatively careful moral economy management of colliery closures between 1947 and 1979. Those earlier experiences conditioned objections that were framed in social contract terms when moral economy norms were abandoned in favour of market logic. Edgerton's and Marquand's accounts also accord with explorations of deindustrialization in Scottish popular culture. Irvine Welsh's 2012 prequel to *Trainspotting*, *Skagboys*, features a dialogue where Davie Renton, the shipyard worker father of the main character, Mark, reconsiders his Unionist constitutional affiliations. Davie finds himself arguing with a retired police officer about events unfolding during the 1984–5 miners' strike in a pub adjacent to the site of his recently closed former workplace, the Henry Robb shipyard in Leith. Welsh portrays Davie's despair that the social fabric of industrial society was 'slowly but irrevocably coming apart'.[10] Scott McCallum relayed a similar perspective in an interview during 2014. Scott reflected on the contentious closure of Cardowan colliery, to the east of Glasgow, during 1983, where his father and brother worked, as well as on his schoolboy experience of the miners' strike. He explained deindustrialization through wilful social violence:

> The Tory government *sighs* They're to blame for it. Politics … Just who was in charge at the time. She [Margaret Thatcher] went out tae put a purpose. She won her purpose. There's no mines now cause of her.[11]

Nicky Wilson, who was an electrician and trade union representative at Cardowan, also rationalized accelerated deindustrialization in terms of cultural values. Nicky singled out Albert Wheeler, the NCB's Scottish Area Director, as having betrayed moral economy obligations by relentlessly pursuing the strategy of a government hostile to the industry: 'He probably suited the government at that time. He'd no social conscience or that. He

---

[8] Edgerton, *Rise and Fall*, p. 498.
[9] D. Marquand, *Mammon's Kingdom: an Essay on Britain Now* (London, 2014), pp. 1–3.
[10] I. Welsh, *Skagboys* (London, 2012), pp. 288–90.
[11] Scott McCallum, interview with author, The Counting House, Dundee, 22 Feb. 2014.

didnae care what happened tae mining communities despite coming fae that originally. Just a ruthless, ruthless, person who had ambition'.[12] A sense of betrayal was encouraged by the nationalized industry's earlier promises of collective economic security. Pat Egan clearly voiced the coalfield moral economy's key tenets. As a young miner, Pat had experienced the NUM's consent to the closure of Bedlay colliery in Lanarkshire during 1982, which led him to relocate to Fife and take up employment at the Longannet complex's drift mines. Pat was prepared to accept that less productive or geologically exhausted pits would close in return for the provision of employment elsewhere:

> *Plan for Coal* wis that you'd hae that big complex at Longannet. You wid have the Fife complex with the Lothians feedin into Cockenzie and there wis a new coalfield which is still a virgin coalfield, the Canonbie coalfield doon in the Borders which runs fae there right doon tae Durham. It's [got] massive seams o coal. That wis never it wis never [exploited], but that was planned where it would go. Cause we used tae talk that we could end up livin in the Borders in a new toon somewhere. Cause that wis the talk that they would build a new toon doon there. Mining's an exhaustive industry. Pits are gonnae shut. They need tae shut, cause once you mine the coal what you gonnae do, y'know? And they weren't just kept open fir the social aspect.[13]

The Coal Board's pursuit of market ends dislocated the social life of coalfield communities and the mining workforce's practices. These developments began long before the pronounced industrial conflict of the 1980s, which has subsequently dominated the public understanding of coalfield deindustrialization.[14] Earlier closures created severe tensions and led to questioning of nationalization's achievements. This was indicated by the response to the closure of Gartshore 9/11 colliery in North Lanarkshire during 1968. Both Board officials and the NUM were 'absolutely flabbergasted' by the decision of sixty miners to refuse transfer and instead pursue a legal appeal for redundancy payments. An NCB report described the 'resistance group' as primarily middle-aged miners who were motivated by the loss of up to a fifth of their earnings through transfer from piecework rates to day wage work.[15] The Board decried this action as 'a mass protest

---

[12] Nicky Wilson, interview with author, John Macintyre Building University of Glasgow, 10 Feb. 2014.

[13] Pat Egan, interview with author, Fife College, Glenrothes, 5 Feb. 2014.

[14] J. Arnold, '"Like being on death row": Britain and the end of coal *c.*1970 to the present', *Contemporary British History*, xxxii (2018), 1–32, at pp. 4–5.

[15] National Records of Scotland, Edinburgh (NRS), Coal Board (CB) 300/14/1, Manpower branch, NCB, Glasgow to Legal Department, NCB, Edinburgh, memorandum:

... directed not only at the NCB but at the National Policy of the NUM'. This choice of 'exit' strategy was influenced by perceptions of economic insecurity, with one man resignedly commenting that 'all I know is pit work – but for the sake of my family I've got to get out now'.[16]

These actions were given a collective nature by the number of men involved, some of whom were union branch officials. Their grievances indicated discontent with the insensitivity of a centrally organized closure programme, which disembedded the workplace from its traditional social order, including the rationing of piecework positions. The mass legal challenge of pit transfers anticipated the subsequent renegotiation of the moral economy through unofficial and official strike action between 1969 and 1974. These struggles produced significant increases in miners' remuneration and a commitment to the industry's future through the *Plan for Coal*, which shaped Pat Egan's understanding of the sector's future during the early 1980s.

Most of the men pursuing redundancy payments at Gartshore rejected NUM representation in favour of a local lawyer 'of mining stock'. He was viewed as a reliable community voice and more combative than the full-time union officials who had agreed unacceptable closure terms. The Board and NUM both felt Gartshore 9/11 was 'the best organized closure ever carried out in their experience'.[17] Divergences between workers at the point of production and union and NCB staff are indicative of the class tensions inherent in experiences of deindustrialization. From their technocratic perspective, the Board's officials had acted to both increase rates of productivity and improve financial performance while protecting employment security. Yet a significant portion of workers felt aggrieved by their experiences of repeated closure, remote administration and a more general view that their industry did not have an optimistic future. Industrial relations in the NCB were typified by 'fluid struggle for control between administrators and administered'.[18] Deindustrialization increased these tensions through adding to the sense of a persistent threat to livelihoods

---

'Background notes: what happened at Gartshore 9/11 closure?' (1968).

[16] NRS, CB 300/14/1, Manpower branch, NCB, Glasgow to Legal Department, NCB, Edinburgh, memorandum: 'Background notes: what happened at Gartshore 9/11 closure?' (1968); T. B. Lawrence and S. L. Robinson, 'Ain't misbehavin: workplace deviance as organizational resistance', *Journal of Management*, xxxiii (2007), 378–94, at pp. 382–3.

[17] NRS, CB 300/14/1, Manpower branch, NCB, Glasgow to Legal Department, NCB, Edinburgh memorandum: 'Background notes: what happened at Gartshore 9/11 closure?' (1968).

[18] J. Krieger, *Undermining Capitalism: State Ownership and the Dialectic of Control in the British Coal Industry* (Princeton, N.J., 1983), p. 26.

and social status that was far more existential than routine trade union concerns over wage and conditions, or even episodic unemployment.

Unlike narratives from younger men such as Scott McCallum, older respondents displayed a stronger awareness of the longer history of colliery closures. Alex Clark left his job at Douglas Castle colliery in South Lanarkshire during the early 1950s to become a full-time Communist party organizer. Four decades later, Alex recalled that when he left there was already a sense that the industry was contracting in the area. Local miners had already transferred to Fife, the Lothians and Clackmannanshire, where Alex would later assist party activities among miners.[19] Alex's widow, Jessie Clark, narrated coal closures in Scotland as a drawn-out process when she was interviewed in 2014:

> I mean it was Thatcher that finished it. I don't have to tell you or anybody else what happened in the end, y'know. But as far as the mines in Scotland were concerned, you know, it was happening quicker [earlier] I think than anywhere else, you know. Because it was having its affect in Ayrshire, as well, and also in Fife and the Lothians. Y'know, was winding down. Because it had been the policy and I think everybody knows this: get the coal out as easily and as quickly as you can, you know. And wi doin it that way it got to a stage that it wasn't very wise, how can I put it. To describe it technically you know, it was quick buck. Let's make a quick buck you know. So, it wasn't scientifically worked oot at all you know. So that they would get the best out the mine for the longest time properly. It was let's make a quick buck you know and get it out as make as much money as you can right now. And as I said to you earlier, according to the guys that worked in the pit, they said there was still plenty coal there.[20]

Jessie's narrative was strongly imbued with moral economy sentiments shaped by class and national dimensions. She juxtaposed the long-term commitment miners made to the industry with the short-term financial priorities which in her view motivated closure decisions. Burying large reserves of coal in the ground, at the cost of jobs and production, stimulated grievances. Jessie's perspective was amplified by the national context because the Scottish coalfields contracted at a faster rate than the industry across the UK. Opencast mining exemplified the distinctions between 'a quick buck' and long-term investment. It became a source of ire when coal

---

[19] A. Clark, 'Personal experience from a lifetime in the communist and labour movements', *Scottish Labour History Review*, x (1996–7), 9–11, at p. 11; A. Clark, 'Personal experience from a lifetime in the communist and labour movements (part 2)', *Scottish Labour History Review*, xi (1997–8), 14–16, at p. 14.

[20] Jessie Clark, interview with author, residence, Broddock, 22 March 2014.

mining experienced its most pronounced employment rundown during the 1960s. Opencast operations were more profitable than deep mining, partially due to their short-term nature and the small number of miners employed, which encouraged opposition on moral economy grounds. Coal Board officials reported to the Scottish Coal Committee in early 1961 'that serious labour trouble would arise if it expanded opencast working while it shut down pits'.[21]

NCB officials recognized they operated between commitments to pursue the most profitable production and obligations towards their workforce and coalfield communities. Ministry of Power officials, who were persistent advocates of liberalized energy markets, continued to pressurize the Board towards opencast production and eschewing social responsibilities during the late 1960s. In preparing the 1967 *Fuel Policy* white paper, a Ministry steering group concluded that 'as other coal producing countries have shown, there can be no question that the right course commercially is to maximize the production of low-cost opencast coal'. These comments anticipated the officials bemoaning the moral economy in the form of 'national conscience', which is detailed above.[22] In November 1969, Michael McGahey contrastingly summarized the NUMSA's role in the countermovement to market liberalization. He underlined that the union would only accept the development of three opencast sites in Scotland on the condition that 'closures in future must of necessity affect opencast workings before deep mine workings'.[23] McGahey reasserted this position in a tripartite discussion during 1977 when he again stated that the price of the NUM accepting opencast development was a guarantee these developments were 'not to supplant deep-mined coal'.[24]

A conviction that 'national conscience' presented a barrier to the necessary reordering of British industry has provided an enduring narrative of deindustrialization and economic 'decline'.[25] Advocates of Thatcherite counter-reforms, such as Corelli Barnett, were influential champions of this assessment. Barnett accused post-Second World War policymakers of

[21] TNA, Coal 31/96, W. E. Fitzsimmons and E. Wright, Scottish Coal Committee Report, 11 Oct. 1961, p. 39.

[22] TNA, Coal 31/154, Steering Committee on Fuel Policy, Opencast coal production: 1967/71, 1967.

[23] TNA, Coal 89/103, Copy of letter from Mr McGahey, NUMSA, to Mr Glass, NCB, 27 Nov. 1969.

[24] TNA, Coal 31/166, Coal industry tripartite discussions meeting held at 2.30pm, Wednesday 31 Jan., Thames House South.

[25] J. Tomlinson, *The Politics of Decline: Understanding Post-War Britain* (London, 2000), p. 16.

operating free from any 'sense of financial limits, but also from any sense of limits on material resources'.[26] However, these characterizations appear poorly evidenced when contrasted with the coal industry's experience under nationalization: the urgent pursuit of production during the first decade of nationalization in the context of coal shortage was followed by a swift pace of colliery closures and a consistent rationalization towards the newest most productive collieries.

Barnett misrecognizes the economic challenges of managing coal's contraction in the context of competition between fuel sources. In 1967, NCB officials in the chairman's office starkly summarized 'a serious conflict of objectives' between the government's pursuit of rapidly converting electricity generating capacity to oil and nuclear, and 'potential damage to the industry and the consequences for the country, both economic and social, of reducing their manpower suddenly at a time of unemployment'.[27] Contradictions between these policy ends were further augmented by the fact that closures were concentrated in the 'peripheral' coalfields, which were already experiencing high rates of unemployment.[28] A year later, Jimmy Hood left the contracting Scottish coalfields for Nottinghamshire after Auchlochan 9 near Coalburn in South Lanarkshire closed. In 2014, when the Labour MP for Lanark and Hamilton East, Jimmy vividly recalled coal closures during this period in terms that were highly critical of energy policies pursued by Labour governments:

> The then government, and it included Tory and Labour because it was in the sixties. In fact, in fairness Wilson came in '64, and the Labour government at that time, they were closing it and going for cheap oil. Cause oil was cheap as chips. And they were being seduced by the cheap oil. It was always going to be there. What a crazy, and look where we are now by doing that. Look where we are now, all the trouble we've caused in the Middle East by relying on their oil. It was, it's crazy. It was crazy. You couldn't sit down and plan to do it any worse if you were given a task how could you destroy people's economies by depriving them of commodities and things like that.[29]

Jimmy's perspective was clearly shaped by a sense of composure that connected his occupational identity and experience with macro developments

[26] C. Barnett, *The Audit of War: the Illusion and Reality of Britain as a Great Nation* (London, 1986), p. 49.

[27] TNA, Coal 31/131, Notes on draft of *Fuel policy* section D: electricity, nuclear power and power station fuel use, F13, 1967.

[28] *Fuel Policy* (Parl. Papers 1967 [Cmnd. 3438]), p. 30.

[29] Jimmy Hood, interview with author, South Lanarkshire Council offices, Lanark, 4 Apr. 2014.

in energy markets and political economy. His views were partly informed by his decision to oppose another Labour premier, Tony Blair, by voting against the invasion of Iraq in 2003.[30] Three decades earlier, the 'oil shock' of 1973–4 had given credence to the NUM's position on the importance of maintaining domestic production to avoid dependency on imported fuels.[31] These events provided renewed impetus for a moral economy viewpoint centred on government obligation to invest in collieries, maintain or expand employment and make use of national resource endowments

Jimmy's reflections demonstrate that deindustrialization in the Scottish coalfields was a long-term process. It was entangled with fundamental changes in the structures of the economy and meeting basic human needs, including the provision of domestic heat and lighting as well as power for commercial and industrial activities. Coal industry restructuring and moves towards final closure over the second half of the twentieth century were not the unmediated outcome of market forces or technological changes. In the context of Britain's nationalized coal industry and the development of policies that aimed to promote a multi-fuel energy economy, colliery closures were a consciously willed process. The logic of pursuing financial performance alone was resisted in a countermovement that emanated from colliery workforces and communities. Trade union representatives articulated a moral economy stance which insisted that closures were negotiated through the nationalized industry's consultation machinery, and that the Board and government undertake responsibility by providing alternative employment for those affected. These measures were at least partially implemented through the provision of transfers for miners upon closure and the direction of manufacturing inward investment towards contracting coalfields.

The moral economy had long roots into the nineteenth century. It was shaped by opposition to unbridled employer power, and the effects that economic fluctuations had in communities often overwhelmingly dependent upon coal mining. Collective memories of the private industry were maintained in families and passed on within coal mining settlements. They were also given institutionalized expression by the labour movement. This historical consciousness underlined the nationalized industry's duty to provide employment by exploiting coal, a valuable national asset over which it had stewardship. Older respondents recalled a long-running

---

[30] 'Obituary: Jimmy Hood', *Times*, 7 Dec. 2017 <https://www.thetimes.co.uk/article/obituary-jimmy-hood-rfgl6sml3> [accessed 22 Dec. 2019].

[31] National Union of Mineworkers (NUM), *National Energy Policy* (London, 1972), pp. 1–8; J. North and D. Spooner, 'The great UK coal rush: a progress report to the end of 1976', *Area*, ix (1977), 15–27, at p. 15.

conflict between government pursuit of financial and energy diversification objectives on the one hand, and the interests of the coal mining workforce and the regions which strongly depended on it on the other. In Scotland, deindustrialization was therefore not 'disorientating' and depoliticizing, which questions the universality of American scholars' conclusions.[32] Other European examples add to this contention, such as Asturias where the Spanish government recently agreed a 'just transition' programme with mining trade unionists who had previously mounted militant strike action against colliery closures.[33] The drawn-out experience of Scotland's postwar deindustrialization proved to be a highly politicizing context. Coalfield contraction informed alternative, class-based, national interests at UK level but also a distinct Scottish identity. The oral testimonies collected for this study also indicate the challenge experienced by historians of deindustrialization. Understanding deindustrialization's temporalities requires discriminating between the trend towards contracting employment in Scottish heavy industries since the early 1920s and individual and collective perceptions often conditioned by comparatively episodic reference points.[34] The sense of betrayal associated with the abandonment of the 1974 *Plan for Coal* exemplifies how generational outlooks coalesce around discrete formative experiences. Differing conceptions of the fusion between class and Scottish national interests also strongly conditioned narratives of economic change.

Class and nation

The mass refusal of transfers at Gartshore 9/11 illustrates the economic substance which shaped collective perceptions of deindustrialization. Willie Doolan recalled that during the twelve years that he was employed at Cardowan colliery, there were persistent rumours of closure: 'there was always that threat hanging over us'.[35] Even earlier than this, coal mining was understood as an ageing and passing activity. For instance, in introducing her poem, 'The newly wed miner', at a reading during 2007, Liz Lochhead explained that she was inspired to write it by a 1950s childhood memory of seeing an elderly opencast worker cycling to and from the housing

---

[32] Cowie and Heathcott, 'Introduction', p. 4.

[33] R. Popp, 'A just transition of European coal regions: assessing the stakeholder positions towards transitions away from coal', *E3G Briefing Paper* (2019) <https://www.e3g.org/showcase/just-transition/> [accessed 20 Nov. 2019].

[34] P. Payne, *Growth and Contraction: Scottish Industry, c.1860–1990* (Glasgow, 1992), p. 19.

[35] Willie Doolan, interview with author, The Pivot Community Centre, Moodiesburn, 12 March 2014, notes.

scheme in Craigneuk, North Lanarkshire, where she grew up. To the young Lochhead, the man embodied a passing way of life within 'a former mining village turned dormitory' in an area strongly affected by colliery closures during the early years of nationalization. Lochhead's poem recasts the miner as a young married man, before he became a representative of the past.[36]

Perceptions of deindustrialization were strongly shaped by localized circumstances. However, these were also mapped on to the Scottish national context. From the late 1950s onwards, a sense of prolonged contractions is visible in commentary on the coal industry. Accounts of mining's past and future often included hostility towards the NCB, which was portrayed as following the dictates of a distant London-based central management, or towards British governments that were uninterested in exploiting viable coal reserves. These were not exclusively drawn from the labour movement or the political left. In February 1959, a businessman, David Murray, wrote a column in the *Scotsman*, entitled 'I want to buy a coal pit' that outlined his frustrations that the NCB held a monopoly on large-scale extraction. Murray objected to the closure of Douglas Castle colliery in South Lanarkshire. He argued it was the product of 'group psychology' within the nationalized industry which secured adherence to central direction from London. Public ownership had resultantly fatally undermined the Scottish coal industry: 'If the iron and steel companies got back the pits that were taken away from them, they could soon show the NCB how to get coal out at much less than its idea of cost'.[37] Murray's criticisms chime with those made by other Scottish authors who have sympathies for the private industry. Halliday's account of his experiences in Scottish divisional management is highly disapproving of the major investment projects undertaken during the 1950s and 1960s. The division was 'chasing its own tail' to meet production targets and achieve extensive reconstruction or new sinkings deep into Limestone coal reserves which would simply have been unfeasible within the remit or localized managerial knowhow of privatized coal companies. However, Halliday was also more complimentary about the 'remarkable' success in concentrating colliery investment towards Upper Hirst coals through the development of the Longannet complex.[38] Such an undertaking was similarly unimaginable without the resources made possible by public ownership.

---

[36] L. Lochhead, 'The newly wed miner' (2007), Seamus Heaney Centre Digital Archive <http://digitalcollections.qub.ac.uk/poetry/recordings/details/108096> [accessed 22 Dec. 2019]; L. Lochhead, *The Colour of Black and White* (Edinburgh, 2003), p. 12.

[37] TNA, POWE/14/857/47, D. Murray, 'I want to buy a coal pit', *Scotsman*, 3 Feb. 1959.

[38] R. Halliday, *The Disappearing Scottish Colliery: a Personal View of Some Aspects of Scotland's Coal Industry since Nationalisation* (Edinburgh, 1990), pp. 41–8.

The class basis of these perspectives should not be understated. However, actors who were otherwise socially and ideologically opposed converged on fundamental details. In 1976, George Montgomery, the NUM's mechanical and electrical safety inspector, referred to the 'wastelands' of 'the old Lanarkshire coalfield' where he had grown up. Montgomery lamented the concentration of colliery closures in Scotland, underlining that 'the present-day miner has been the victim of continual closure and redundancy'. These injustices were blamed on 'the blunders and bungling of the planner'.[39] Montgomery was writing in the pages of *New Edinburgh Review*, and further conferred the distinctive Scottish context by objecting to coal's absence from contemporary energy discussions, which he saw as unduly concentrated on North Sea oil.[40] In the same publication, the Scottish nationalist poet, T. S. Law, stated that 'my father and my grandfather saw the heart of Lanarkshire cleared out'.[41] Law's narrative at this point departs far from Murray and Montgomery's in other respects. He underlined the history of miners' struggles for social justice. Law was scathing in comments on 1970s coal strikes, underlining the revulsion he held towards middle-class opponents of the NUM's mobilization for improved living standards. They left him 'indignant and quite appalled at the utter impertinence of the continuing strictures against the miners by people who have not only never done a day's hard manual work in their lives, but are incapable of sustained work of any kind at all'.[42] Unlike Montgomery, Law drew more explicit attention to a fusion of class and national interests at stake in these confrontations, underlining the role of British government and capital in appropriating Scottish natural resources, which he saw as continuing in the North Sea.[43]

Through these readings, a shared feeling of hostility towards centralization within the nationalized industry are apparent alongside a view that British government policy had negative consequences in the Scottish coalfields. In each case, elements of Scottish national identity and class interest merge and shape a countermovement framed around a protective role for Scottish nationhood against the threats of market forces and remote administration. The remedies put forward were incompatible: free enterprise was promoted by opponents of nationalization; the NUM sought a government commitment to coordinated energy policymaking

---

[39] G. Montgomery, 'Introduction', *New Edinburgh Review*, xxxii (1976), 1–3, at p. 1.

[40] Montgomery, 'Introduction', p. 1.

[41] T. S. Law, 'A Wilson memorial', *New Edinburgh Review*, xxxii (1976), 22–8, at p. 24.

[42] Law, 'A Wilson memorial', p. 24.

[43] Law, 'A Wilson memorial', p. 24.

and for political devolution to ensure decentralization within a Unionist framework; and Scottish nationalists advocated for independence and sovereignty over natural resources.[44] These stances stemmed from markedly different political positions and were roughly approximate to the three major ideological traditions that held sway, to varying degrees, within the Scottish coalfields during this period. While viable alternatives to Labour party dominance of Scottish coalfield politics were marginalized by the late 1930s, Conservative or Unionist representation remained significant for at least two further decades.[45] Jimmy Hood's early political memories included the victory of his predecessor, Labour's Judith Hart, in the Lanark constituency at the 1959 general election: 'I remember ma mother swinging me round the room because it was just announced that Judith Hart had won the '59 election. Labour lost the election, but we beat Patrick Maitland, and Judith Hart was elected after years of Tories being MPs'.[46]

Maitland's politics exemplified how dimensions of Unionist–Nationalism shaped perceptions of deindustrialization. Although his appeal rested on an accentuated Britishness, he also affirmed a politics of distinct local and Scottish national interest. Maitland's responses to colliery closures included calling on the NCB to 'make available for purchase by private enterprise any collieries which the Board finds uneconomic to keep open'.[47] He made this argument from the floor of the House of Commons during the same month that David Murray outlined a similar position in the press considered above. Maitland also made a representation to Lord Mills, the minister of power, over the closure of Douglas Castle, a major employer in an isolated part of his constituency. His letter was sent on the same day as the Coal Board met NUM representatives and expressed concern over the impact on the area's long-term prospects. Maitland employed distinctly moral economy terms, discussing the wrongs of what he considered viable reserves 'being flooded and left to waste' by the Board while local consumers struggled to find coal.[48]

Maitland's ultimately doomed representations parallel the struggle to save Blackhill colliery in Northumbria around the same time. In each case, trade union representatives and community campaigners were joined by Tory grandees in campaigning against closure but faced with indifference

---

[44] S. Maxwell, *The Case for Left-Wing Nationalism* (Edinburgh, 1981).

[45] A. Campbell, *The Scottish Miners, 1874–1939*, ii: *Trade Unions and Politics* (Aldershot, 2000), p. 1.

[46] Jimmy Hood, interview.

[47] TNA, POWE 14/587, Hansard column 103, Friday 20 Feb. 1959.

[48] TNA, POWE 14/587, Letter from P. Maitland to Lord Mills, 26 Feb. 1959.

by the government and the NCB's centralized bureaucratic hierarchy. These forces broadly fit with Polanyi's conception of the coalitions that often make up countermovements to the imposition of market forces, combining worker interests with those of local social elites also troubled by threats to communal institutions.[49] Jack Parsons captured the events just south of the Anglo-Scottish border in *The Blackhill Campaign*.[50] His film profiles the struggles of a small-scale community in the context of mounting critiques of organized capitalism signified by C. Wright Mills' power elite thesis.[51] New left academics and activists influenced by similar perspectives have been formative to the study of deindustrialization, most influentially in the United States. As a result, the literature has often privileged instances of local activists struggling against 'global' conglomerates.[52]

Although the shared dimensions of the 'community versus capital' thesis has been central to readings of deindustrialization across international contexts, the influence of larger 'imagined' communities and political relationships should not be understated.[53] In the Scottish coalfields, these dimensions were given further significance by the presence of a nationalized coal industry and the role of both British and Scottish national identities in shaping the experience and framing of workplace closures. Maitland's Unionist–Nationalist view of the nationalized industry's obligations to Scottish communities was to some extent shared by Labour politicians, including sharp opponents of political devolution or 'home rule'. These tendencies were strongly present during the third reading of the Electricity (Borrowing Powers) (Scotland) Bill in December 1962. Opposition members used this opportunity to advocate in favour of extending government funding to fulfil the recently published report of the Scottish Coal Committee, which had met to discuss the industry's future following the onset of contraction during the late 1950s. It recommended investigating the possibility of expanding coal-fired power generation in Scotland.[54]

These developments demonstrate the importance of national contexts: the case for further investment was propelled by a sense of obligation

---

[49] K. Polanyi, *The Great Transformation: the Political and Economic Origins of Our Time* (Boston, Mass., 2001), p. 101.

[50] *The Blackhill Campaign*, J. Parsons, 1963, 50 mins <https://player.bfi.org.uk/free/film/watch-blackhill-campaign-1963-online> [accessed 22 Dec. 2019].

[51] C. Wright Mills, *The Power Elite* (New York, 1956).

[52] S. High and D. Lewis, *Corporate Wasteland: the Legacy and Memory of Deindustrialization* (New York, 2009), pp. 65–85.

[53] Cowie and Heathcott, 'Introduction', p. 13.

[54] *Electricity in Scotland: Report of the Committee on the Generation and Distribution of Electricity in Scotland* (Parl. Papers 1962 [Cmnd. 1859]).

towards the Scottish coalfield and sustained by an argument that the UK government was failing to adhere to the social and political partnership that Union entailed. Margaret Herbison led the charge in condemning government inaction. She was the MP for North Lanarkshire and the daughter of a miner who was killed in a colliery accident.[55] Herbison asserted that 'more and more Scottish pits' faced closure without power station investment. She rhetorically asked: 'can we take it that the Secretary of State has resigned himself to the fact that Scotland must always have a high rate of unemployment?'.[56] Herbison's comments were echoed by Willie Ross, the shadow secretary of state for Scotland and MP for Kilmarnock who numbered Ayrshire miners among his constituents. Ross labelled the government's bill 'a confession of failure', arguing that only further investment in coal-fired capacity could 'bring hope' to the coal industry.[57]

Power stations became a politically contentious issue as they came to dictate employment in the surrounding coalfields during episodes of mounting colliery closures. These dynamics implicated both devolved agencies such as the South of Scotland Electricity Board, but also centrally administered organizations, including the NCB as well as UK government. Colliery closures were pivotal in anticipating the more extensive discussion of the constitution and Scottish nationhood which would follow over the proceeding decades. Key actors included – but were far from limited to – entrepreneurs, civil servants, Coal Board and Electricity Board officials as well as elected politicians and trade unionists. All were responding in a variegated manner to pressure from below, which was marked by expectations of workforce consultation and the provision of economic security. These priorities were shaped by collective memories of the first half of the twentieth century and earlier time periods, but also by a fusion of class and national consciousness which was informed by more recent industrial experiences. The pursuit of increasingly centrally directed closures eroded faith in the nationalized industry's capacity to deliver secure employment and stimulated support for Scottish political autonomy, especially within the NUMSA, but also across the wider labour movement. Serving distinct Scottish interests was not, however, exclusively a Scottish nationalist aim and was often contextualized within a Unionist–Nationalist framework that underlined the economic obligations which political partnership entailed.

---

[55] T. Dalyell, 'Margaret Herbison', *ODNB* <https://doi.org/10.1093/ref:odnb/64016> [accessed 20 Oct. 2011].

[56] TNA, POWE 14/1495, Hansard Parliamentary Debates, dclxix, no. 32, Wednesday 12 Dec. 1962: third reading of Electricity (Borrowing Powers) (Scotland) Bill.

[57] TNA, POWE 14/1495, Hansard, dclxix, no. 32, Wednesday 12 Dec. 1962: third reading of Electricity (Borrowing Powers) (Scotland) Bill.

These developments lend weight to Hobsbawm's view that expressions of nationalism incorporate 'the assumptions, hopes, needs, longings and interests of ordinary people, which are not necessarily national and still less nationalist'.[58] Hobsbawm was reflecting on the nineteenth-century experience of European industrialization, but episodes of deindustrialization also lend themselves to the reshaping of territorial politics and solidaristic affiliations through capitalism's uneven development. These were embodied in the debate over power stations by emotive appeals to Scotland's past and the perceived wanton neglect of its future. Parliamentary advocates of further investment in a coal-fired generation gave voice to less technical arguments than they did ones regarding economic welfare and obligations towards coalfield communities, which were given national rather than merely local importance.

Economic emigration became a spectre which haunted these discussions. The power of this trope partly came from the memory of interwar economic distress when large numbers of Scots emigrated overseas and to a lesser extent southwards, including through schemes such as the Stewart and Lloyds steelworks investment at Corby in the English Midlands.[59] After the Second Wold War, migratory trends continued and objections to southwards emigration were given more prominence as it exceptionally outstripped overseas departures during the 1960s.[60] This was partly due to incentives provided by the NCB. Scottish miners were granted the opportunity to take up well-paid industrial employment in England, while Scotland experienced higher rates of closures and divestment. Thomas Fraser, the Labour MP for Hamilton, explained during the 1962 debate that 'All we are saying to the government is that it would be better to export electricity from the North than to export people. And that is the choice which has to be made'.[61] Fraser's juxtaposition between people and goods is illustrative of the countermovement to market logic and opposition to economic upheaval. The form of objection raised by Fraser inserts a crucial national content to these deliberations. Localized objections to closure were

---

[58] E. Hobsbawm, *Nations and Nationalism since 1780: Programme, Myth, Reality* (Cambridge, 1990), p. 10, also quoted in J. Phillips, *The Industrial Politics of Devolution* (Manchester, 2008), p. 9.

[59] A. Mycock, 'Invisible and inaudible? England's Scottish diaspora and the politics of the Union', in *The Modern Scottish Diaspora: Contemporary Debates and Perspectives*, ed. M. Leith and D. Sim (Edinburgh, 2014), pp. 99–117, at p. 105.

[60] T. M. Devine, *To the Ends of the Earth: Scotland's Diaspora, 1750–2010* (London, 2011), p. 270.

[61] TNA, POWE 14/1495, Hansard, dclxix, no. 32, Wednesday 12 Dec. 1962: third reading of Electricity (Borrowing Powers) (Scotland) Bill.

given the standing of national crises where they reached a critical mass. In the complex Scottish situation, objections to closures were voiced within the framework of the Union and stimulated demands for increased political autonomy.

Margaret Herbison portrayed the policies of Sir Alex Douglas-Holme's Conservative government as driving an exodus of miners from Scotland. She underlined that this far from voluntary movement was to the detriment of Scotland, and the workers affected:

> Last night I listened to an account of the social difficulties caused by the almost enforced migration of our miners to coalfields in England. I was horrified to find that some of the houses being built by the Coal Board will be in completely isolated communities. It is bad enough to lose one's job in Scotland, but it is far worse to be taken to an area where one has to live in an isolated community, as too many of our miners and their families have to do. All these things worry those of us who are deeply concerned about the kind of lives which our Scottish people want to live.[62]

In his best-selling history of modern Scotland, Tom Devine argues that deindustrialization in the late twentieth century brought about 'a deep crisis of national identity' that stemmed from 'a collective psyche' invested in heavy industries.[63] The extent of socioeconomic transformation which took place over the second half of the twentieth century is indisputably the greatest since the first industrial revolution. It was a longer process than Devine allows for though. Rather than the sudden outcome of Thatcherite policymaking, deindustrialization in the coalfields can be traced back to the reorganization of coal mining that began under nationalization in 1947. A narrative based around threats to the Scottish nation was already being mobilized by the NUMSA and the broader labour movement during the early 1960s. Objections to colliery closures were given national significance in Scotland due to their swift pace, fears of increasing unemployment and the prominent role of state policy in determining the coal industry's future, especially through power station investment. Discontent was magnified by the growth of emigration, which both triggered fears of a return to the interwar depression and appeared to signify Scotland's inequitable treatment vis-à-vis the rest of the UK. Moral economy expectations of just procedure during colliery closures had origins in earlier experiences of dislocation within localized coalfield communities. But the presence of a nationalized industry and

---

[62] TNA, POWE 14/1495, Hansard, dclxix, no. 32, Wednesday 12 Dec. 1962: third reading of Electricity (Borrowing Powers) (Scotland) Bill.

[63] T. Devine, *The Scottish Nation, 1707–2007* (London, 2007), p. 643.

the contours of government policymaking, as well as the establishment of a relatively centralized Scottish miners' union, all encouraged a national expression of miners' discontents. Political forces broadly described as Unionist, sometimes with a capital U, were part of these expressions. Objections to closures, including those mounted by explicit Scottish Nationalists, were also often characterized by critiquing NCB policy in terms of a Unionist partnership framework.[64]

Narratives of deindustrialization

Personal accounts of deindustrialization are shaped by a complex range of factors besides predominant political narratives. Oral history provides an important corrective by giving voice to individual as well as collective memories. Life stories are inevitably 'deeply personal' and shaped by contemporary contexts as well as the imperative of constructing a 'composed' version of events.[65] Interviewees spelt out their current political position in dialogue with composed memories of past injustices. The pain associated with workplace closure and community fragmentation was a widespread feature in the oral testimonies collected for this study. Other common inclusions were tensions between personal narratives of comparative success and economic comfort with collective stories of job loss. In further cases, there was contravention from dominant accounts. Although Coal Board transfer programmes were condemned by Labour parliamentarians from Scottish coalfield constituencies, and viewed with suspicion by union leaders, they were far from universally deplored by miners.[66]

During 1968, both the NUMSA and the secretary of state for Scotland, Willie Ross, heavily lobbied the NCB to reconsider the decision to close Auchlochan 9 colliery in South Lanarkshire.[67] The colliery's remote status was also alluded to by Scottish Office officials. W. K. Fraser of the regional development department informed Ministry of Power officials in London that they should not falsely assume that miners could easily find other jobs:

> I hope that the Ministry's advice and the Sub-Committee's views are not simply to be set aside by reference to a map to the area which suggests, quite wrongly,

[64] TNA, Coal 31/96, B. Woolfe, SNP, to A. Robens, NCB, London, 1963.

[65] A. Portelli, *The Battle of Valle Giulia: Oral History and the Art of Dialogue* (Maddison, Wis., 1997), pp. 57–8.

[66] L. Daily, 'Scotland on the Dole', *New Left Review*, xvii (1962), 17.

[67] TNA, POWE 52/85, Note for the record: secretary of state's meeting with the Scottish Area NUM on Friday 19 Jan. 1968; W. Ross, Scottish Office, to R. Marsh, Ministry of Power, 9 Feb. 1968.

that if the men at Auchlochan were prepared to travel a short distance, they could find employment without difficulty.[68]

The agglomeration of voices arguing for the retention of Scottish coal employment indicates the forces behind the slowing of colliery closures which followed. However, the perspective they relay also jarred with the memories of two interviewees who saw the closure that followed – and the Coal Board's subsequent offer of a transfer to Nottingham – as an opportunity. Gilbert Dobby was a recently qualified twenty-two-year-old engineer who recalled that the offer of an NCB subsidized house was a significant incentive:

> I dare say that the pit closin as well, maybe benefited me a little. I don't know. I got married in July in '68 and we decided to move we got the chance tae move anywhere in Britain … We moved doon there. Coal Board paid fir oor flittin as well. Didnae cost us a penny tae move doon there. The first year you were there they paid the difference in yer rent. The next year they paid half the difference, the next year a quarter, and it went on like that. Ah wis livin in a lovely house paying less rent than what ma neighbours were paying. As I say, got ma furniture moved doon and everything, which felt really good. So, if you like, I think I was sorta loyal. That was one of the things that made me sort of loyal to the Coal Board and I enjoyed workin in the pits.[69]

Jimmy Hood similarly recalled the closure of Auchlochan 9 and the offer of transfer in starkly more optimistic terms than the official coalfield representatives tasked with opposing it:

> Seven years' free rent, a new hoose and what, fifty per cent mare wages? It wisnae much of a dilemma to be honest. The tragedy was I was leaving where I was born and bred. But I was young. That wasnae a big issue for me. It's more difficult to make them sort of decisions when you're a wee bit older or a lot or a bit older … I think I was twenty. It wisnae a decision at aw really.[70]

Both Jimmy and Gilbert described a markedly different, more diverse environment in Nottinghamshire than they had left behind in Lanarkshire. Jimmy recalled a 'cosmopolitan' environment characterized by miners who had migrated from across the UK including large contingents from the north-east of England, Wales and other parts of England.[71] Gilbert

---

[68] TNA, POWE 52/85, W. K. Fraser, Regional Development Department, St Andrews House, Edinburgh to J. R. Jenkins, Ministry of Power, London, 1968.

[69] Gilbert Dobby, interview with author, Coalburn Miners' Welfare, 11 Feb. 2014.

[70] Jimmy Hood, interview.

[71] Jimmy Hood, interview.

remembered a multicultural workforce including colleagues from the Soviet Union, Poland and Biafra.[72] Jimmy only returned to Scotland after being a leader of the striking minority within the Nottinghamshire coalfield in the 1984–5 dispute.[73] Gilbert returned to Scotland earlier, to work at Killoch colliery in Ayrshire and then Bedlay and Polkemmet collieries in Lanarkshire and West Lothian respectively. He cited distance from family as leading him to ultimately give up the higher earnings he obtained in Nottinghamshire. When Gilbert returned to Scotland, he not only brought his young family into proximity with grandparents, but also worked alongside his father at Bedlay.[74] Willie Hamilton also left the Lanarkshire collieries. He went on to work at Cummins engine factory in Shotts and later moved to Ford's plant at Langley, to the west of London. Like Gilbert, he explained that his wife's inability to settle and being far from family led him to return to Shotts and the NCB during the early 1960s. Willie's life-story narrative emphasized he became a miner because 'there was nowhere else to go' in Shotts as a fifteen-year-old school leaver in 1951. His decision to return to mining, as well as his description of a social infrastructure and sense of belonging provided by collieries, perhaps suggests a greater sense of attachment.[75]

Despite Gilbert Dobby's view of the opportunities provided at Auchlocan 9's closure, his testimony also exemplifies Steven High's understanding that deindustrialization incorporates 'wilful acts of violence perpetrated against working people'.[76] Gilbert described the loyalty he felt the NCB had shown him and that he had reciprocated as an employee for over twenty years. He also discussed the mutual respect which characterized the nationalized industry. This was demonstrated by his father insisting that colliery management referred to him as 'Mr Dobby' when partaking in meetings as a union representative, which summarized the hard-fought-for status miners were granted under public ownership. The social violence of deindustrialization was epitomized by the closure of Polkemmet at the behest of the Coal Board following the pit's pumps being switched off during the 1984–5 strike.[77] After Polkemmet closed, despite his years of service, Gilbert was made redundant and spent a year unemployed. The incremental logic

---

[72] Gilbert Dobby, interview.
[73] Jimmy Hood, interview.
[74] Gilbert Dobby, interview.
[75] Marian and Willie Hamilton, interview, residence, Shotts, 14 March 2014.
[76] S. High, 'Beyond aesthetics: visibility and invisibility in the aftermath of deindustrialization', *International Labor and Working-Class History*, lxxxiv (2013), 140–53, at p. 141.
[77] High, 'Beyond aesthetics', p. 141.

of deindustrialization was summarized when he subsequently became a driving instructor. After the Ravenscraig steelworks closed in 1992, a flood of new entrants into the sector led Gilbert to change occupations once more by becoming a school caretaker.[78] Polkemmet's closure was recalled by other respondents such as Peter Downie who worked as an overman at the pit alongside his three sons. Peter emphasized the cost in terms of public resources, the personal economic cost of lost jobs and the social damage of dispersed social bonds.[79]

A sense of injustice and enforced removal from anticipated life courses and communal expectations was reiterated in several interviews, with many respondents exclaiming they would still be working in collieries if they had not closed. Mick McGahey, who lost his job at Bilston Glen colliery following his arrest while picketing during 1984, stated that:

> If the pits had still been open, I'd have probably still been working in the pit. I'd have been retiring in the pits. So, it destroyed communities and there was nothing replaced the loss o they jobs in the industry, or the social fabric ae the communities that people lived in. Destroyed.[80]

The collective and individual costs of deindustrialization were mentioned in several testimonies, with parallels to Mick's story of displacement. Opportunities created through economic change were also discussed, which created a lacuna in some narratives. Bill McCabe, who was a participant in the Caterpillar factory occupation during 1987, explained that 'I'd have been quite happy on the truck tul I was sixty-five y'know!' The factory was a homely environment for Bill where he was tutored by older shop stewards and worked alongside members of his immediate and extended family. Following the factory's closure, and subsequently working in the North Sea oil industry, Bill took up white-collar employment in insurance. He reflected that Caterpillar's closure 'made me look at having tae dae a lot of different things in life, which is good in some ways and bad in others'.[81] Factories and collieries were sites of collective social interaction and identity formation. Bill McCabe's recollections suggest the importance of reckoning with social mobility in narratives of deindustrialization while maintaining an awareness of the capacity for critical nostalgia in comprehending the

---

[78] Gilbert Dobby, interview.

[79] Moodiesburn focus group, retired miners' group, The Pivot Community Centre, Moodiesburn, 25 March 2014.

[80] Mick McGahey, interview with author, Royal Edinburgh Hospital, 31 March 2014.

[81] Bill McCabe, interview with author, Tannochside Miners' Welfare, Tannochside, 20 Jan. 2017.

losses associated with economic restructuring. The breakup of community and weakening of trade union power were not fully compensated for by unevenly distributed access to alternative employment or financial gains. Workplace democracy and collectivity were the most significant and enduring losses.

Coal's status as a nationalized industry, and the centrality of the sector's fortunes to modern British history, strongly prefigures the collective memory of deindustrialization. Mining, especially under the NCB, had the character of 'a civilization', with distinct norms and values and a culture of internal governance.[82] In 1976, George Montgomery celebrated this by commenting that 'with few exceptions all positions of authority are held by people who have spent a life time in the industry'.[83] The nationalized industry's social democratic infrastructure gave form to a distinctive occupational standing. Gatherings such as the Scottish miners' gala epitomized these national connections, as well as elements of competition or rivalry. In 1967 John Hamilton was an apprentice blacksmith at Ponfeigh colliery in South Lanarkshire. During an interview in 2016, John emphasized the paternal role played by the men he worked alongside, some of whom took him to watch Scotland play international football matches at Hampden in Glasgow. Almost half a century later, John enthusiastically recalled himself playing for Scotland in the form of a select miners' team in a football competition against a Derbyshire side at that year's gala:

> It doesnae matter whether you're a kid or you're seventy years old. If you're representing your country, you're quite proud. At that time, I was playing junior football and getting trials for professional teams. So, it was a big thing for me, playing against England. And the icing on the cake was that we won one-nothing![84]

As a member of the winning team, John was presented with an engraved Ronsonol lighter. The importance of these events and similar competitions was also recalled by former management employees. Even during the early 1980s, under Albert Wheeler's austere anti-trade union direction, the NCB's Scottish Area continued to enthusiastically support participation in UK-wide health and safety competitions.[85] Ian Hogarth, who worked at the

---

[82] A. Perchard, '"Broken men" and "Thatcher's children": memory and legacy in Scotland's coalfields', *International Labor and Working-Class History*, lxxxiv (2013), 78–98, at p. 94.

[83] Montgomery, 'Introduction', p. 4.

[84] John Hamilton, interview with author, South Lanarkshire Integrated Children's Services office, Larkhall, 26 Apr. 2016.

[85] John Hamilton, interview.

Board's Scottish headquarters at Green Park in Edinburgh, recalled they were viewed as a means to bring prestige to 'satellite' coalfields such as Scotland. As a result, Wheeler personally ensured participants were given generous subsidies to attend events in other parts of the UK.[86]

While civilization narratives powerfully shape coal mining discourses, they also compete with emphases on class struggle. Especially when viewed in relation to the long-term impact of deindustrialization, the events of 1984–5 and the sudden acceleration of colliery closures under the Thatcher government loom large in predominant public and personal memories.[87] However, the conflict at Gartshore 9/11 and Gilbert Dobby's memory of his father's assertion of equitable status between management and miners are important reminders that class tensions persisted across the nationalized industry's lifespan. This was bluntly expressed by J. Wormald, a National Association of Colliery Overmen Deputies and Shotfirers agent, at a meeting with Coal Board officials in 1974, when he 'commented that absence from a pit to go to Doncaster races was recorded, but absence from the office to go to Ascot was not'.[88] Within the oral testimonies, there was a strong tendency for narratives to blend dimensions of class and Scottish nationhood. In part this reflected the influence of contemporary politics, especially as most testimonies were collected shortly before the 2014 referendum on Scottish independence. The testimonies reflected the tendency for reflections on major social changes to present 'myriad narratives', spanning past and present in dialogue with a range of perspectives drawn from varied ideological sources.[89] Tommy Canavan, who had formerly been an NUM representative at Cardowan colliery, exemplified such trends when he discussed the approaching referendum through analogies with the interwar coal industry:

> Aw I'm Scottish aw right, and I'll be voting Yes because it cannae be any worse than they Tory arseholes we've got in the now ... There's nobody entitled tae come fae another country and lay doon the law, and that's the law and you will just follow it. Cause that's what the coal owners done. That's what the coal owners done for years: 'if you don't do it you'll be oot a job oot a hoose', and noo they're saying, 'you'll no get a currency you're no getting intae Europe'. It's the same thing the very same! On a bigger scale, it's the same thing.[90]

---

[86] Ian Hogarth, interview with author, National Mining Museum, Newtongrange, 28 Aug. 2014.

[87] Arnold, '"Like being on death row"', p. 9.

[88] TNA, Coal 31/168, Coal Industry Examination: Demand and Supply Working Group 2nd meeting, Thames House South, 22 May 1974.

[89] J. Kirk, *Class, Culture and Social Change: on the Trail of the Working Class* (London, 2007), p. 7.

[90] Tommy Canavan, interview with author, residence, Kilsyth, 19 Feb. 2014.

Earlier in the same interview, Tommy had made another comment which appeared to counteract his inclination towards Scottish independence in class terms: 'Mining politics are the same in Kilsyth as they are in Croy as they are in Yorkshire as it in Nottingham as it in Kent as it in Lanarkshire, Lancashire, same politics'.[91] Tommy's remarks indicate the tension between Scottish national responses to deindustrialization and inclinations towards class-based UK-wide solidarity. Both trends persisted within the NUMSA between the 1940s and 1980s, and they are now marked in historical narratives of deindustrialization. These collective memories rest on the fusion of class and national consciousness which were marked in responses to colliery closures from the late 1950s. The closure of coal mines, steelworks and factories was an imposition of class power and social violence that created a sense of crisis, removal and social redundancy. Moral economy perspectives persist decades after the last major closures, demonstrating the strength of an outlook deeply embedded in Scotland's industrial fabric. The retention of moral economy outlooks reinforces the overwhelming significance of deindustrialization as a societal transformation whose substance is central to contemporary politics and culture.

[91] Tommy Canavan, interview.

Conclusion:
the meaning and memory of deindustrialization

On 26 July 2019, workers left the Caledonian railway works in Springburn, Glasgow, for the last time. Over 200 jobs were lost as the yard shut its doors after the German owners deemed the site surplus to requirement, ending over 160 years of railway engineering. Following a ceremony to mark the departure of the works' final production – a refurbished train – the workforce marched out together. They were led by a piper, and accompanied by industrial chaplains, trade union officials and local politicians. After the procession was cheered by residents of adjacent housing schemes, the union convenor, Les Ashton, delivered a speech. Ashton underlined that employment at 'the Caley' had been inherited by the present workers from their forebearers. He rejected the notion that jobs at the railway works were the property of the current workforce which could be legitimately traded for redundancy payments. They were custodians who wished to pass on employment at the works to a further rising cohort of future employees.[1]

The closure in Springburn marked another departure of (in this case much modernized) infrastructure with origins in the coal and steam era from Scotland's landscape. Ashton's collectivist sense of intergenerational solidarity clearly resonated with the coalfield moral economy. During an interview conducted in 2014, the National Union of Mineworkers (NUM) Scotland president, Nicky Wilson, angrily recalled the National Coal Board's (NCB's) conduct during contentious closures in the 1980s. Board officials enticed men 'tae sell their jobs' through the offer of large redundancy packages and enhanced pension benefits.[2] These attempts to undermine workforce solidarity were a double transgression of the moral economy's foundation in an understanding of colliery employment as a community resource, and an attack upon the trade union representation through which it was mediated. In a thoughtful reflection on the great strike for jobs as it concluded, Michael Jacobs had recognized twenty-nine years

---

[1] E. Gibbs, 'The "Caley" and Scotland's "invisible" workers', *Conter*, 26 July 2019 <https://www.conter.co.uk/blog/2019/7/26/the-caley-and-scotlands-invisible-workers> [accessed 23 Dec. 2019].

[2] Nicky Wilson, interview with author, John Macintyre Building University of Glasgow, 10 Feb. 2014.

'Conclusion: the meaning and memory of deindustrialization', in E. Gibbs, *Coal Country: The Meaning and Memory of Deindustrialization in Postwar Scotland* (London, 2021), pp. 251–58. License: CC-BY-NC-ND 4.0.

previously that one of its remarkable features was the basis upon which it was fought. Striking miners were primarily motivated by the maintenance of the industry and passing jobs on to their descendants.[3]

The ceremonial nature of the last day of work at the Caledonian works demonstrates continuity in the politics of deindustrialization across late twentieth- and early twenty-first-century Scotland. Scott McCallum recollected that 'there were quite a few tears shed' when the Cardowan colliery chimney was demolished after the pit's closure in 1983. This was a community occasion, with Scott and other schoolmates from mining households attending.[4] There was a cathartic dimension to the 'ritualistic' practices enacted at the Caley and Cardowan.[5] These developments strongly evidence the refusal of industrial workers to be rendered 'invisible'.[6] One effect of deindustrialization is to cast remaining industrial workers as anachronistic historical leftovers. Just as workers resisted industrialization by mobilizing their traditions of craft customs, responses to the long experience of deindustrialization are characterized by recourse to the collectivist culture of industrial workplaces. While E. P. Thompson memorably insisted that the industrial working class was 'present at its own making', episodes of deindustrialization reveal that it is also present at its own unmaking.[7] When the last train left the Caledonian works, it was workers and their union officials who marked the occasion, while management were conspicuous by their absence. The construction of a memorial for the Auchengeich disaster by striking miners in 1984 (discussed in chapter three) is another example of the impulse to collectively mark industrial heritage when its future is threatened.

A larger array of industrial monuments now populates the Scottish coalfields. Memorials erected as a result of community campaigns tend towards the commemorations of specific events, while industrial monuments built as part of commercial regeneration efforts often provide a more generic tribute to the industrial era.[8] These distinctions underline that

[3] M. Jacobs, 'End of the coal strike', *Economic and Political Weekly*, xx (1985), 443–4.

[4] Scott McCallum, interview with author, The Counting House, Dundee, 22 Feb. 2014.

[5] S. High and D. Lewis, *Corporate Wasteland: the Legacy and Memory of Deindustrialization* (New York, 2009), pp. 9–11.

[6] J. Clarke, 'Closing Moulinex: thoughts on the visibility and invisibility of industrial labour in contemporary France', *Modern and Contemporary France*, xix (2011), 443–58, at p. 444.

[7] T. Strangleman, 'Deindustrialisation and the historical sociological imagination: making sense of work and industrial change', *Sociology*, li (2017), 466–82, at p. 466; E. P. Thompson, *The Making of the English Working Class* (Middlesex, 1968) p. 9.

[8] A. Clark and E. Gibbs, 'Voices of social dislocation, lost work and economic

deindustrialization is a temporal challenge to historians. It requires analysis of a *longue durée* structural change that evolved over more than half a century, without losing empathy for the strongly ingrained cultural practices that instilled industrial activities with a sense of durability. Coalfields are in some senses an extreme example given the tendency for paternal inheritance of occupational identities and the trend for coal mining settlements to be strongly dependent on the industry. Generational cohorts are an effective unit of analysis which reveal how distinct formative experiences and labour market conditions shaped different political and cultural outlooks. Comparatively short and episodic events related to the socioeconomic climate and experiences of industrial relations provided important reference points that were retained across lifetimes. But successive generations were also strongly shaped by intergenerational processes of socialization. A strong sense of historical and familial attachment was formative to Scottish coalfield politics and underpinned moral economy perspectives.

Deindustrialization has been a central feature of the Scottish coalfields since the mid twentieth century. It was initially experienced through local crises, especially in the Shotts area in eastern Lanarkshire. These contested experiences of community abandonment contributed to a growing sense of nationwide coal crisis during the late 1950s and early 1960s. This developed through the onset of falling coal employment levels and the relationship between coal production and power station investment. Responses to deindustrialization were historically conditioned by the collective memory of industrialization and the political culture of the coalfields that developed through the industry's nineteenth-century expansion and the conflicts of the interwar years. Objections to deindustrialization hinged on the juxtaposition of communitarian interest with the dictates of market logic. Michael McGahey summarized his understanding of the forces that determined colliery closure in characteristic humanist terms within the pages of *Scottish Marxist* during the early 1970s: 'Inanimate matter does not compete – it is the pecuniary forces behind various fuels which are in competition'.[9]

McGahey's emphasis on resistance to commodification of coal and coal miners' labour exemplifies the conflict between market logic and a protective countermovement that shaped the double movement of deindustrialization in postwar Scotland. While McGahey was writing in the pages of a

---

restructuring: narratives from marginalised localities in the 'new Scotland', *Memory Studies*, xiii (2020), 39–59.

[9] M. McGahey, 'The coal industry and the miners', *Scottish Marxist*, i (1972), 16–19, at p. 16.

Communist Party of Great Britain (CPGB) journal on this occasion, the NUM Scottish Area (NUMSA) leadership was able to speak for a broad coalition when it articulated moral economy sensibilities. This reflected their foundations in historical experience. Within the testimonies collected for this book, former miners with anti-communist views and former colliery managerial staff concurred with the perspective of communists over the nationalized industry's obligations towards coalfield communities. These perspectives were often rooted in an understanding of the abuses committed under private ownership. The moral economy centred on a view that collieries and the employment they sustained were a collective resource. Those inclinations are far from uniquely Scottish and follow an international pattern. When Scottish miners commenced stay-down strikes to oppose colliery closures at Devon, Alloa, in 1959 and at Kinneil, West Lothian, during 1982 they deployed similar protest repertoires to those used in South Wales in the 1930s and in Occitanie, western France, during the 1960s.[10] In all of these examples, pit closures within areas marked by the long-term rundown in colliery employment were met by communal assertions of rights to economic resources. Collieries were understood as sources of wealth and nourishment which had been paid for at a high cost of deaths and injuries.

Although these commonalities are important in theorizing deindustrialization, this volume has also analysed the importance of historically specific societal and political contexts. The translation of localized developments onto a national plain, and the form of politicization that deindustrialization provokes, can both be relatively malleable. As prolonged contraction hit the Scottish coalfields, responses to closure became enmeshed with views of Scotland's distinct interests within the Union and debates over the constitution. Through the relatively decentralized NCB before 1967, the Scottish Office, and the South of Scotland Electricity Board, as well as the institutional voices of labour and business, the Scottish Trades Union Congress and Scottish Council (Development and Industry), there were significant devolved dimensions to these deliberations. One major outcome of accelerated coalfield contraction under an increasingly centralized NCB during the 1960s was that the NUMSA's support for home rule became more pronounced.

As events in Springburn demonstrated, deindustrialization continues to be mediated by national framings. During the 2010s, Scottish workers

[10] D. Reid, *The Miners of Decazeville: a Genealogy of Deindustrialization* (Cambridge, Mass., 1985), p. 204; J. Jenkins, 'Hands not wanted: closure, and the moral economy of protest, Treorchy, South Wales', *Historical Studies in Industrial Relations*, xxxviii (2017), 1–36, at pp. 3–7.

continued articulating their interests in a moral economy language directed at policymakers and governments in Edinburgh and London. The same week as the workers left the Caley for the last time, an argument about the (mis)use of natural resources brewed elsewhere within Scottish industry on terms which were strongly redolent of the coalfield moral economy. Pat Rafferty and Gary Smith, Scottish secretaries of the Unite and GMB unions respectively, denounced the 'paltry return' of only eight jacket sleeves awarded to the BiFab fabrication yards from the large Neart Na Gaoithe offshore wind development ten miles from the Fife coast. BiFab's largest yard in Methil, eastern Fife, sits atop the former site of Wellesley colliery that closed in 1967. The area has experienced considerable job losses in mining, manufacturing, shipbuilding and dock work in recent decades.[11]

The logic deployed by Rafferty and Smith has strong parallels with arguments made for power station investment at Longannet analysed in chapter one. The building of Longannet and the drift mines which supply it secured work for Fife miners half a century ago, persisting into the early twenty first century. During the 1960s, coal's competitors were alternative fuel sources including imported oil. In the context of renewables, it is Scotland's place in multinational supply chains which is the height of controversy. However, in each case, workers, communities and trade union representatives have argued for the exploitation of natural resources in a manner which develops local employment and economic security. Alongside the importance of national sovereignty over resource deposits, labour movement arguments have also rested on the central role of public investment. This was embodied by the nationalized coal industry and power generators during the 1960s, and by the occupiers of Caterpillar's factory in Tannochside in 1987 who stated that closure was an injustice in the context of the regional assistance payments that their plant had received. Contemporary arguments for economic justice in renewables manufacturing relate to the significant public subsidies provided for renewables as well as the Scottish Government support given to BiFab's Canadian owners.[12] The reference points of coal miners therefore remain highly relevant to discussion over the meaning of an environmentally and socially 'just transition'.[13]

---

[11] B. Wray, '"A paltry return": unions criticise reported 200 NnG manufacturing jobs for BiFab', *Common Space*, 24 July 2019 <https://www.commonspace.scot/articles/14522/paltry-return-unions-criticise-reported-200-nng-manufacturing-jobs-bifab> [accessed 23 Dec. 2019].

[12] Wray, '"A paltry return"'.

[13] Petrocultures Research Group, *After Oil* (2016) <http://afteroil.ca/wp-content/uploads/2016/02/AfterOil_fulldocument.pdf> [accessed 23 Dec, 2019].

A sense of finality is key to explaining the sense of loss associated with coalfield deindustrialization. Narratives are strongly conditioned by the way in which colliery closures were managed and the economic cost to workers and communities. The less tangible cultural cost of removal from heritage and a strongly invested occupational identity also strongly colours accounts of closure and job loss. Mick McGahey's testimony was characterized by a strong familial embeddedness in the industry. Mick recalled his grandfather taking him for walks in Lanarkshire when he was a young boy and telling him about collieries that he had worked in with Mick's father.[14] Miners' sense of a lost occupational working culture is often accompanied by an understanding of the role their industry played in transforming Scotland, Britain and the international economy. Sam Purdie underlined miners' contribution by recalling the industrial innovations that had been pioneered in the area surrounding the now depopulated mining village of Glenbuck in Ayrshire where he grew up:

> Because you've lost it. If you think about the history we've lost. If you look at places like Lugar, there's hardly a mention of people like Murdoch who I consider one of the founding fathers of the petrochemical industry. Nut. McAdam, who's got a cairn somewhere on the outside of Muirkirk. And the Katrine mills were started in conjunction with Robert Owen. They're important. Well I think the miners are important. If it hadn't been for us there wouldn't have been an industrial revolution. It took coal and iron and the labour force to make the industrial revolution happen. Are we just gonna forget that? I don't think so. I don't think so.[15]

Sam's rhetorical framing, listing the miners' contribution after innovative industrialists and their endeavours, indicates their central importance, but also underlines a fear of being forgotten along with Glenbuck itself. These comments underline the importance of esteem and the continuity of a struggle for recognition which is central to coalfield politics, history and memory.[16] Drawing a balance sheet of this nature necessarily entails a form of critical nostalgia that can both identify the real losses associated with deindustrialization but also critique the past and right historical wrongs. Alongside noting coal miners' achievements in industrial, social and democratic advancement, the testimonies underlined the dangers of the industry and ambivalence towards its health effects, as well as painful memories of victimization and displacement. Narratives of loss and removal

[14] Mick McGahey, interview with author, Royal Edinburgh Hospital, 31 March 2014.

[15] Sam Purdie, interview with author, UWS Hamilton campus, 3 May 2018.

[16] A. Honneth, *The Struggle for Recognition: the Moral Grammar of Social Conflicts* (Cambridge, Mass., 1995), p. 113.

are given further prescience by the sense of continuing injustices associated with colliery closures. Scottish coalfield history shares an affinity with American miners for whom 'memory, is indeed the final site of conflict'.[17]

During 2018, the Scottish Government began a review into the policing of the 1984–5 miners' strike.[18] This undertaking was the outcome of a decades-long campaign fought by victimized miners after the strike concluded in March 1985.[19] The review included a series of evidence-gathering meetings in coalfield locations which were attended by local miners. At these meetings, witnesses recalled individual and collective abuses including unjust arrests and the targeted victimization and sacking of union activists. These memories were accompanied by reflections on the purpose of the government's prosecution of the strike which was rationalized in terms of the disposal of viable economic assets and the undermining of the trade union voice. The Lanarkshire meeting was held at the Auchengeich Miners' Club on 6 December 2018. Several former strikers gave evidence relating to experiences of policing during the strike and underlined the social and economic costs incurred by victimized men in navigating an increasingly adverse labour market after the dispute's conclusion.[20]

A devolved Scottish government reviewing the injustices associated with coal industry conflicts can be understood as a success for the politics which developed through the long experience of deindustrialization. In an emotional tribute to Michael McGahey, written in 2002, John McAllion, who was then the Labour MSP for Dundee East, attributed the achievement of a Scottish parliament in no small part to his efforts.[21] As this volume has demonstrated, McGahey's contribution was highly significant. However, the expression of Scottish national discontent with deindustrialization – and the resultant emboldening of commitments to Scottish political autonomy – originated under an earlier generation of communist miners' leaders. Politics in Scotland under devolution bares the mark of these origins. This was personified by

---

[17] A. Portelli, *They Say in Harlan County: an Oral History* (Oxford 2010), p. 192.

[18] 'Policing during miners' strike: independent review', *Scottish Government* (2018) <https://www.gov.scot/groups/independent-review-policing-miners-strike/> [accessed 23 Dec. 2019].

[19] S. McGrail and V. Paterson, *Cowie Miners, Polmaise Colliery and the 1984–85 Miners' Strike* (Glasgow, 2017).

[20] Observation notes from meeting at Auchengeich Miners' Club, 6 Dec. 2018; J. Phillips 'Containing, isolating and defeating the miners: the UK cabinet ministerial group on coal and the three phases of the 1984–85 strike', *Historical Studies in Industrial Relations*, xxxv (2014), 117–41.

[21] J. McAllion, 'Rose like a lion', *Scottish Review* (2002) <http://www.scottishreview.net/JohnMcAllion422a.html> [accessed 23 Dec. 2019].

Nicola Sturgeon's response to the death of Margaret Thatcher in 2013. The deputy first minister at the time recalled, 'The brutal deindustrialization she presided over, and the unemployment that resulted, [which] has left deep scars in every community in this city [Glasgow] and across Scotland'.[22] Interpretations of deindustrialization and coalfield politics in Scotland are not homogeneous. Politically influential accounts have tended towards an understanding of accelerated closures and falling industrial employment as 'an external attack' imposed on Scotland from London, ignoring dynamics of class conflict within Scotland itself. However, the influence of the moral economy's stress on government responsibility towards communities is also evident. This has played a crucial role in shaping a 'social justice' discourse of collective partnership and a shared national interest which predominates in contemporary Scottish politics.[23]

Experiences of job losses in Springburn and precariousness in Methil demonstrate that deindustrialization is not confined to the past. The dynamics of the double movement continue to unfold in Scotland. Renewing industrial employment and embedding production within communitarian concerns has been given heightened importance by the context of climate change and the urgent necessity behind another major transition in energy generation. In this context, political outlooks grounded in the sensibilities of industrial society remain central. Conceptions of justice which understand natural resources and employment as community resources imbued with significance for national sovereignty continue to generate resistance to the logic of capital accumulation. Deindustrialization is a long process, with profound impacts that span across generations. Cultural associations with origins in coal mining, including familial and local identities and the memorialization and commemoration of industrial history, remain present across much of central Scotland. Less overt but nevertheless profound implications of industrialization in patterns of settlement also reverberate, including in the marked socioeconomic disadvantages and health inequalities experienced within deindustrialized areas.[24] The legacy of employment in Scotland's coalfields, and the memories of the closure of collieries, steel mills and factories are formative to the political and economic situation in the 2020s.

[22] 'Thatcher remembered: the Scottish Nationalist, Nicola Sturgeon', *STV* (2013) <http://stv.tv/news/west-central/221629-margaret-thatcher-remembered-by-nicola-sturgeon-msp-for-political-division/> [accessed 23 Dec. 2019].

[23] J. Phillips, 'Contested memories: the Scottish parliament and the 1984–5 miners' strike', *Scottish Affairs*, xxvi (2015), 187–206 ; E. Gibbs, '"Civic Scotland" versus communities on Clydeside: poll tax non-payment, *c.* 1987–1990', *Scottish Labour History*, xlix (2014), 86–106.

[24] C. Beatty, S. Fothergill and T. Gore, *The State of the Coalfields 2019: Economic and Social Conditions in the Former Coalfields of England, Scotland and Wales* (Sheffield, 2019), pp. 5–7.

Appendix: biographies of oral history participants

Alan Blades grew up during the 1960s and 1970s in Greengairs, a mining village adjacent to Airdrie in North Lanarkshire. Alan worked at Bedlay colliery between 1979 and 1982, following in his father's and brother's footsteps. He transferred to Solsgirth upon closure and was a participant in the 1984–5 strike. After taking redundancy in 1997, Alan later worked at the Chungwha factory at the Eurocentral industrial estate near Airdrie, where he now lives.

John Brannan was born in 1948 and grew up in Viewpark, North Lanarkshire. His father worked as a miner in the Lanarkshire coalfield before later taking a job at the Tunnock's factory in Uddingston. John started work at the Caterpillar factory as an eighteen-year-old, and subsequently became the engineering union covenor at the plant. He was involved in several major strikes at the plant before leading the occupation against its closure during 1987. John latterly entered the building trade and was also a leading member of the Caterpillar Workers Legacy Group which marked the occupation's thirtieth anniversary.

Tommy Canavan grew up in a mining family in Croy, North Lanarkshire, between the late 1940s and mid 1960s. He followed his father and grandfathers into the mining industry by entering Cardowan colliery during the 1960s and subsequently transferred to Solsgirth, Clackmannan, following Carodwan's closure in 1983. Tommy was a National Union of Mineworkers (NUM) representative at Cardowan and was active in Lanarkshire during the 1984–5 strike.

Jessie Clark grew up in Douglas Water, South Lanarkshire, during the 1920s and 1930s. Her father was a victimized miner, and a member of the Independent Labour party and later a Communist. Jessie shared his political convictions and went on to marry another Communist miner, Alex Clark. She was first employed in domestic service during the 1930s before taking a job in the Douglas Castle colliery canteen during the 1940s. Jessie subsequently worked in local government.

Gilbert Dobby was born in Coalburn in 1946 where he grew up in a mining family. He entered local collieries as an apprentice engineer during the 1960s before transferring to Nottinghamshire after major closures in the South

'Appendix: biographies of oral history participants', in E. Gibbs, *Coal Country: The Meaning and Memory of Deindustrialization in Postwar Scotland* (London, 2021), pp. 259–65. License: CC-BY-NC-ND 4.0.

Lanarkshire coalfield during the late 1960s. Gilbert subsequently returned to Scotland during the early 1970s, working at collieries in Ayrshire, Lanarkshire and West Lothian. Gilbert took redundancy after Polkemmet was closed after it flooded during the 1984–5 miners' strike, which he supported throughout the year. Gilbert subsequently became a driving instructor and then a school caretaker before retiring.

Willie Doolan grew up in a mining family in Moodiesburn during the 1950s and 1960s. He began work at Cardowan colliery during the early 1970s where he was active in both the NUM and the Communist Party of Great Britain (CPGB). After Cardowan closed he transferred to the Longannet complex. Willie has continued his connection with coalfield culture through the Moodiesburn Miners Memorial Committee's efforts to commemorate the memory of the Auchengeich pit disaster of 1959. In 2017 he was elected as a Labour councillor for Gartcosh, Glenboig and Moodiesburn.

Pat Egan grew up in a mining family in Twechar, North Lanarkshire, during the 1960s and 1970s. He followed in his father's footsteps by entering Bedlay colliery in the late 1970s. After Bedlay closed, Pat transferred to the Longannet complex in Fife, and later moved to Glenrothes, but he was also active in Lanarkshire during the 1984–5 strike. Pat then worked at the Longannet complex until Castlebridge colliery closed in 2002. He is now employed by Unite the Union.

Billy Ferns was born in 1936 and grew up in Glasgow. His father worked at Cardowan colliery where Billy later found work. Billy moved into National Coal Board (NCB) housing in Bishopbriggs and was a highly active picketer during the 1984–5 dispute. He transferred to the Longannet complex after Cardowan closed, before retiring after taking redundancy during the late 1980s.

Barbara Goldie grew up in Cambuslang during the 1930s and 1940s. Her father worked in local collieries and was an active trade unionist. Barbara's sister and two brothers worked at the local Hoover factory. Barbara had a range of jobs including work at the Templeton's carpet factory in Glasgow.

George Greenshields grew up in Coalburn during the 1960s and 1970s. His father and brothers had worked in local collieries and continued to commute to collieries further afield after the final closures in the area during the late 1960s. George worked at the large opencast site at Dalquhandy from 1978 until the early 1990s. He is now a Labour councillor for Clydesdale South.

John Hamilton was born in Kirkmuirhill, South Lanarkshire in 1949. He grew up in a mining family in Lesmehagow and entered local collieries during the mid 1960s. John subsequently migrated to Canada in 1969 but returned to Lanarkshire during the early 1970s. He re-entered mining in 1980 when he started work at Bedlay and then transferred to Polkemmet in West Lothian but left the coal industry before the 1984–5 strike.

Marian Hamilton grew up in Shotts during the 1940s and 1950s. Her father was an iron moulder, and her grandfather had been a miner. She worked at the Hartwood hospital and married Willie Hamilton, a local miner. They briefly migrated to Windsor, West London, during the 1960s but subsequently returned to Shotts.

Willie Hamilton grew up in Shotts between the mid 1930s and early 1950s. He entered the coal mining industry, following his father and grandfather, working in the Shotts area at Stane colliery and then Kingshill 3 before transferring to Polemmet in 1974. He was trained as a shotfirer and subsequently rose to the rank of overman. Willie interspersed his coal mining work with jobs in manufacturing and eventually left the coal mining industry during the late 1970s.

Ian Hogarth was born in Springboig, Glasgow, in 1928 where he grew up adjacent to the Lanarkshire coalfield. His father was an accountant for Bairds and Scottish Steel. Ian entered the NCB's management training during the early 1950s. He was the ventilation officer at Cardowan before being made responsible for ventilation across the Central West Area. In 1959, Ian transferred to the NCB's Scottish headquarters at Green Park in Edinburgh where he remained until retiring in 1987.

Jimmy Hood was born in Lesmahagow, South Lanarkshire, during 1948. His family on both sides were strongly connected to the mining industry. Jimmy trained as an engineer in local collieries before transferring to Nottingham following the closure of Auchlochan 9 in 1968. Jimmy was a leader among the minority of striking miners in Nottinghamshire during the 1984–5 strike. He subsequently returned to stand as the Labour candidate for Clydesdale in 1987 and remained the local MP until 2015. Jimmy died after suffering a heart attack in December 2017.

John Kay was born in Glasgow in 1925 and worked in engineering factories before migrating to New Zealand, where he lived from 1949 to 1957. He joined the Communist Party there and became increasingly active upon his return to Glasgow. John became a full-time organizer for the CPGB during the 1960s, first as Glasgow secretary and then as Scottish industrial organizer, which was a post he retained until retiring in 1990. John died following illness in November 2019.

Margaret Keena grew up in Newton Rows, Cambuslang, South Lanarkshire, during the 1930s and 1940s. Her father was a miner at Newton colliery as were many of her other male relatives. Margaret's sister went on to work at the Hoover factory, while Margaret was employed in a range of jobs which included working in a textile factory and as a bus conductor.

Duncan Macleod's father and both grandfathers were miners from the Carluke area of South Lanarkshire. His father subsequently moved to Derbyshire in order to join the police force. Duncan was born there in 1953. The family later moved back to Carluke in the 1970s and Duncan stayed in the town while working as a telecommunications engineer before retiring.

Marian Macleod is from Law near Carluke, South Lanarkshire, and still lives in the Carluke area. Her grandfather worked at local collieries before major closures affected the area during the 1950s. Marian found work at Honeywell's factory in Newhouse before she later became a purchasing director for Motherwell Bridge.

Peter Mansell-Mullen grew up in the south of England during the 1930s and 1940s. After graduating from Oxford with a PPE degree, Peter joined the NCB as a manager during the early 1950s, where he met his wife, who was also a management trainee at the time. Peter trained in Nottinghamshire and then became an NCB Area secretary in Cannock, Staffordshire. Peter later moved to NCB headquarters at Hobart House, London, where he became director of manpower.

Jennifer McCarey grew up in Mossend, North Lanarkshire during the 1970s and 1980s. Her mother and father were both active trade unionists. Jennifer's father was the convenor for non-manual workers at Ravenscraig, having followed his father into the industry. Jennifer was an active Labour party member and supporter of the 1984–5 miners' strike. She subsequently became a professional trade union organizer in Birmingham during the late 1980s. Jennifer now works for Unison in Scotland.

Bill McCabe was born in Viewpark, North Lanarkshire during the early 1960s. His grandfather had been a coal miner in the area while his father and brother both worked at the nearby Caterpillar factory in Tannochside. Bill followed them into the factory during the 1980s and became a shop steward before being highly involved in the occupation against its closure. Bill latterly worked in North Sea oil and then insurance but has remained a committed trade unionist.

Scott McCallum grew up in a mining family in Cardowan village. His great-grandfather, grandfather, father and brother were all coal miners. Scott was at primary school during the 1984–5 strike and was taken on demonstrations with his family. He subsequently became a joiner in the Stepps area before moving to Dundee.

Mick McGahey was a third-generation communist miner. His grandfather was jailed for his activities during the 1926 general strike and miners' lockout and forced out of the Lanarkshire coalfield, but returned during the 1930s. Mick's father became president of the NUM Scottish Area in 1967, which led the family to relocate to Liberton on the outskirts of Edinburgh. Mick worked at Bilston Glen colliery in Midlothian from the early 1970s until he was victimized following his arrest during the 1984–5 strike. He now works at the Edinburgh Royal Hospital where he is the Unison convenor.

Michael McMahon grew up in Newarthill, North Lanarkshire. His grandfather was a miner, while his father was employed at the Terex factory in Holytown where Michael also worked as a welder. Michael became an active trade unionist at the factory and was chair of the Scottish Trades Union Congress (STUC) youth committee during the mid 1980s. He also joined the Labour party. Michael left Terex to study politics and sociology at Glasgow Caledonian University during the 1990s and and represented Hamilton North and Bellshill and then Uddingston and Bellshill as an MSP between 1999 and 2016.

Siobhan McMahon grew up in Bellshill, North Lanarkshire, during the 1980s and 1990s. Her father and paternal grandfather both worked at the Terex factory in Holytown while her maternal grandfather was a miner. Siobhan worked for her father as an MSP's researcher before becoming a Labour Central Scotland list MSP herself between 2011 and 2016.

Angela Moohan grew up in Livingston in West Lothian. She joined the Labour party Young Socialists and the Militant tendency as a seventeen-year-old during 1984 and subsequently became involved in miners' strike solidarity activities. Angela has retained a strong involvement with the labour movement and later married and started a family with Brendan, a striking miner she met through the strike.

Brendan Moohan grew up in Musselburgh, East Lothian during the 1960s and 1970s. His grandfather had been a Communist activist who was blacklisted out of the Lanarkshire coalfields and migrated across central Scotland following the 1926 general strike and miners' lockout. Brendan followed his father into employment at Monktonhall colliery in Midlothian. He was arrested during the 1984–5 strike and then sacked. Brendan subsequently studied community education at Edinburgh University and is now a youth worker with West Lothian council. He was active in the Militant tendency during the 1980s and remains a member of the Labour party.

Sam Purdie was born in the Ayrshire mining village of Glenbuck in 1936. His family moved to Muirkirk when Glenbuck was depopulated in 1954. Sam worked as an engineer at Kames colliery and then Cairnhill mine before leaving mining to study at Ruskin College, Oxford. He subsequently worked for Marathon, which built oil rigs at Clydebank, and then Bechtel internationally. Sam left the Labour party for the Scottish National Party and stood for the Nationalists in South Ayrshire during the 1970 by-election and subsequently general election. Sam is presently involved in campaigning to preserve the memory of Glenbuck.

Anthony Rooney was born in Bellshill in 1938 where he grew up in miners' rows. His father and both grandfathers worked in local collieries. Anthony found work at the Caterpillar factory in Tannochside where he was a shop steward. He has also been a longstanding Labour activist in the Bellshill area of North Lanarkshire.

John Slaven grew up in Birkenshaw, North Lanarkshire, between the mid 1960s and mid 1980s. His parents had moved from Glasgow to Tannochside where his father took a job at the Caterpillar factory, which opened in 1959. John's mother also subsequently found a job at the factory where she was involved in the 1987 occupation against its closure, which took place after his father had taken redundancy in the early 1980s. John had been involved in the local Labour party but left Lanarkshire for London after leaving school in 1985. While in London, he took a job on the railways and became an active trade unionist. He now works for the STUC in Glasgow.

Mary Spence was born in Hamilton in 1944 but moved to Hampshire shortly afterwards. She returned to Lanarkshire with her father in 1959. Mary's father came from a coal mining family but had entered the civil service. Mary felt that this established a large social distinction and created tension with her grandmother who looked after her father's father and brothers when they suffered from coal mining related illnesses. Mary later moved to East Kilbride where she worked as a teacher.

Margaret Wegg was born in 1941. She grew up in Cardowan village in Scottish Special Housing Association housing, which was secured through her father's employment at Cardowan colliery. Margaret had several jobs including working as a typist before starting work at the Cardowan colliery canteen. She was made redundant when the pit closed in 1983 and subsequently became a leading Women Against Pit Closures activist in the area during the 1984–5 miners' strike. Her husband Jerry was a miner at Cardowan colliery, and he subsequently transferred to Castlebridge colliery in Clackmannan.

Appendix: biographies of oral history participants

Rhona Wilkinson was born in 1968 and is from a mining family background in Breich, West Lothian. Her paternal grandfather worked at local collieries in the West Lothian area. Rhona's mother's family were also of local mining heritage. Her father worked as an engineer in local foundries. Rhona works in the public sector and lives in Fauldhouse.

Nicky Wilson grew up in Easterhouse on the eastern outskirts of Glasgow during the 1950s and 1960s. He entered Cardowan colliery as an apprentice electrician during the mid 1960s and subsequently became a Scottish Colliery, Enginemen, Boilermen and Tradesmen's Association (SCEBTA) representative at the pit. Nicky transferred to the Longannet complex after Cardowan closed and was active during the 1984–5 miners' strike. He is now the NUM's Scottish president.

Bibliography

Observation notes

Observation notes from Auchengeich colliery memorial service, 16 Sept. 2018.

Observation notes from Auchengeich colliery memorial service, 15 Sept. 2019.

Observation notes from meeting at Auchengeich Miners' Club, 6 Dec. 2018.

Oral history interviews

Alan Blades, interview with author, residence, Airdrie, 26 Feb. 2014.

John Brannan, interview with author, UWS Hamilton campus, 21 Feb. 2017.

Tommy Canavan, interview with author, residence, Kilsyth, 19 Feb. 2014.

Jessie Clark, interview with author, residence, Broddock, 22 March 2014.

Gilbert Dobby, interview with author, Coalburn Miners' Welfare, 11 Feb. 2014.

Willie Doolan, interview with author, The Pivot Community Centre, Moodiesburn, 12 March 2014.

Willie Doolan, interview with author, The Pivot Community Centre, Moodiesburn, 14 June 2019.

Pat Egan, interview with author, Fife College, Glenrothes, 5 Feb. 2014.

Billy Ferns, interview with author, residence, Bishopbriggs, 17 March 2014.

Barbara Goldie and Margaret Keena, interview with author, Whitehall Bowling Club, Cambuslang, 8 Dec. 2014.

George Greenshields, interview with author, Coalburn Miners' Welfare, 11 Feb. 2014.

John Hamilton, interview with author, South Lanarkshire Council Integrated Children's Services office, Larkhall, 26 Apr. 2016.

'Bibliography', in E. Gibbs, *Coal Country: The Meaning and Memory of Deindustrialization in Postwar Scotland* (London, 2021), pp. 267–92. License: CC-BY-NC-ND 4.0.

Marian and Willie Hamilton, interview with author, residence, Shotts, 19 March 2014.

Ian Hogarth, interview with author, National Mining Museum, Newtongrange, 28 Aug. 2014.

John Kay, interview with author, residence, Bishopbriggs, 11 Aug. 2014.

Duncan and Marian Macleod, interview with author, residence, Carluke, 1 March 2014.

Peter Mansell-Mullen, interview with author, residence, Strathaven, 3 Oct. 2014.

Bill McCabe, interview with author, Tannochside Miners' Welfare, 20 Jan. 2017.

Scott McCallum, interview with author, The Counting House, Dundee, 22 Feb. 2014.

Jennifer McCarey, interview with author, iCafe, Woodlands, Glasgow, 9 Oct. 2014.

Mick McGahey, interview with author, Royal Edinburgh Hospital, 31 March 2014.

Michael McMahon, interview with author, constituency office, Bellshill, 21 Feb. 2014.

Siobhan McMahon, interview with author, Central Scotland Regional List MSPs Office, Coatbridge, 28 March 2014.

Moodiesburn focus group, retired miners' group, The Pivot Community Centre, Moodiesburn, 25 March 2014.

Angela Moohan, interview with author, residence, Livingston, 5 Feb. 2015.

Brendan Moohan, interview with author, residence, Livingston, 5 Feb. 2015.

Sam Purdie, interview with author, UWS Hamilton campus, 3 May 2018.

Anthony Rooney, interview with author, Morrisons café, Bellshill, 24 Apr. 2014.

Shotts focus group, Shotts history group, including former miners and respondents from mining family backgrounds, Nithsdale Sheltered Housing Complex, Shotts, 4 March 2014.

John Slaven, interview with author, STUC Building Woodlands, Glasgow, 5 June 2014.

Mary Spence, interview with author, The Terraces café, Olympia shopping centre, East Kilbride, 11 Aug. 2014.

Margaret Wegg, interview with author, residence, Stepps, 17 Nov. 2014.

Rhona Wilkinson, interview with author, residence, Fauldhouse, 7 Nov. 2014.

Nicky Wilson, interview with author, John Macintyre Building, University of Glasgow, 10 Feb. 2014.

Manuscripts

National Mining Museum Scotland

National Union of Mineworkers Scottish Area

Executive Committee Minutes, July 1982 to June 1983.

Minutes of Executive Committee and Special Conferences, 8 July 1946 to 11 June 1947.

Minutes of Executive Committee and Special Conferences, 23 June 1947 to 8 June 1948.

Minutes of Executive Committee and Special Conferences, 20 June 1949 to 2 June 1950.

Minutes of Executive Committee and Special Conferences, 18 June 1951 to 20 June 1952.

Minutes of Executive Committee and Special Conferences from 18 June 1956 to 5 to 7 June 1957.

Minutes of Executive Committee and Special Conferences, 12 June 1961 to 6/8 June 1962.

Minutes of Executive Committee and Special Conferences from 27 June 1966 to 14/16 June 1967.

Minutes of Executive Committee and Special Conferences from 24 June 1968 to 18/20 June 1969.

Minutes of Executive Committee and Special Conferences, June 1969 to 15/16 June 1970.

Minutes of Executive Committee and Special Conferences from 28 June 1971 to 14/16 June 1972.

Minutes of Executive Committee and Special Conferences, 27 June 1977 to 14/16 June 1978.

Closure records
NMMS, FC/3/2/3/2 Cardowan

National Records of Scotland
National Coal Board (all references proceeded by CB)
207/14/3 Wester Auchengeich
207/14/4 Wester Auchengeich
207/14/5 Auchincruive
207/24/1 Auchengeich
210/14/3 Auchlochan
210/25/1 Auldton
222/14/1 Baton
223/14/3 Bedlay
256/14/1 Cardowan
256/33/2 Cardowan
280/30/1 Douglas Castle
298/6/1 Garscube
295/14/1 Fortisat
300/14/1 Gartshore 9/11
300/14/2 Gartshore 9/11
321/14/1 Hillhouserigg
327/14/1 Kennox
334/19/2 Kingshill 1
334/19/3 Kingshill 3
410/14/1 Stane
483/24/1 Broomside

Scottish Economic Planning (all references proceeded by SEP)
4/13 Investigation of sites for individual firms: Hoover Ltd
4/567 Individual Areas LO area file: East Kilbride
4/568 Individual areas: East Kilbride
4/690 Distribution of industry
4/762 Research studies
4/781 Statistics and records
4/784 Statistics and records
4/1199 Unemployment in Scotland
4/1629 Location of Industry, Lanark County: Honeywell Controls Ltd

4/2337 Electronics industry
4/3550 Unemployment and redundancies
4/3791 Burroughs Corporation, Detroit, USA
4/4070 Burroughs Machines Ltd, Cumbernauld
4/4251 Individual areas: Cumbernauld
15/437 East Kilbride New Town
17/56 Scottish Economic Planning Board
17/70 Central Scotland Growth Areas

The National Archives
National Coal Board (all files preceded by Coal)
30/629 Report on Britain's energy supply
31/96 Closure of collieries: Scotland
31/120 Chairman's office
31/123 Fuel policy: nuclear power
31/130 Social Costs
31/135 Plan for Coal
31/138 Long-term inquiry
31/166 Coal industry examination: tripartite
31/168 Coal Industry Examination Working Party on Supply and Demand
31/433 Plan for Coal: background information
74/1287 Headquarters
101/488 Scottish manpower
101/580 Operational Research Executive
Ministry of Fuel and Power and successor departments (all files preceded by POWE)
14/857 Fuel supply to the British Electricity Authority: policy on coal
14/1495 Fuel supplies to the Central Electricity Generating Board
14/2501 Alternatives to oil dependency
33/2156 Coal/Oil Conversion
37/481 NCB Reorganisation and Development Programmes: Closure of High Cost Pits-1959 Scotland
52/17 Choice of fuel for the next Scottish Power Station after Cockenzie to be built at Longannet

52/85 Colliery closure programme for Scotland, Aug. 1967–March 1969
52/278 Brief for the minister of power's meeting with the STUC
52/305 Colliery closure programme for Scotland, Aug. 1967 onwards
Prime Minister's Office (all files preceded by PREM)
13/1610 Manpower: appeal to slowdown colliery closures
15/1144 Power, 1971–2

Official publications

Abercrombie, P. and R. H. Matthew, *The Clyde Valley Regional Plan, 1946* (Edinburgh, 1949).

Cardowan Colliery (Accident) HC Deb 27 Jan. 1982 vol.16 cc889-91, *Hansard* <http://hansard.millbanksystems.com/commons/1982/jan/27/cardowan-colliery-accident> [accessed 2 Dec. 209].

Census 1951 Scotland vol. iv: Occupation and Industries (Edinburgh, 1956).

Census 1961 Scotland Occupation and Industry County Tables: Glasgow and Lanark, Leaflet no. 15 (Edinburgh, 1966).

Census 1971 Scotland Economic Activity County Tables part 2 (Edinburgh, 1976).

Census 1981 Scotland Economic Activity 10%: Strathclyde Region (microfiche, Edinburgh, 1983).

Electricity in Scotland: Report of the Committee on the Generation and Distribution of Electricity in Scotland (Parl. Papers 1962 [Cmnd. 1859]).

Fuel Policy (Parl. Papers 1967 [Cmnd. 3438]).

NCB Scottish Division, *Scotland's Coal Plan* (Edinburgh, 1955).

Sample Census 1966 Scotland Economic Activity County Tables Leaflet no. 3: Glasgow and Lanark (Edinburgh, 1968).

Business reports

Scottish Council Research Institute, *US Investment in Scotland* (Edinburgh, 1974).

Toothill, J. N., *Inquiry into the Scottish Economy, 1960–1961: Report of a Committee appointed by the Scottish Council (Development and Industry) under the Chairmanship of J. N. Toothill* (Edinburgh, 1961).

Trade union reports

STUC, *Annual Report 1950–1951*, liv (1951).

STUC, *Annual Report 1951–1952*, lv (1952).
STUC, *Annual Report 1952–1953*, lvi (1953).
STUC, *Annual Report 1957–1958*, lxi (1958).
STUC, *Annual Report 1967–1968*, lxxi (1968).
STUC, *Annual Report 1972–1973*, lxxvi (1973).
STUC, *Annual Report 1978–1979*, lxxxii (1979).

Films and documentaries

The Blackhill Campaign, J. Parsons, 1963, 50 mins <https://player.bfi.org.uk/free/film/watch-blackhill-campaign-1963-online> [accessed 22 Dec. 2019].

'Busby, Stein and Shankly: the football men', *Arena*, episode one, H. McIlvanney, BBC, UK, originally broadcast on 28 Mar. 1997, 55 mins.

Here We Go: Women Living the Strike, TV2day, M. Wright, 2009.

Newspapers and media

Aldridge, J., 'Labour to act over Monklands scandal', *Independent*, 2 July 1994 <http://www.independent.co.uk/news/uk/labour-to-act-over-monklands-council-scandal-mp-to-heal-wounds-caused-by-allegations-against-local-authority-john-arlidge-reports-1410968.html> [accessed 2 Dec. 2019].

'Bothwellhaugh ex-residents committee', *Bothwellhaugh* <http://www.bothwellhaugh.com/> [accessed 2 Dec. 2019].

'EU referendum results by region: Scotland', *Electoral Commission*, 29 Sept. 2019 <https://www.electoralcommission.org.uk/who-we-are-and-what-we-do/elections-and-referendums/past-elections-and-referendums/eu-referendum/results-and-turnout-eu-referendum/eu-referendum-results-region-scotland> [accessed 19 Nov. 2019].

'Future of Sunbeam plant in jeopardy', *Herald*, 15 Sept. 1978, p. 3.

Gibbs, E. 'The "Caley" and Scotland's "invisible workers"', *Conter*, 26 July 2019 <https://www.conter.co.uk/blog/2019/7/26/the-caley-and-scotlands-invisible-workers> [accessed 23 Dec. 2019].

Hoskyns, J. and N. Strauss, 'Stepping Stones' (1977), available via *Centre for Policy Studies* <https://www.cps.org.uk/files/reports/original/111026104730-5B6518B5823043FE9D7C54846CC7FE31.pdf> [accessed 21 Nov. 2019].

Kay, D., 'Liverpool legend Bill Shankly's spirit rekindled in village he first kicked a ball', *Liverpool Echo*, 3 Sept. 2019 <https://www.liverpoolecho.co.uk/sport/football/football-news/liverpool-most-important-date-history-16843384> [accessed 2 Dec. 2019].

'Last reminders of a dying bread', *Herald*, 31 May 1994 <https://www.heraldscotland.com/news/12694321.last-reminder-of-a-vanishing-breed/> [accessed 21 Dec. 2019].

Little, A. 'Scotland's decision', *BBC News*, 4 Sept. 2014 <https://www.bbc.co.uk/news/special/2014/newsspec_8699/> [accessed 1 Dec. 2019].

Lochhead, L., 'The newly wed miner' (2007), *Seamus Heaney Centre Digital Archive* <http://digitalcollections.qub.ac.uk/poetry/recordings/details/108096> [accessed 2 Dec. 2019].

'Obituary: Jimmy Hood', *Times*, 7 Dec. 2017 <https://www.thetimes.co.uk/article/obituary-jimmy-hood-rfgl6sml3> [accessed 22 Dec. 2019].

'Policing during miners' strike: independent review', *Scottish Government* (2018) <https://www.gov.scot/groups/independent-review-policing-miners-strike/> [accessed 23 Dec. 2019].

Reid, J., 'Mick McGahey', *Herald*, 2 Feb. 1999 <http://www.heraldscotland.com/sport/spl/aberdeen/mick-mcgahey-1.307647> [accessed 21 Dec. 2019].

Roberts, M., 'Annotated copy of *Employment Policy* (1944)', *Margaret Thatcher Foundation* <http://fc95d419f4478b3b6e5f3f71d0fe2b653c4f00f32175760e96e7.r87.cf1.rackcdn.com/2312B65342E04F2B8107131C635023BD.pdf> [accessed 22 Nov. 2019].

Robertson, R., 'Rob Roberston meets the union leader on the coal face to save Monktonhall', *Herald*, 16 May 1997 <http://www.heraldscotland.com/news/12324728.Rob_Robertson_meets_the_union_leader_working_at_the_coal_face_in_the_fight_to_save_Monktonhall_Digging_deep_for_survival/> [accessed 22 Nov. 2019].

Swarbrick, S., 'Breathing fresh life into the story of forgotten Lanarkshire mining village Bothwellhaugh', *Herald*, 11 Feb. 2017 <http://www.heraldscotland.com/life_style/pictures/15084711.display/> [accessed 2 Dec. 2019].

'Thatcher remembered: the Scottish Nationalist, Nicola Sturgeon', *STV*, 17 Apr. 2013 <http://stv.tv/news/west-central/221629-margaret-thatcher-remembered-by-nicola-sturgeon-msp-for-political-division/> [accessed 23 Dec. 2019].

'The struggle to build Vietnam', *Labour Herald*, 29 Apr. 1983, pp. 6–7.

Wray, B., '"A paltry return": unions criticise reported 200 NnG manufacturing jobs for BiFab', *Common Space*, 24 July 2019 < https://www.commonspace.scot/articles/14522/paltry-return-unions-criticise-reported-200-nng-manufacturing-jobs-bifab> [accessed 23 Dec. 2019].

Published secondary literature

Abrams, L., *Oral History Theory* (London, 2010).

Ackers, P., 'Review essay: life after death: mining history without a coal industry', *Historical Studies in Industrial Relations*, i (1996), 159–70.

— 'On paternalism: seven observations on the uses and abuses of the concept in industrial relations, past and present', *Historical Studies in Industrial Relations*, v (1998), 173–93.

— 'Gramsci at the miners' strike: remembering the 1984–1985 Eurocommunist alternative industrial strategy', *Labor History*, lv (2014), 151–72.

Allen, V., *The Militancy of British Miners* (Shipley, 1982).

— 'The year-long miners' strike, March 1984–March 1985: a memoir', *Industrial Relations Journal*, xl (2009), 278–91.

Anderson, B., *Imagined Communities: Reflections on the Origin and Spread of Nationalism* (London, 2009).

Andrews, G., *Endgames and New Times: the Final Years of British Communism, 1964–1991* (London, 2004).

Arnold, J., '"The death of sympathy": coal mining, workplace hazards, and the politics of risk in Britain, c. 1970–1990', *Historical Social Research*, xli (2016), 91–110.

— '"Like being on death row": Britain and the end of coal c. 1970 to the present', *Contemporary British History*, xxxii (2018), 1–32.

— '"That rather sinful city of London": the coal miner, the city and the country in the British cultural imagination, c. 1969–2014', *Urban History*, xlvii (2019), 292–310.

Arnot, R. P., *A History of the Scottish Miners: from the Earliest Times* (London, 1955).

Ashworth, W., *The History of the British Coal Industry*, v: *1946–1982: the Nationalized Industry* (Oxford, 1986).

Barron, H., 'Women of the Durham coalfield and their reaction to the 1926 miners' lockout', *Historical Studies in Industrial Relations*, xxii (2006), 53–83.

— *The 1926 Miners' Lockout: Meanings of Community in the Durham Coalfield* (Oxford, 2009).

Beatty, S. and S. Fothergill, 'Labour market adjustment in areas of chronic industrial decline: the case of the UK coalfields', *Regional Studies*, xxx (1996), 627–40.

Beatty, C., S. Fothergill and T. Gore, *The State of the Coalfields 2019: Economic and Social Conditions in the Former Coalfields of England, Scotland and Wales* (Sheffield, 2019).

Benn, T., *Conflicts of Interest: Diaries, 1977–80* (London, 1991).

Benson, J., *Affluence and Authority: a Social History of Twentieth-Century Britain* (London, 2005).

Beynon, H. and T. Austrin, 'The performance of power: Sam Watson, a miners' leader on many stages', *Journal of Historical Sociology*, xxviii (2015), 458–90.

Beynon, H., R. Davies and S. Davies, 'Sources of variation in trade union membership across the UK: the case of Wales', *Industrial Relations Journal*, xliii (2012), 200–21.

Bluestone, B. and B. Harrison, *The Deindustrialization of America: Plant Closings, Community Abandonment, and the Dismantling of Basic Industry* (New York, 1982).

Bonnett, A., *Left in the Past: Radicalism and the Politics of Nostalgia* (New York, 2010).

Boyle, J., B. Knox and A. McKinlay, '"A sort of fear and run place": Unionising BSR, East Kilbride, 1969', *Scottish Labour History*, liv (2019), 103–25.

Braudel, F., *The Mediterranean and the Mediterranean World in the Age of Philip II* (2 vols., London, 1972).

Brotherstone, T., 'Energy workers against Thatcherite neoliberalism: Scottish coal miners and North Sea offshore workers: revisiting the class struggle in the UK in the 1980s', *International Journal on Strikes and Social Conflict*, i (2013), 135–54.

Bruley, S., 'The politics of food: gender, family, community and collective feeding in South Wales in the general strike and miners' lockout of 1926', *Twentieth Century British History*, xviii (2007), 54–77.

Cameron, E. A., 'The stateless nation and the British State since 1918', in *The Oxford Handbook of Modern Scottish History*, ed. T. M. Devine and J. Wormald (Oxford, 2013), pp. 620–34.

Cameron, G. C., *Industrial Movement and the Regional Problem* (Edinburgh, 1966).

Campbell, A., 'Exploring miners' militancy, 1889–1966', *Historical Studies in Industrial Relations*, vii (1999), 147–64.

—— *The Scottish Miners, 1874–1939*, i: *Industry, Work and Community* (Aldershot, 1999).

—— *The Scottish Miners, 1874–1939*, ii: *Trade Unions and Politics* (Aldershot, 1999).

—— 'Scotland', in *Industrial Politics and the 1926 Mining Lockout: the Struggle for Dignity*, ed. J. McIlroy, A. Campbell and K. Gildart (Cardiff, 2009), pp. 173–89.

—— 'Traditions and generational change in Scots miners' unions, 1874–1929', in *Generations in Labour History: Papers Presented to the Sixth British–Dutch Conference on Labour History, Oxford 1988*, ed. A. Blok, D. Damsma, H. Diedriks and L. H. van Voss (Amsterdam, 1989), pp. 23–37.

Campbell, A. and J. McIlroy, 'Miner heroes: three communist trade union leaders', in *Party People, Community Lives: Explorations in Biography*, ed. J. McIlroy, K. Morgan and A. Campbell (London, 2001), pp. 143–68.

Chick, M., *Electricity and Energy Policy in Britain, France and the United States since 1945* (Cheltenham, 2007).

Church, R., 'Employers, trade unions and the State, 1889–1987: the origins and decline of tripartism in the British coal industry', in *Workers, Owners and Politics in Coal Mining: an International Comparison of Industrial Relations*, ed. G. D. Feldman and K. Tenfelde (New York, 1990), pp. 12–73.

Clark, A., 'Personal experience from a lifetime in the communist and labour movements', *Scottish Labour History Review*, x (1996–7), 9–11.

—— 'Personal experience from a lifetime in the communist and labour movements (part 2)', *Scottish Labour History Review*, xi (1997–8), 14–16.

—— 'Stealing our identity and taking it over to Ireland: deindustrialization, resistance and gender in Scotland', in *The Deindustrialized World: Confronting Ruination in Postindustrial Places*, ed. S. High, L. MacKinnnon and A. Perchard (Vancouver, 2017), pp. 331–47.

Clark, A. and E. Gibbs, 'Voices of social dislocation, lost work and economic restructuring: narratives from marginalised localities in the "new Scotland"', *Memory Studies*, xiii (2020), 39–59.

Clarke, J., 'Closing Moulinex: thoughts on the visibility and invisibility of industrial labour in contemporary France', *Modern and Contemporary France*, xix (2011), 443–58.

Condratto, S. and E Gibbs, 'Afterindustrial citizenship: adapting to precarious employment in the Lanarkshire coalfield, Scotland and Sudbury hardrock mining, Canada', *Labour/Le Travail*, lxxxi (2018), 213–39.

Connell, R. W., *Masculinities* (Cambridge, 2005).

Connolly, S., 'Women and work since 1970', in *Work and Pay in 20th-century Britain*, ed. N. Crafts, I. Gazeley and A. Newell (Oxford, 2007), pp. 142–75.

Cowie, J., *Capital Moves: RCA's Seventy-Year Quest for Cheap Labour* (New York, 2001).

— *Stayin' Alive: the 1970s and the Last Days of the Working Class* (New York, 2010).

Cowie, J. and J. Heathcott, 'Introduction: the meanings of deindustrialization', in *Beyond the Ruins: the Meanings of Deindustrialization*, ed. J. Cowie and J. Heathcott (Ithaca, N.Y., 2003), pp. 1–15.

Craig, F. W. S. (ed.), *British General Election Manifestos, 1900–1974* (London, 1975).

Crane, S. A., 'Writing the individual back into collective memory', *American Historical Review*, cii (1997), 1372–85.

Curtis, B., 'A tradition of radicalism: the politics of the South Wales miners, 1964–1985', *Labour History Review*, lxxvi (2011), 34–50.

Daly, L., 'Scotland on the dole', *New Left Review*, xvii (1962), 17–23.

Dalyell, T., 'Margaret Herbison', *ODNB* <https://doi.org/10.1093/ref:odnb/64016> [accessed 20 Oct. 2011].

Davidson, N., 'Gramsci's reception in Scotland', *Scottish Labour History*, xxxviii (2010), 37–58.

Dennis, N., F. Henriques and C. Slaughter, *Coal is Our Life: an Analysis of a Yorkshire Mining Community* (London, 1956).

Devine, T., *The Scottish Nation, 1700–2007* (London, 2007).

— *To the Ends of the Earth: Scotland's Diaspora, 1750–2010* (London, 2011).

Duncan, R., *The Mineworkers* (Edinburgh, 2005).

Ebke, M., 'The decline of the mining industry and the debate about Britishness of the 1990s and early 2000s', *Contemporary British History*, xxxii (2018), 121–41.

Edgerton, D., *The Rise and Fall of the British Nation: a Twentieth Century History* (London, 2018).

Elliot, G., *Labourism and the English Genius* (London, 1993).

Emery, J. 'Belonging, memory and history in the North Nottinghamshire coalfield', *Journal of Historical Geography*, lviii (2018), 77–89.

Featherstone, D. and D. Kelliher, *'There was just this Enormous Sense of Solidarity': London and the 1984–5 Miners' Strike* (London, 2018).

Fielding, N. and H. Thomas, 'Qualitative interviewing', in *Researching Social Life*, ed. N. Gilbert (London, 2008), pp. 245–65.

Finch, H. and J. Lewis, 'Focus groups', in *Qualitative Research Practice: a Guide for Social Science Students and Researchers*, ed. J. Ritchie and J. Lewis (London, 2003), pp. 170–98.

Finlay, R. J., *Modern Scotland, 1914–2000* (London, 2004).

Findlay, P., 'Resistance, restructuring and gender: the Plessey occupation', in *The Politics of Industrial Closure*, ed. T. Dickson and D. Judge (Basingstoke, 1986), pp. 70–95.

Fisher, M., *Capitalist Realism: is There No Alternative?* (Winchester, 2009).

Foden, M., S. Fothergill and T. Gore, *The State of the Coalfields: Economic and Social Conditions in the Former Mining Communities of England, Scotland and Wales* (Sheffield, 2014).

Fothergill, S., 'The new alliance of mining areas', in *Restructuring the Local Economy*, ed. M. Geddes and J. Benington (Exeter, 1992), pp. 51–77.

Foster, J., 'The twentieth century', in *The New Penguin History of Scotland: from the Earliest Times to Present Day*, ed. R. A Housting and W. W. Knox (London, 2001), pp. 417–96.

Foster, J. and C. Woolfson, *The Politics of the UCS Work-In: Class Alliances and the Right to Work* (London, 1986).

— 'How workers on the Clyde gained the capacity for class struggle: the Upper Clyde Shipbuilders' work-in, 1971–2', in *British Trade Unions and Industrial Politics*, ii: *the High Tide of Trade Unionism, 1964–1979*, ed. J. McIlroy, N. Fishman and A. Campbell (Aldershot, 1999), pp. 297–325.

Francis, H. and G. Rees, '"No surrender in the Valleys": the 1984–85 miners' strike in South Wales', *Llafur*, v (1989), 41–71.

Fyrth, J., 'Introduction: in the thirties', in *Britain, Fascism and the Popular Front*, ed. J. Fyrth (London, 1985), pp. 9–29.

Geoghegan, P., *The People's Referendum: Why Scotland will Never be the Same Again* (Glasgow, 2015).

Gibbs, E., '"Civic Scotland" versus communities on Clydeside: poll tax non-payment, *c.* 1987–1990', *Scottish Labour History*, xlix (2014), 86–106.

Gibbs, E. and J. Phillips, 'Who owns a factory? Caterpillar tractors in Uddingston, 1956–1987', *Historical Studies in Industrial Relations*, xxxviii (2018), 111–37.

— 'Remembering Auchengeich: the largest fatal accident in Scottish coal mining in the nationalised era', *Scottish Labour History*, liv (2019), 47–57.

Gildart, K., *North Wales Miners: a Fragile Unity, 1945–1996* (Cardiff, 2001).

— 'Mining memories: reading coalfield autobiographies', *Labor History*, i (2009), 139–61.

Goldthorpe, J., D. Lockwood, F. Bechhofer and J. Platt, *The Affluent Worker: Industrial Attitudes and Behaviour* (London, 1968).

Graham, H. and P. Preston, 'The Popular Front and the struggle against fascism', in *The Popular Front in Europe*, ed. H. Graham and P. Preston (Basingstoke, 1987), pp. 1–19.

Halliday, R. S., *The Disappearing Scottish Colliery: a Personal View of some Aspects of Scotland's Coal Industry since Nationalisation* (Edinburgh, 1990).

Harvie, C., *No Gods and Precious Few Heroes: Twentieth Century Scotland* (Edinburgh, 1998).

Hassan, G., 'Back to the future: exploring twenty years of Scotland's journey, stories and politics', in *The Story of the Scottish Parliament: the First Two Decades Explained*, ed. G. Hassan (Edinburgh, 2019), pp. 1–27.

Hassan, G. and E. Shaw, *The Strange Death of Labour Scotland* (Edinburgh, 2012).

Haywood, S. and M. Mac An Ghaill, *Men and Masculinities: Theory, Research and Practice* (Buckingham, 2003).

Hechter, M., *Internal Colonialism: the Celtic Fringe in British National Development, 1536–1966* (London, 1975).

Heughan, H. E., *Pit Closures at Shotts and the Migration of Miners* (Edinburgh, 1953).

High, S., *Industrial Sunset: the Making of North America's Rustbelt* (Toronto, 2003).

— 'Beyond aesthetics: visibility and invisibility in the aftermath of deindustrialization', *International Labor and Working-Class History*, lxxxiv (2013), 140–53.

High, S. and D. Lewis, *Corporate Wasteland: the Legacy and Memory of Deindustrialization* (New York, 2009).

High, S., L. MacKinnnon and A. Perchard, 'Introduction', in *The Deindustrialized World: Ruination in Post-Industrial Places*, ed. S. High, L. MacKinnnon and A. Perchard (Vancouver, 2017), pp. 3–22.

Hilwig, S. J., '"Are you calling me a fascist?": a contribution to the oral history of the 1968 Italian student rebellions', *Journal of Contemporary History*, xxxi (2001), 581–97.

Hobsbawm, E., 'Introduction: inventing traditions', in *The Invention of Tradition*, ed. E. Hobsbawm and T. Ranger (Cambridge, 1983), pp. 1–14.

— *Nations and Nationalism since 1780: Programme, Myth, Reality* (Cambridge, 1990).

Holland, S., *The Regional Problem* (London, 1976).

Honneth, A., *The Struggle for Recognition: the Moral Grammar of Social Conflicts* (Cambridge, Mass., 1995).

Hood, N. and S. Young, 'US investment in Scotland: aspects of the branch factory syndrome', *Scottish Journal of Political Economy*, xxiv (1976), 279–94.

Horwood, S. N., *Strikebreaking and Intimidation: Mercenaries and Masculinity in Twentieth-Century America* (Chapel Hill, N.C., 2002).

Horrocks, R., *Masculinity in Crisis: Myths, Fantasies and Realities* (Basingstoke, 1994).

Hospers, G., 'Restructuring Europe's rustbelt: the case of the German Ruhrgebiet', *Intereconomics*, xxxix (2004), 147–56.

Hughes, A., *Gender and Political Identities in Scotland, 1919–1939* (Edinburgh, 2010).

Hutton, G., *Coal Not Dole: Memories of the 1984/85 Miners' Strike* (Glasgow, 2005).

Hyman, R., 'Inequality, ideology and industrial relations', *British Journal of Industrial Relations*, ii (1974), 171–91.

Isenberg, N., *White Trash: the 400-year Untold History of Class in America* (New York, 2016).

Ives, M., *Reform, Revolution and Direct Action Amongst British Miners: the Struggle for the Charter in 1919* (Chicago, Ill., 2016).

Jacobs, M., 'End of the coal strike', *Economic and Political Weekly*, xx (1985), 443–4.

Jenkins, J., 'Hands not wanted: closure, and the moral economy of protest, Treorchy, South Wales', *Historical Studies in Industrial Relations*, xxxviii (2017), 1–36.

Johnston, R. and A. McIvor, 'Dangerous work, hard men and broken bodies: masculinity in the Clydeside heavy industries, *c.* 1930–1970s', *Labor History Review*, lxix (2004), 135–51.

Jones, B., *The Working Class in Mid-Twentieth Century England* (Manchester, 2012).

Jones, B., B. Roberts and C. Williams, '"Going from darkness into light": South Wales miners' attitudes towards nationalisation', *Llafur*, vii (1996), 96–110.

Kaufman, G., 'Varley, Eric Graham, Baron Varley', *ODNB* <https://doi.org/10.1093/ref:odnb/100192> [accessed 20 Oct. 2019].

Keating, M. and D. Beliman, *Labour and Scottish Nationalism* (Edinburgh, 1979).

Kelliher, D., 'Constructing a culture of solidarity: London and the British coalfields in the long 1970s', *Antipode*, xlix (2017), 106–24.

Kelly, E., 'Review essay: sectarianism, bigotry and ethnicity – the gulf in understanding', *Scottish Affairs*, i (2005), 106–17.

Kenny, M., *The First New Left: British Intellectuals after Stalin* (London, 1995).

Kirk, J., *Class, Culture and Social Change: on the Trails of the Working Class* (Basingstoke, 2007).

Knight, P. T., *Small-Scale Research: Pragmatic Inquiry in Social Science and the Caring Professions* (London, 2008).

Knotter, A., '"Little Moscows" in western Europe: the ecology of small-place Communisms', *International Review of Social History*, lxi (2011), 475–510.

Knox, W., *Industrial Nation: Work, Culture and Society in Scotland, 1800–Present* (Edinburgh, 1999).

Knox, W. and A. McKinlay, 'American multinationals and British trade unions, *c.* 1945–1974', *Labor History*, li (2010), 211–29.

— 'The union makes us strong? Work and trade unionism in Timex, 1948–1983', in *Jute No More: Transforming Dundee*, ed. J. Tomlinson and C. A. Whatley (Dundee, 2011), pp. 266–90.

Krieger, J., *Undermining Capitalism: State Ownership and the Dialectic of Control in the British Coal Industry* (Princeton, N.J., 1983).

Law, T. S., 'A Wilson memorial', *New Edinburgh Review*, xxxii (1976), 22–8.

Lawrence, T. B. and S. L. Robinson, 'Ain't misbehavin: workplace deviance as organizational resistance', *Journal of Management*, xxxiii (2007), 378–94.

Leeworthy, D. 'The secret life of us: 1984, the miners' strike and the place of biography in writing history "from below"', *European Review of History: Revue europeenne d'histoire*, xix (2012), 825–46.

Levitt, I. 'The origins of the Scottish Development Department, 1943–1962', *Scottish Affairs*, xiv (1996), 42–63.

McCormack, J. (with S. Pirani), *Polmaise: the Fight for a Pit* (WordPress version, 2015) <https://polmaisebook.wordpress.com/> [accessed 21 Dec. 2019].

McCrone, D. *The New Sociology of Scotland* (London, 2017).

McDermott, M. C., *Multinationals: Foreign Divestment and Disclosure* (Maidenhead, 1989).

MacDonald, C. M., 'Gender and nationhood in modern Scottish historiography', in *The Oxford Handbook of Modern Scottish History*, ed. T. M. Devine and J. M. Wormald (Oxford, 2012) pp. 602–19.

MacDougall, I., 'Reminiscences of John MacArthur, Fife militant', *Scottish Marxist*, vi (1974), 13–23.

— (ed.), *Militant Miners: Recollections of John McArthur, Buckhaven; and Letters, 1924–6, of David Proudfoot, Methil, to G. Allen Hunt* (Edinburgh, 1981).

McGahey, M., 'The coal industry and the miners', *Scottish Marxist*, i (1972), 16–19.

McGrail, S. and V. Paterson, *Cowie Miners, Polmaise Colliery and the 1984–85 Miners' Strike* (Glasgow, 2017).

McIlroy, J., '"Every factory our fortress": Communist party workplace branches in a time of militancy, 1956–79, part 1: history, politics, topography', *Historical Studies in Industrial Relations*, x (2000), 99–139.

McIlroy, J. and A. Campbell, 'Beyond Betteshanger: Order 1305 in the Scottish coalfields during the Second World War, part 1: politics prosecutions and protest', *Historical Studies in Industrial Relations*, xv (2003), 27–72.

— 'Beyond Betteshanger: Order 1305 in the Scottish coalfields during the Second World War, part 2: the Cardowan story', *Historical Studies in Industrial Relations*, xvi (2003), 39–80.

— 'Coalfield leaders, trade unionism and communist politics: exploring Arthur Horner and Abe Moffat', in *Towards a Comparative History of Coalfield Societies*, ed. S. Berger, A. Croll and N. La Porte (Aldershot, 2005), pp. 267–83.

McIlvanney, W., *Surviving the Shipwreck* (Edinburgh, 1991).

Macintyre, S., *Little Moscows: Communism and Working-Class Militancy in Inter-War Britain* (London, 1980).

McIvor, A., 'Women and work in twentieth century Scotland', in *People and Society in Scotland*, iii: *1914–1990*, ed. T. Dickson and J. H. Treble (Edinburgh, 1992), pp. 138–73.

— 'Gender apartheid? Women in Scottish society in the twentieth century', in *Scotland in the Twentieth Century*, ed. T. M. Devine and R. J. Finlay (Edinburgh, 1996), pp. 188–209.

McIvor, A. and R. Johnston, *Miners' Lung: a History of Dust Disease in British Coal Mining* (Aldershot, 2006).

McKibbin, R., *Classes and Cultures: England, 1918–1951* (Oxford, 1998).

McLean, B., 'The 1974 struggle', *Scottish Marxist*, vi (1974), 25–35.

McShane, H., 'The march: the story of the historic Scottish hunger march', *Variant*, xv (2002), 30–4.

Maxwell, S., *The Case for Left-Wing Nationalism* (Edinburgh, 1981).

Mannheim, K., *Essays on the Sociology of Knowledge* (London, 1952).

Marquand, D., *Mammon's Kingdom: an Essay on Britain Now* (London, 2014).

Meadhurst, J., *That Option No Longer Exists: Britain, 1974–76* (Arelsford, 2014).

Miliband, R., *Parliamentary Socialism: a Study in the Politics of Labour* (London, 1973).

Mills, C., *Regulating Health and Safety in the British Mining Industries, 1800–1914* (Famham, 2010).

Mills, C. W., *The Power Elite* (New York, 1956).

Milne, S., *The Enemy Within: the Secret War against the Miners* (London, 2012).

Mitchell, J., *Devolution in the United Kingdom* (Manchester, 2009).

— *Hamilton 1967: the Byelection that Transformed Scotland* (Gosport, 2017).

Mitchell, T., *Carbon Democracy: Political Power in the Age of Oil* (London, 2013).

Moffat, A., *My Life with the Miners* (London, 1965).

Montgomery, G., 'Introduction', *New Edinburgh Review*, xxxii (1976), 1–3.

Morgan, D., 'Class and masculinity', in *Handbook of Studies on Men and Masculinities*, ed. M. S. Kimmel, J. Hearn and R. W. Connell (London, 2005), pp. 165–77.

Morton, G., *Unionist-Nationalism: Governing Urban Scotland, 1830–1860* (East Linton, 1999).

Murray, A. and F. Hart, 'The Scottish economy', *Scottish Marxist*, i (1972), 23–35.

Mycock, A., 'Invisible and inaudible? England's Scottish diaspora and the politics of the Union', in *The Modern Scottish Diaspora: Contemporary Debates and Perspectives*, ed. M. S. Leith and D. Sim (Edinburgh, 2014), pp. 99–117.

Nairn, T., 'Three dreams of Scottish nationalism', *New Left Review*, xlix (1968), 3–18.

— *The Break-Up of Britain: Crisis and Neo-Nationalism* (London, 1977).

Nettleingham, D., 'Canonical generations and the British left: narrative construction of the British miners' strike, 1984–85', *Sociology*, li (2017), 850–64.

North, J. and D. Spooner, 'The great UK coal rush: a progress report to the end of 1976', *Area*, ix (1977), 15–27.

NUM, *National Energy Policy* (London, 1972).

Oglethorpe, M. K., *Scottish Collieries: an Inventory of the Scottish Coal Industry in the Nationalised Era* (Edinburgh, 2006).

— 'The Scottish coal mining industry since 1945', *Scottish Business and Industrial History*, xxvi (2011), 77–98.

Parker, M., *Thatcherism and the Fall of Coal* (Oxford, 2000).

Passerini, L., 'Work ideology and consensus under Italian fascism', *History Workshop*, viii (1979), 82–108.

Paterson, R., 'The pulpit and the ballot box: Catholic assimilation and the decline of church influence', in *Scotland's Shame?: Bigotry and Sectarianism in Modern Scotland*, ed. T. M. Devine (Edinburgh, 2000), pp. 219–30.

Payne, P. L., *Growth and Contraction: Scottish Industry, c. 1860–1990* (Glasgow, 1992).

— 'The end of steelmaking in Scotland *c.* 1967–1993', *Scottish Economic and Social History*, xv (1995), 66–84.

Perchard, A., *The Mine Management Professions in the Twentieth-Century Scottish Coal Mining Industry* (New York, 2007).

— '"Broken men" and "Thatcher's children": memory and legacy in Scotland's coalfields', *International Labor and Working-Class History*, lxxxiv (2013), 78–98.

Perchard A. and K. Gildart, '"Run with the foxes and hunt with the hounds": managerial trade-unionism and the British Association of Colliery Management, 1947–1994', *Historical Studies in Industrial Relations*, xxxix (2018), 79–110.

Perchard, A. and J. Phillips, 'Transgressing the moral economy: Wheelerism and management of the nationalised coal industry in Scotland', *Contemporary British History*, xxv (2011), 387–405.

Petrocultures Research Group, *After Oil* (2016) <http://afteroil.ca/wp-content/uploads/2016/02/AfterOil_fulldocument.pdf> [accessed 23 Dec. 2019].

Phillips, J. *The Industrial Politics of Devolution: Scotland in the 1960s and 1970s* (Manchester, 2008).

— 'The "retreat" to Scotland: the Tay Road Bridge and Dundee's post-1945 development', in *Jute No More: Transforming Dundee*, ed. J. Tomlinson and C. A. Whatley (Dundee, 2011), pp. 246–65.

— *Collieries, Communities and the Miners' strike in Scotland, 1984–85* (Manchester, 2012).

— 'Material and moral resources: the 1984–5 miners' strike in Scotland', *Economic History Review* lxv (2012), 256–76.

— 'Deindustrialization and the moral economy of the Scottish coalfields, 1947 to 1991', *International Labor and Working-Class History*, lxxxiv (2013), 99–115.

— 'Containing, isolating and defeating the miners: the UK cabinet ministerial group on coal and the three phases of the 1984–85 strike', *Historical Studies in Industrial Relations*, xxxv (2014), 117–41.

— 'Contested memories: the Scottish parliament and the 1984–5 miners' strike', *Scottish Affairs*, xxvi (2015), 187–206.

— 'The closure of Michael colliery in 1967 and the politics of deindustrialization in Scotland', *Twentieth Century British History*, xxvi (2015), 551–72.

— 'Economic direction and generational change in twentieth century Britain: the case of the Scottish coalfield', *English Historical Review*, cxxxii (2017), 885–911.

— 'The moral economy of deindustrialization in post-1945 Scotland', in *The Deindustrialized World: Ruination in Post-Industrial Places*, ed. S. High, L. MacKinnnon and A. Perchard (Vancouver, 2017), pp. 313–30.

— 'The meanings of coal community in Britain since 1947', *Contemporary British History*, xxxii (2018), 39–59.

— *Scottish Coal Miners in the Twentieth Century* (Edinburgh, 2019).

Phillips, J., V. Wright and J. Tomlinson, 'Deindustrialization, the Linwood car plant and Scotland's political divergence from England in the 1960s and 1970s', *Twentieth Century British History*, 30 (2019): 399–423, doi: 10.1093/tcbh/hwz005 [accessed 10 Jan. 2020].

Polanyi, K., *The Great Transformation: the Political and Economic Origins of our Time* (Boston, Mass., 2001).

— *The Present Age of Transformation: Five Lectures by Karl Polanyi, Bennington College, 1940* (Northbrook, Ill., 2007).

Polanyi Levitt, K., *From the Great Transformation to the Great Financialization: on Karl Polanyi and Other Essays* (London, 2013).

Popp, R., 'A just transition of European coal regions: assessing the stakeholder positions towards transitions away from coal', *E3G Briefing Paper* (2019) <https://www.e3g.org/showcase/just-transition/> [accessed 20 Nov. 2019].

Popular Memory Group, 'Popular memory: theory, politics, methodology', in *The Oral History Reader*, ed. R. Perks and A. Thomson (London, 2006), pp. 43–53.

Portelli, A., *The Death of Luigi Trastulli and Other Stories: Form and Meaning in Oral History* (Albany, N.Y., 1991).

— *The Battle of Valle Giulia: Oral History and the Art of Dialogue* (Maddison, Wis., 1997).

— '"This mill won't run no more": oral history and deindustrialization', in *New Working-Class Studies*, ed. R. Russo and S. L. Linkon (Ithaca, N.Y., 2005), pp. 54–9.

— 'What makes oral history different', in *The Oral History Reader*, ed. R. Perks and A. Thomson (Oxford, 2006), pp. 32–42.

— *They say in Harlan County: an Oral History* (Oxford, 2010).

Priscott, D., 'The miners' strike assessed', *Marxism Today*, Apr. 1985, pp. 21–7.

Rafeek, N. C., *Communist Women in Scotland: Red Clydeside from the Russian Revolution to the End of the Soviet Union* (London, 2008).

Randall, J. N., 'New towns and new industries', in *The Economic Development of Modern Scotland, 1950–1980*, ed R. Saville (Edinburgh, 1985), pp. 245–69.

Reid, D., *The Miners of Decazeville: a Genealogy of Deindustrialization* (Cambridge, Mass., 1985).

Reid, J., *Reflections of a Clyde-Built Man* (London, 1976).

Richards, A., *Miners on Strike: Class Solidarity and Division in Britain* (Oxford, 1996).

Robens, A., *Ten Year Stint* (London, 1972).

Rogan, T., *The Moral Economists: R. H. Tawney, Karl Polanyi, E. P. Thompson and the Critique of Capitalism* (Princeton, N.J., 2017).

Roper, M. and Tosh, J., 'Historians and the politics of masculinity', in *Men and Masculinities: Critical Concepts in Sociology*, i: *Politics and Power*, ed. S. M. Whitehead (Oxford, 2006), pp. 79–99.

Samuel, R., 'Introduction', in *The Enemy Within: Pit Villages and the Miners' Strike of 1984–5*, ed. R. Samuel, B. Bloomfield and G. Boanas (London, 1986), pp. 1–38.

— *The Lost World of British Communism* (London, 2006).

— *Theatres of Memory: Past and Present in Contemporary Culture* (London, 2012).

Saunders, J., *Assembling Cultures: Workplace Activism, Labour Militancy and Cultural Change in Britain's Car Factories, 1945–82* (Manchester, 2019).

Saville, J., 'Labourism and the Labour government', *Socialist Register*, iv (1967), 43–71.

Saville, R., 'The coal business', *Scottish Economic and Social History*, viii (1988), 93–6.

Schenk, C., *International Economic Relations since 1945* (London, 2011).

Scherger, S., 'Concepts of generation and their empirical application: from social formations to narratives – a critical appraisal and some suggestions', *CRESC Working Paper*, cxvii (2012).

Scott, P., 'Regional development and policy', in *The Cambridge Economic History of Modern Britain*, iii: *Structural Change and Growth, 1939–2000*, ed. R. Floud and P. Johnson (Cambridge, 2004), pp. 332–67.

— 'The household economy since 1870', in *The Cambridge Economic*

History of Modern Britain, ii: *1870 to the Present*, ed. R. Floud, R. Humphries and P. Johnson (Cambridge, 2014), pp. 382–86.

Seifert, R. V. and T. Sibley, *Revolutionary Communist at Work: a Political Biography of Bert Ramelson* (London, 2012).

Smith, P. and J. Brown, 'Economic crisis, foreign capital and working-class response, 1945–1979', in *Scottish Capitalism: Class, State and Nation from before the Union to the Present*, ed. T. Dickson (Edinburgh, 1985), pp. 287–320.

Smith, R., 'New towns for Scottish miners: the rise and fall of a social ideal (1945–1948)', *Scottish Economic and Social History*, ix (1989), 71–9.

Spence, J. and C. Stephenson, '"Side by side with our men?" Women's activism, community, and gender in the 1984–1985 British miners' strike', *International Labor and Working-Class History*, lxxv (2009), 68–84.

Stephenson, J. D. and C. G. Brown, 'The view from the workplace: women's memories of work in Stirling, *c*. 1910–*c*. 1950', in *The World is Ill Divided: Women's Work in Scotland in the Nineteenth Century and Early Twentieth Century*, ed. E. Gordon and E. Breitenbach (Edinburgh, 1990), pp. 7–28.

South Africa Democracy Education Trust, *The Road to Democracy in South Africa*, ii: *1970 to 1980* (Pretoria, 2006).

Stizia, L., 'Telling Arthur's story: oral history relationships and shared authority', *Oral History*, xxvii (1999), 58–67.

Stokes, R., *Opting for Oil: the Political Economy of Technological Change in the West German Chemical Industry, 1945–1961* (New York, 1994).

Strangleman, T., 'Networks, place and identities in post-industrial communities', *International Journal of Urban and Regional Research*, xxv (2001), 253–66.

— '"Smokestack nostalgia", "ruin porn" or working-class obituary: the role and meaning of deindustrial representation', *International Labor and Working-Class History*, lxxxiv (2013), 23–37.

— 'Deindustrialisation and the historical sociological imagination: making sense of work and industrial change', *Sociology*, li (2017), 466–82.

Summerfield, P., *Reconstructing Women's Wartime Lives: Discourse and Subjectivity in Oral Histories of the Second World War* (Manchester, 1998).

— 'Culture and composure: creating narratives of the gendered self in oral history interviews', *Cultural and Social History*, i (2004), 65–93.

Sutcliffe-Braithwaite, F. and N. Thomlinson, 'National Women Against Pit Closures: gender, trade unionism and community activism in the miners' strike, 1984–5', *Contemporary British History*, xxxii (2018), 78–100.

Taylor, A., *The NUM and British Politics*, i: *1944–1968* (Aldershot, 2003).

— *The NUM and British Politics*, ii: *1969–1995* (Aldershot, 2006).

Taylor, S., *The Fall and Rise of Nuclear Power in Britain: a History* (Cambridge, 2016).

Terris, I., *Twenty Years Down the Mines* (Ochiltree, 2001).

Thereborne, G., 'Twilight of Swedish social democracy', *New Left Review*, cxiii (2018), 5–26.

Thompson, E. P., *The Making of the English Working Class* (Middlesex, 1968).

— 'The moral economy of the English crowd in the eighteenth century', *Past and Present*, i (1971), 76–136.

— *Customs in Common* (London, 1991).

Thomson, A., 'Anzac memories: putting popular memory theory into practice in Australia', in *The Oral History Reader*, ed. R. Perks and A. Thomson (London, 2006), pp. 300–10.

Tomlinson, J., *The Politics of Decline: Understanding Post-War Britain* (London, 2000).

— 'A "failed experiment"? Public ownership and the narratives of post-war Britain', *Labour History Review*, lxxiii (2008), 228–43.

— 'De-industrialisation not decline: a new meta-narrative for post-war British history', *Twentieth Century British History*, xxvii (2016), 76–99.

— *Managing the Economy, Managing the People: Narratives of Economic Life in Britain from Beveridge to Brexit* (Oxford, 2017).

Tomlinson, J. and E. Gibbs, 'Planning the new industrial nation: Scotland, 1931 to 1979', *Contemporary British History*, xxx (2016), 584–606.

Torrance, D., *The Battle for Britain: Scotland and the Independence Referendum* (London, 2013).

Walker, G., 'The Orange Order in Scotland between the wars', *International Review of Social History*, xxxvii (1992), 277–306.

— 'Sectarian tensions in Scotland: social and cultural dynamics and the politics of perception', in *Scotland's Shame? Bigotry and Sectarianism in Modern Scotland*, ed. T. M. Devine (Edinburgh, 2000), pp. 125–34.

Walker, J., 'The road to Sizewell: the origins of the UK nuclear power programme', *Contemporary Record*, i (1987), 33–50.

Walkerdine, G. and L. Jimenez, *Gender, Work and Community after De-Industrialization: a Psychosocial Approach to Affect* (Basingstoke, 2012).

Ward, D., *Unionism in the United Kingdom, 1918–1974* (Basingstoke, 2005).

Whatley, C., 'Some historical pointers: miners in 18th century Ayrshire', *Scottish Marxist*, vi (1974), 5–12.

Wight, D., *Workers Not Wasters: Masculine Respectability, Consumption and Unemployment in Central Scotland: a Community Study* (Edinburgh, 1993).

Williams, C., *Capitalism, Community and Conflict: the South Wales Coalfield, 1898–1947* (Cardiff, 1998).

Williams, R., *Culture and Materialism: Selected Essays* (London, 2005)

Wilson, G., '"Our chronic and desperate situation": anthracite communities and the emergence of redevelopment policy in Pennsylvania and the United States, 1945–1965', in *De-industrialization: Social, Cultural and Political Aspects*, ed. B. Altena and M. van der Linden (Cambridge, 2002), pp. 137–58.

Winterton, J. and R. Winterton, *Coal, Crisis and Conflict: the 1984–85 Miners' Strike in Yorkshire* (Manchester, 1989).

Woolfson, C. and J. Foster, *Track Record: the Story of the Caterpillar Occupation* (London, 1988).

Wrigley, C., 'Women in the labour market and the unions', in *British Trade Unions and Industrial Politics*, ii: *the High Tide of Trade Unionism, 1964–79*, ed. J. McIlroy, N. Fishman and A. Campbell (Aldershot, 1999), pp. 43–69.

Young, J. D., 'The British miners, 1867–1976', *New Edinburgh Review*, xxxii (1976), 5–9.

Young, N. and S. Young, *Multinationals in Retreat: the Scottish Experience* (Edinburgh, 1982).

Zweiniger-Bargielowska, I., 'South Wales miners' attitudes towards nationalization: an essay in oral history', *Llafur*, vi (1993), 70–84.

Fiction, poetry and songs

Gaughan, D., 'Why old men cry', *Redwood Cathedral* (Greentrax, 1998).

Kay, J., 'Last room in operations' (2014), unpublished poem used with permission of the author.

Lochhead, L., *The Colour of Black and White* (Edinburgh, 2003).

McGrath, J., *Six-Pack: Plays for Scotland* (Edinburgh, 1996).

McIlvanney. W., *Docherty* (London, 1975).

Monaghan. J., 'United colours of Cumnock', *Concept*, viii (2017) <http://concept.lib.ed.ac.uk/issue/view/216> [accessed 20 Dec. 2019].

Moohan, B., *The Enemy Within* (Livingston, 2012).

Welsh, D. *Skagboys* (London, 2012).

Index

accidents, 58, 62, 105–6, 109, 131, 134, 170, 178, 240
   Cardowan gas explosion 1982, 105–6
affluence, 6, 14, 121, 142, 153, 157–8, 169, 172
Affluent Worker study by Goldthorpe et al, 6, 177
Afghanistan, 197
Airdrie, North Lanarkshire, 98, 103, 116, 118, 135, 159
alcohol abuse, 121, 132–5
Allison, Willie, Boglea miner, 158
Allen, Vic, industrial relations scholar, 192
Anglo-Scottish Union, *see* the Union
anti-communism, 80, 112–13, 194–5
APEX trade union, 141
Appalachian miners, 71
Arnot, Robert Page, historian of the Scottish miners, 58, 190
Auchengeich colliery, North Lanarkshire, 1–5, 18, 74–6, 106–9, 118, 126, 163, 257
Auchinleck, East Ayrshire, 38
Auchlocan 9 colliery, South Lanarkshire, 71–2, 122, 233, 242–5
Auldton colliery, South Lanarkshire, 72

Baird, R. Cardowan miner and NUM activist, 181
Bairds, coal and steel owners, 77, 111–12, 163
Bardykes colliery, South Lanarkshire, 174
Barlinnie prison, Glasgow, 161, 179
Barony colliery, East Ayrshire, 35, 37–8, 51, 70, 208, 210
Baton colliery, North Lanarkshire, 63–7, 161
Bedlay colliery, North Lanarkshire, 3, 41–2, 56–8, 62, 68, 73–4, 76–7, 80–1, 84–6, 110–15, 167, 175–6, 199–200, 229, 245
Bellshill, North Lanarkshire 33, 97–9, 107, 114, 142, 149, 173, 218
Benn, Tony, Labour secretary of state for energy, 1975–9, 47–50, 165, 192, 211, 213, 218–19
Berg air brake factory, Cumbernauld, North Lanarkshire, 144
Betteshanger colliery, Kent, 45
Billy Elliot by Stephen Daldry, 91
Bilston Glen colliery, Midlothian, 127, 137, 189, 202, 246
Bings, *see* coal bings
Birkenshaw, North Lanarkshire, 170–1
Birmingham, 100
Birmingham Sound Reproducers factory, East Kilbride, 145
Bishopbriggs, East Dunbartonshire, 130
blacklisting, 11, 103, 110, 127, 130, 137, 160, *see also* victimization
Blades, Alan, Bedlay and Solsgirth miner, 103, 116, 137–8, 160
Blantyre, South Lanarkshire, 93, 196
Blood Red Roses by John McGrath, 145
Board of Trade, 14, 17, 41, 64, 138, 143
Boglea colliery, North Lanarkshire, 75, 161

293

Bogside colliery, Fife, 82
Bolton, George, NUMSA president, 1987–96, 13, 178, 191
Bolton, Guy, Devon colliery miner and NUM activist, 179
Bothwell Constituency Labour Party, 205
Bothwellhaugh, North Lanarkshire, 102, 118
Brannan, John, leader of the Caterpillar tractor factory occupation, 1987, 113–14
Brassed Off by Mark Hernan, 91
breadwinner wage, 119–21, 123, 125, 130–1, 153
Breich, West Lothian, 101, 125
Brexit, 6
British Telecom, 142
Broad Left strategy, 196–8
Broomside colliery, North Lanarkshire, 67
Burroughs office machinery factory, Cumbernauld, 42, 52, 84, 147

Cairnhill mine, South Ayrshire, 113, 211
Caledonian railway works, Glasgow, 251–2
Callaghan, Jim, Labour prime minister, 1976–9, 44–9, 62, 196, 210
Cambuslang, South Lanarkshire, 43, 101, 111, 169–71, 174, 189
Cameron, Veronica, convener of engineering workers at Burroughs, 147
Canavan, Tommy, Cardowan miner and NUM activist, 77, 83–5, 192, 248–9
Cardowan colliery, North Lanarkshire, 1, 3, 41–2, 52–63, 68–88, 96–7, 105–7, 126, 133–4, 142, 150, 161, 164–7, 173, 181, 192, 200–2, 228, 235, 248, 252

Carluke, South Lanarkshire, 96–8
Castlebridge colliery, Clackmannanshire, 48, 106
Caterpillar tractor factory, Tannochside, North Lanarkshire, 52, 113–14, 131, 135, 146–7, 167–73, 184, 218, 246, 255
Catholic Action, 111–13
Catholicism, 13, 80, 110–18, 159, *see also* sectarianism
central coalfields, 40, *see also* Nottinghamshire
Central Energy Generating Board, 47, *see also* energy policy; power stations
Central Policy Review Staff, 47
Chapel mine, North Lanarkshire, 66
Chrysler, 78–9, *see also* Linwood
Chungwha television factory, Mossend, North Lanarkshire, 135
Clark, Alex, Lanarkshire miner and communist, 60, 69, 184, 231
Clark, Jessie, Douglas Castle pit canteen worker and communist, 59–60, 69, 94–5, 110, 124, 130, 141–2, 159–60, 184, 199, 231
Clyde Valley Regional Plan, 28, 33, 138
Clydesdale steelworks, Bellshill, 100
coal bings, 98, 130
Coal Industry Act 1980, 49–51, 81
Coal is Our Life by Dennis et al, 91, 127
Coalburn, South Lanarkshire, 69–72, 121, 233,
Coatbridge, North Lanarkshire, 43, 98, 111, 118,
Cockenzie power station, East Lothian, 38, 229
Cold War, 194–7, 223
collective memory, 1, 11, 17, 55, 58, 83, 119, 186, 188, 226, 234, 240, 243, 249, 253, *see also* oral history
Colliery Officials and Staff Association, 85, 208–9

Index

Coltness Iron Company, 59, 130
Communist Party of Great Britain, 12–13, 60, 66, 108–12, 127, 161, 179, 184, 188, 191–9, 202–4, 211–12, 220–3, 254
community
    coalfield communities, 2, 6–18, 42, 46, 61–89, 91–94, 118, 184–5, 199–200, 226, 230, 239
    fragmentation, 56, 95–99, 103, 243, 247
    new postwar industrial communities, 18, 96–97
    Scottish national coal community, 2, 18, 57, 187–90, 216–23
commuting distances, 2, 41, 64, 68–71, 86, 88, 91, 95, 97–8, 139–42, *see also* suburbanization
Connolly, James, Irish revolutionary, 214
Conservative party, 7, 29, 32–4, 44, 53, 71, 85, 99, 198, 211, 226–8, 238–9, 241–3, 248
Cooperative movement, 72–3, 205
Corby, English Midlands, 241
Cosmopolitan collieries, 3, 41, 69, 80, 122, 177, 199, 202, 244
council housing, 69, 93, 98–9, 101, *see also* public housing
Craigneuk, North Lanarkshire, 238
Critical nostalgia, *see* nostalgia
Croy, North Lanarkshire, 111, 116, 249
Cullen, T, miner and NUM representative at Gartshore 9/11 colliery, 181
cultural circuit, 17, 57–9, 84, 106, 111, *see also* oral history; collective memory
Cumbernauld, North Lanarkshire, 28, 42, 52, 84, 144–8
Cummins engine factory, Shotts, 167, 245
Cumnock, East Ayrshire, 38, 185–6

Daily Worker, CPGB newspaper, 179
Daly, Lawrence, NUM general secretary, 1968–84, 73, 77, 196–8, 209–10
Day, Alex, engineering trade unionist and STUC general council member, 191
deindustrialization
    conceptualization, 2–8, 14–19, 25, 172, 225–35, 243–9
    in Scottish national consciousness, 11, 35, 53–4, 99, 109, 127, 196, 199–214, 223, 235–43, 251, 258
    loss and dislocation, 103–4, 118, 131, 137, 151, 171, 201, 204, 207, 211–12, 225, 252–7
    temporal phases, 22–6, 30, 49, 63, 84, 87, 91–2, 97, 109, 119, 133, 147–8, 153, 155–6, 166–70, 186
Department of Energy, *see* Ministry of Fuel and Power and successors
Derbyshire, English Midlands, 98, 125, 247
devolution, 7, 13, 47, 53, 92, 99, 192, 196–8, 212–13, 222, 238–9, *see also* home rule
Devon colliery, Clackmannanshire, 35–6, 178–9, 254
disability, 11, 55, 62, 68, 93, 128, 173–4
disasters, 16–17, 58, 118, 163
    Auchengeich disaster 1959 and commemoration, 1–5, 16–17, 106–8, 118, 128, 165, 252
    Barony disaster 1962, 37–8
    Kames disaster 1957, 126
    Michael disaster 1967, 75, 109, 165
Dobby, Gilbert, Auchlocan 9 apprenticed engineer, 71–2, 86–7, 121, 244–6, 248

Docherty by William McIlvanney, 177–8
domestic service, 161
Donagher, Ed, Cairnhill mine miner and Catholic Action activist, 112
Doolan, Willie, Cardowan miner and communist activist, 2, 77, 84, 86, 106–9, 128–9, 165, 199, 205, 237
double movement, 8–9, 14–15, 30, 56, 60–2, 65, 73, 88, 122, 228, 130, 161, 181, 253, 258, *see also* Karl Polanyi
Dougan, Tom, engineering union official and STUC General Council member, 113, 203
Douglas Castle colliery, South Lanarkshire, 60, 62, 69–71, 160, 231, 236, 238
Douglas Water, South Lanarkshire, 59, 69, 72–3, 95, 130, 199
Douglas Water Cooperative Society, 72
Downie, Peter, Bedlay and Polkemmet colliery official, 58, 68, 78–9, 87, 151, 178, 200, 246
Dunoon, Argyll and Bute, 191
Durham, north-east England, 91, 133, 216, 229
Durham Miners' Association, 133
Durham miners' gala, 216

Eadie, Alex, Labour MP for Midlothian, 1966–92, 48, 221
East Kilbride, South Lanarkshire, 28, 32, 73, 94, 139, 143, 145–6, 193
East Germany, 193–4
Edinburgh People's Festival, 190
Egan, Pat, Bedlay and Longannet miner, 3, 62, 80, 84, 110–12, 114, 175, 199–200, 229–30
Employment Policy white paper 1944 (cmnd 6527), 64, 71
energy policy, 4, 14, 29, 44–52, 79–81, 181, 187, 204, 225, 233, 238, *see also* fuel competition; power stations
Ewing, Winifred, SNP MP for Hamilton, 1967–70, 210
Ezra, Derek, Chairman of the NCB, 1972–83, 38, 50–1, 56

Fauldhouse, West Lothian, 96, 101, 134
Feminism, 141, 152
Ferns, Billy, Cardowan miner and NUM activist, 97, 124, 129, 134
Ferranti, 31
Fife Socialist League, 196
First World War, 160, 164
Firth of Forth, 30, 48
Fishcross, Clackmannanshire, 180
Focus groups, 15–17
folk revival, 212
Ford Langley factory, Berkshire, 140, 245
Ford, Sidney, NUM president, 1960–71, 206–8
Fox, Alan, industrial sociologist, 177
Frame, John, Cardowan colliery manager, 62, 86
Frances colliery, Fife, 83, 184
Fraser, Thomas, Labour MP for Hamilton, 1943–67, 161, 241–2
French miners, 193
fuel competition, 2, 14–15, 21–3, 29–34, 44–8, 59, 77, 109, 179–82, 225–34, 253–5, *see also* energy policy; power stations
Fuel Policy white paper 1967 (cmnd 3438), 33–4, 77, 232

Gaetsewe, Jon, South African trade unionist and anti-Apartheid activist, 193
gala days, 17, 98, 101, 104, 218, 225, *see also* Scottish miners' gala

Index

Gallacher, Willie, Communist MP for West Fife, 1935–50, 192, 196–8
Garscube colliery, Glasgow, 62
Gartshore 3/2 colliery, North Lanarkshire, 194
Gartshore 9/11 colliery, North Lanarkshire, 68, 74–6, 181, 200, 209, 229–30, 235, 248
Gateside colliery, South Lanarkshire, 162
Gaughan, Dick, folk musician, 175–6
gender relations, 14, 15–16, 18, 41, 95, 106, 119–27, 132, 134, 137–43, 148–53, 160
generations, 3, 13, 15–16, 18, 57, 77, 103, 113, 117, 129, 155–86, 225, 235, 251, 253, 257–8
and gender, 131, 134–7, 139, 141–3, 149, 151–3
Glenbuck, East Ayrshire, 101–12, 118, 256
Glenochil colliery, Clackmannanshire, 27, 35–6, 45
Glenrothes, Fife, 27
Glentore colliery, North Lanarkshire, 58, 68
Goldie, Barbara, Cambuslang resident and miners' daughter, 101, 169–70
Gormill, Frank, SCEBTA General Secretary, 200
Gormley, Joe, NUM president, 1971–82, 47, 51, 56
Green Park, Edinburgh, NCB Scottish headquarter, 164, 248
Greengairs, North Lanarkshire, 58, 68, 79, 102–3, 116, 159
Greenshields, George, Coalburn resident and opencast mineworkers, 151

Hamilton, South Lanarkshire, 71, 96, 161, 210, 233, 241
Hamilton by-election, 1943, 163
Hamilton by-election, 1967, 210
Hamilton Palace colliery, North Lanarkshire, 102
Hamilton, David, NUM activist at Monktonhall colliery, Midlothian, 53, 201
Hamilton, John, blacksmith at Ponfeigh colliery, 110–11, 247–8
Hamilton, Marian, Shotts resident, iron moulder's daughter and miner's wife, 16, 100, 140–1
Hamilton, Willie, colliery official at Kingshill 3 and Polkemmet, 16, 61, 96, 117, 140, 166–7, 174–5, 245
Hampden Park football stadium, Glasgow, 247
Harlan County, Kentucky, United States, 140
Hart, Judith, Labour MP for Lanark, 1959–83, 72, 238
health and safety, 9, 11, 61, 109, 162–5, 207, 248
Heart of Midlothian Football Club, 117
Heath, Ted, prime minister, 1970–4, 44–7, 79, 210
Henderson, Jimmy, Cardowan colliery undermanager, 166
Herbison, Margaret MP for North Lanarkshire, 1945–70, 75, 242–4
Here We Go: Women Living the Strike by Maggie Wright, 151
Herriot, Chris, Monktonhall colliery NUM youth delegate, 185
Hibernian Football Club, 117
Hillhouserigg colliery, North Lanarkshire, 65–7
Hirst coal, 39, 236
Hobart House, London, NCB headquarters, 34, 37, 187
Hobsbawm, Eric, historian, 214, 241
Hogarth, Ian, NCB manager, 126, 163–5, 248

Hogg, Mick, Lothians miner, victimized during the 1984–5 strike, 109
Holyrood Park, Edinburgh, 216–18
Holytown, North Lanarkshire, 114, 135
home rule, 7, 19, 53, 187, 198, 210, 212, 219, 222–3, 239, 254, *see also* devolution
Honeywell office machinery factory, Newhouse, 33, 142–3, 167
Hood, Jimmy, Lanarkshire and Nottinghamshire miner, 122, 233–4, 238, 244–5
Hoover appliances factory, Cambuslang, South Lanarkshire, 43, 170–1
Horner, Arthur, NUM general secretary, 1946–59, 27
Howell, David, secretary of state for energy, 1979–81, 50–1
Hungary, 193, 196–7
Hunterston ore terminal, North Ayrshire, 129
Hunterston power station, North Ayrshire, 40, 48
Independent review: policing during the 1984–5 miners' strike, 257

independence, *see* Scottish independence
industrial injuries, 3, 62, 67, 106, 131, 176, 254
inter-colliery transfers, 3, 10–11, 17, 27, 36–40, 55–6, 61–78, 82–8, 110–11, 122–5, 161, 163–5, 192, 199–200, 208–9, 211, 229–31, 234–5, 243–4
internationalism, 13, 190, 192, 194–5, 197, 205–6, 216–25
interwar coal industry, 11, 26, 28, 55, 91, 95, 101, 103, 109, 112, 140, 155–64, 169–70, 178–80, 186–8, 213, 226, 241–2, 248, 253,

Invergordon aluminium smelter, Easter Ross, 52
Inverkip power station, Inverclyde, 48
inward investment, 6, 14, 17, 21–2, 28–9, 33, 41–4, 114–18, 139, 227, 234, *see also* multinational enterprises

just transition, 4, 235, 255
just-in-time production, 135

Kay, Jackie, poet, 198
Kay, John, CPGB Scottish industrial organizer, 198, 214, 219
Keena, Barbara, Cambuslang resident and miners' daughter, 169–70
Kennox colliery, South Lanarkshire, 73–4
Kent coalfields, 45, 140, 189, 192, 249
Kenya, 192–4
Kerrigan, Peter, communist activist and International Brigade volunteer, 204
Killoch colliery, East Ayrshire, 51, 70–2, 150, 211, 245
Kilmarnock, East Ayrshire, 43, 240
Kilsyth, North Lanarkshire, 41–2, 76–7, 97, 249
Kincardine power station, Fife, 38
Kingshill 1 colliery, North Lanarkshire, 67
Kingshill 2 colliery, North Lanarkshire, 68
Kingshill 3 colliery, North Lanarkshire, 68, 166, 180–1
Kinneil colliery, West Lothian, 48, 83–4, 20–2, 254
Kirkintilloch, East Dunbartonshire, 97, 129
Knights of St Columba, 80, 110

Labour party, 13, 53, 99, 110–11, 117, 136, 174, 183, 187–98, 205, 210–13, 218–20, 238

labourism, 12–14, 188, 191, 195–7, 220
Lanark, South Lanarkshire, 69, 72, 124
Lancashire, 249
Larkhall, South Lanarkshire, 69, 73, 121, 206
Law, South Lanarkshire, 142
Law, T. S., poet, 129, 237
Lee Jeans factory, Greenock, Inverclyde, 147
Leith, Edinburgh, 103, 176, 228, 230,
Leith Links, Edinburgh, 216
Lesmahagow, South Lanarkshire, 69, 184
Lesmahagow male voice choir, 184
Lewis, John L., American miners' leader, 66
Liberal Party, 198
Liddell, Helen, MP for Monklands East and then Airdrie and Shotts, 1994–2005, 118
Limestone coal, 27, 39, 236
Linwood car factory, Renfrewshire, 78–9,
Little, Alan, journalist, 92
Little Moscows, mining villages with a significant communist presence, 199
Liverpool Football Club, 102
Lloyd, George, Conservative Minister of Fuel and Power, 29–30
Lochhead, Liz, poet, 237–8
Loganlea colliery, West Lothian, 64
Longannet
  drift mines complex, 3, 39–40, 48, 68, 108, 124, 135, 167, 199–200, 206, 229, 236
  power station, 38–9, 201, 255
Lovable Lingerie factory, Cumbernauld, 148
Lumphinnans, Fife, 199

MacColl, Ewan, folk musician, 190
MacDougall, Ian, oral historian, 204
MacGregor, Ian, chairman of the NCB, 1983–6, 50–1, 85
Macleod, Duncan, son of a Law colliery blacksmith, 98, 101, 125
Macleod, Marian, miners' granddaughter and Honeywell and Motherwell Bridge employee, 142–3
Maitland, Patrick, Conservative MP for Lanark, 1951–9, 238–9
Mannheim, Karl, social theorist, 156
Mansell-Mullen, Peter, senior NCB official, 122–3, 164
Marathon oil rig yard, Clydebank, 165
Marshall, Liz, canteen worker at Killoch, 150
Marxism, 12, 190–1, 199, 204, 208
Masculinity, 119–21, 125–37, 155, 183
Masonic Lodge, 60, 80, 110–11, 113
mass production industries, 6, 21–3, 31, 143, 227
Maxwell, Billy, Cardowan miner, 59, 79
McAllion, John, MP for Dundee East 1987–2001, MSP for Dundee East, 1999–2007, 257
McCabe, Bill, Caterpillar shop steward and factory occupier, 184, 246–7
McCallum, Scott, son, brother and grandson of Cardowan miners, 105, 133, 151, 167–8, 173–4, 228–31, 252
McCarey, Jennifer, daughter of Ravenscraig shop steward and Unison union organizer, 112–14, 132, 141–2, 145, 152
McCormack, John, Polmaise NUM delegate, 202
McDougall, William, Devon colliery miner and stay-down strike leader, 179

McGahey, Jimmy, miner, communist activist and father of Michael, 127, 191, 199, 256
McGahey, Michael, president of the NUMSA, 1967–87, 47–8, 51, 53, 73–80, 82, 85, 101, 111, 128, 157–8, 162, 167, 178–81, 189–93, 197–204, 207–18, 220, 232, 253
McGahey, Mick, son of Michael and Bilston Glen surface worker, 100–1, 111, 127, 137, 162–5, 191–3, 246, 256
McGrath, John, playwright, 145
McGuigan, James, Devon colliery stay-down striker, 179
McIlvanney, William, writer, 175–6
McLean, Bill, NUMSA general secretary, 1969–77, 46, 204
McMahon, Michael, Terex engineer and Labour MSP for Hamilton North and Bellshill and then Uddingston and Bellshill, 1999–2016, 93, 114–16, 135–6
McMahon, Siobhan, daughter of Michael and Labour Central List MSP, 2011–16, 97–8, 106–7,
mechanization, 3, 22, 27, 106–9, 177
memorials to industry, 1–2, 5, 102, 107–8, 252
Meta colliery, Clackmannanshire, 35
Michael colliery, Fife, 75, 109, 163
Migration upon colliery closure, 1
  between Scottish coalfields, 28, 36, 56, 63
  reluctance to migrate, 64–5, 68, 241–4
  to English coalfields, 37, 98, 226
Militant Miners by Ian MacDougall, 204
Militant tendency, 136, 183–4
Mills, C. Wright, social theorist, 239
Mills, Percy, Conservative minister of fuel and power, 1957–8, 71, 238
miners' lockouts, 1921 and 1926, 4, 11, 57, 127, 130, 148, 157, 160, 162, 188–9
miners' rows, 91, 94–5, 101
miners' strike, Yorkshire, 1955, 29
miners' strike, 1972 and 1974, 44, 46, 56, 78–9, 129–30, 158, 166, 177–8, 200, 204, 207–8
miners' strike, 1984–5
  historical context, 4–5, 11, 57
  build up, 49, 50–2, 55, 81–8, 201
  consequences, 55, 87, 105
  activism, victimization and memories, 91, 105, 107, 109, 121, 129, 134, 148–50, 167–8, 183–5, 189, 202, 228, 245–7, 252, 257
Miners' Welfares, 77, 92, 104, 124, 151, 183
  Auchengeich, 259
  Bellshill, 149
  Douglas, 71, 92
  Kilsyth, 95
  Kirkintilloch, 95, 129
  Twechar, 95
Mines Act, 1842, 9, 120
Ministry of Fuel and Power and successors, 14, 23, 29–32, 35, 37–8, 40, 64–5, 181, 213, 225, 232, 243–4
Ministry of Labour, 32, 138
Ministry of Technology, 33
Mitchell, John, Frances colliery miner and NUM activist, 182
modernization, 21–35, 39–44, 69, 171, 227–8
Moffat, Abe, president of the NUMSA, 1942–61, 12–13, 27–8, 63–7, 113, 160–2, 164, 179–80, 188–92, 194–5, 198–9
Moffat Abe (SCEBTA), son of Abe

Moffat and SCEBTA president, 85
Moffat, Alex, brother of Abe and president of the NUMSA, 1961–7, 33, 47, 80, 113, 128, 158, 188–92, 198–9, 208
Monaghan, Jim, Ayrshire poet, 185
Monklands East by-election, 1994, 117–18
Monktonhall colliery, Midlothian, 53, 168, 173, 183–4, 202
Montgomery, George, NUM mechanical and electrical safety inspector, 205, 237, 247
Moodiesburn, North Lanarkshire, 1–3, 16, 59, 116, 133, 151, 163, 177
Moodiesburn retired miners' group, 16, 59, 78, 133, 151
Moohan, Angela, Livingston resident and supporter of the 1984–5 strike, 168
Moohan, Brendan, Monktonhall miner, Militant tendency activist and poet, 92, 103–4, 134–7, 168, 173, 182–5, 217–18
moral economy
  abandonment by NCB and government, 80–9, 99
  extension to other industries, 136, 137, 144–5, 147, 153, 155–6, 165, 169, 251, 253–5, 258
  during pit closures, 17–18, 21, 30, 41–2, 48, 53–4, 63–5, 68–78
  in the nationalized coal industry, 60–3, 107, 111, 117, 127, 129, 163, 166, 182, 186, 199, 207, 209, 228–34
  link to Scottish nationalism, 190, 228, 240, 242–3, 249
  theoretical framework, 7–14, 55–60
  see also E. P. Thompson
Mossend, North Lanarkshire, 99–100, 114

Motherwell, North Lanarkshire, 32, 52, 71, 98
Motherwell Bridge, 142–3
Muirhead, North Lanarkshire, 59
Muirkirk, East Ayrshire, 101, 256
multinational enterprises, 6, 22, 42–3, 52, 113, 143–5, 255, see also inward investment
Murray, David, businessman, 236–8
Musselburgh, East Lothian, 48, 103, 183

Nairn, Tom, nationalism theorist, 219–20
National Association of Colliery Overmen Deputies and Shotfirers, 82, 248
National Coal Board
  and the moral economy, 10, 18, 55–63, 64–89, 128, 146, 173, 179, 190, 202, 232–3, 244
  centralization, 22, 34–7, 70, 76, 124, 188, 205, 209, 225, 227–9, 235, 236–44, 254
  closure programmes, 2–5, 26–7, 72, 159, 179, 201–2, 207, 230, 251
  housing, 94–7
  management, 15, 45, 94, 122, 162–7, 177
  relations with government, 23, 30, 34, 38–40, 47–54, 85
  Scottish Division, 13–14, 187, 216
  see also nationalized coal industry
National Union of Mineworkers and National Union of Mineworkers Scottish Area, 2, 5, 11–15, 19, 189–92
  and Scottish nationhood, 53–4, 94, 203–5, 211–24, 240–3, 254
  engagement with UK energy policymaking, 18, 27, 33–6, 45–8, 232–47

internal politics, 108–113, 149–50, 158–63, 176–84, 187–98, 206–9
responses to local pit closures, 45, 52, 63–89, 229–30, 238
see also internationalism
National Union of Mineworkers South Wales Area, 204, 222
nationalized coal industry, 2–11, 14
in energy policymaking, 21–31, 36, 45, 49–53, 235
industrial and community relations, 94–8, 106, 109–11, 118, 155–65, 178, 182–8, 219–23, 231–48, 254–5
pit closures under, 55–63, 77, 82–4, 88, 225, 229, 251
see also health and safety; National Coal Board
Nationalism, 11, 210–13, 219–23, 241, *see also* Unionist-Nationalism
NCB houses, 95–7, 244
New Edinburgh Review, 199, 205, 237
New left, 12, 157, 196, 209, 239
New Stevenson, North Lanarkshire, 114
new towns, 27–8, 42, 73, 94, 139
Newarthill, North Lanarkshire, 93, 112
Newhouse industrial estate, North Lanarkshire, 33, 114, 142–3, 167
Newton colliery, South Lanarkshire, 128
Newtongrange, Midlothian, 176
National Health Service, 137
nineteenth-century coal industry, 9, 22, 120, 129, 172, 212, 234
North Atlantic Treaty Organization, 195–6
Noble, Michael, Conservative secretary of state for Scotland, 1962–4, 32
nostalgia, 16, 18, 92, 100–4, 108–9, 185, 226, 247, 256

Nottinghamshire coalfield, 34, 122, 129, 164, 233, 244–5, 249
nuclear disarmament, 192, 195, 198, 222–3, *see also* Cold War
nuclear power, 2, 14, 29–34, 37, 40, 44–8, 109, 198, 227, 233

Occitan miners, 71, 254
occupational identities, 6–7, 11–18, 91–5, 98–104, 119, 127, 129–31, 143, 153, 155, 164, 187, 193, 200–2, 213, 216, 222, 234, 247, 253–6
O'Grady, Frances, Trade Union Congress general secretary, 14–31
Oil
power station fuel, 14, 29–34, 50–1, 109, 235, 255
industrial uses, 29
imports and price fluctuations, 34–5, 44–6, 50, 78–80, 235–6
North Sea, 46–9, 136, 165, 233–7, 246
opencast mining, 73, 88, 153, 165, 227, 233–4, 238
oral history, 15, 17, 80, 92, 103, 116, 204, 243
Orange Order, 15, 59, 112, 116
Organon pharmaceuticals plant, Newhouse, 167
Orgreave coking plant, South Yorkshire, 129
overspill, 28
Owens, Joe, Polkemmet colliery NUM delegate, 183

Paris, Bill, Shotts miner, 61, 95
Parker, Ronald, chair of the NCB Scottish Division, 39, 68, 74–5
Parsons, Jack, director, 241
paternalism, 9, 94, 114, 120, 130, 164, 171
Paynter, Will, NUM general secretary, 1959–69, 177

Pearson, Bill, NUMSA, general secretary, 1945–1956, 161–2
Percy Mills, Conservative Minister of Fuel and Power, 1957–8, 71, 238
Perth, 160, 190
Peterhead power station, 48
pit baths, 62, 94–5, 125–6, 140
Plan 2000, 45
Plan for Coal (1974), 44, 48–52, 79, 81, 88, 229–30, 235
Plessey factory, Bathgate, West Lothian, 148
Pneumoconiosis, 164, 173, *see also* disability
Poland, 41, 194, 221, 245
Polanyi, Karl, social theorist, 8–11, 120, 163, 225, 239, *see also* double movement
Policy colliery, Falkirk, 35
political education, 192–3
Polmaise colliery, Stirlingshire, 82, 202
Ponfeigh colliery, South Lanarkshire, 247
Popular Front, 198, 212
Prague Spring, 197
Priory colliery, South Lanarkshire, 196
private housing, 18, 91, 97–8
privately owned coal industry, 9–11, 17–18, 27, 55–7, 61, 77, 94, 110–17, 124, 158–9, 164–5, 169–70, 175, 234–8
Privatization, 49–54, 86–7
Protestantism, 13, 110–18, *see also* sectarianism
public housing, 6, 14, 18, 91–4, 99, 117–18, 226, *see also* council housing
pubs, 97, 122, 140, 184, *see also* Miners' Welfares
Purdie, Sam, Kames mining engineer and South Ayrshire SNP candidate, 101, 113, 126, 165, 211, 256

Ragged Trousered Philanthropist by Robert Tressell, 184
Rangers Football Club, 78, 117
Ravenscraig steelworks, North Lanarkshire, 29, 31–2, 52, 72, 82, 88, 99–100, 132, 141, 147, 166, 171, 246
RCA, 144
redundancies, 37–8, 41–4, 56, 62–8, 71–5, 81–4, 119, 124, 127, 131–5, 143, 146–7, 150, 153, 172, 181–2, 229–30, 237, 246, 251
regional policy, 6, 14, 17–18, 21–6, 41–2, 51–4, 64–5, 138, 144–7, 171, 227
Reid, Jimmy, leader of the UCS work-in, 128, 165, 197, 212–13
Rhondda Valley, South Wales, 91
Rigside, South Lanarkshire, 69, 73, 94
Robens, Alf, NCB chairman, 1960–71, 37–40, 45, 76–7, 205–6, 210
Robeson, Paul, American singer, 190
Rooney, Anthony, miners' son and Caterpillar shop steward, 173–4, 218–19
Ross, Willie, Labour secretary of state for Scotland, 1964–70 and 1974–6, 43, 240, 243
Rothes colliery, Fife, 27, 35–7, 45, 165–6
Royal Musselburgh Golf Club, 183
Rust belt, 226

Salutation Hotel, Perth, 190
Scargill, Arthur, NUM president, 1982–2002, 52, 82, 191–2
Scottish coal lobby, 34, 38, 48, 73
Scottish Colliery Enginemen, Boilermen and Tradesmen's Association, 82, 85, 191, 200–1
Scottish Council (Development and Industry), 23, 31, 37, 42, 139, 254
Scottish Covenant campaign, 211

Scottish independence, 92, 238, 248–9
Scottish Marxist, CPGB journal, 203–8, 253
Scottish Miner, NUMSA newspaper, 190, 192, 196
Scottish miners' gala, 58, 187, 198, 204, 216–25, 249
Scottish National Party, 7, 37, 118, 200, 214–16, 222–3, 247
Scottish Office, 14, 17, 23, 28–32, 41–3, 47, 138, 243, 254
Scottish Special Housing Association, 36, 94, 96, 161
Scottish Steel, 29, 63, 77, 163
Scottish Trades Union Congress, 13, 15, 37, 44, 46, 80, 136, 184, 190, 193–8, 201, 205, 212–13, 222–3, 254
Seafield colliery, Fife, 109
Second World War, 31, 160–1
sectarianism, ethno-religious, 13–14, 92, 109–18
self-rescuer, 109, 162–3
Shankly, Bill, Ayrshire miner and Liverpool Football Club manager, 102
shift working, 115, 135–6, 159, 201
shipbuilding, 22, 31, 33, 44, 53, 128, 255
Shotts, North Lanarkshire, 16, 18, 55–72, 95–6, 100, 117, 124, 142, 140, 166–7, 245, 253
Shotts ironworks, North Lanarkshire, 100
Sillars, Jim, Labour then Scottish Labour Party MP for South Ayrshire, 1970–9, 210
Silences, 127, *see also* oral history
Slaven, John, son of Caterpillar workers, 131–3, 141, 146–7, 170–2
social conservatism, 13, 73, 92, 102, 104, 118, 183, 228

social democracy, 12, 49, 58, 68, 135, 138, 155–8, 171, 187–8, 195–8, 206, 219, 223, 228, 247
socialism, 104, 183
solidarity,
    class and workplace, 3, 6, 82, 166, 168, 190, 200, 202, 219, 226, 249, 251
    community, 102
    international, 192, 197, 221
    see also internationalism; mining disasters, 105, 107
Solovyev, G., Soviet miners' leader, 189
Solsgirth colliery, Clackmannanshire, 3, 68, 124
South Africa, 193, 197, 222
South Ayrshire by-election, 1970, 211
South of Scotland Electricity Board, 23, 37, 47–8, 240, 254
South Wales coalfield, 107, 125, 192, 202–3, 220, 254
Soviet Union, 112, 160, 189, 193, 195–7, 245
Spain, 112, 193, 204, 235
Spence, Mary, East Kilbride teacher from a mining family, 93–4, 99
Springboig, Glasgow, 163
Springburn, Glasgow, 251–8
Stalinism, 183, 188–197
steel industry, 1, 22–31, 44, 51–3, 63, 72, 77, 81, 88, 92, 96–101, 111, 117–18, 127, 132, 137, 155, 163, 166–7, 170–2, 174, 203, 236, 241, 246
Stepps, North Lanarkshire, 151
Stewart and Lloyds, 241
Strathclyde Technical College, Glasgow, 165
Sturgeon, Nicola, first minister of Scotland, 259–60
suburbanization, 18, 98, *see also* commuting distances

Index

Sunbeam appliances factory, East Kilbride, South Lanarkshire, 143–4
super pits, 4, 91, 118, 227

Tannahill, E., Kingshill 3 miner and NUM activist, 180–1
Tannochside, North Lanarkshire, 52, 113, 131, 146, 167–72, 218, 255
Taylor, John, Labour MP for West Lothian, 1951–62, 211–13
Terex factory, Holytown, 93, 114–18, 135–6
Terris, Ian, Cardowan and Rothes shotfirer, 165–6
Thatcher, Margaret, Conservative prime minister, 1979–90, 4, 15, 49–54, 81, 83, 167, 192, 203, 210, 228–9, 232, 242, 248, 258
Thompson, E. P., labour historian, 8–11, 252, *see also* moral economy
Timex watch and electronics factory, Dundee, 148
Timpany, Alex, Barony miner and NUM activist, 208
Tombs, Frank, Scottish and Southern Electricity Board chairman, 47
Toothill, John, Ferranti chairman and author of the Toothill report, 31
Toothill report, 31–3, 41, 138
Torness power station, East Lothian, 48
Tory party, *see* Conservative party
Turnwood, Betty, Shotts resident and former British Telecom employee, 95, 140
Twechar, North Lanarkshire, 84, 97, 111, 175, 200

Uddingston, South Lanarkshire, 33
Unemployment, 11, 16, 26, 29, 31, 37–43, 49, 52, 55, 60–4, 84, 119, 130–2, 138–9, 153, 155–65, 170–2, 182, 186, 210, 231–3, 240–2, 246, 258

Union, the, 1, 11, 242, 254
Unionism, 15, 59, 187, 226–8, 243
Unionist-Nationalism, 219, 238
Unionist party, *see* Conservative party
Unison, 137, 152
United Mine Workers of America, 66
United Mineworkers of Scotland, 188
Unity Theatre, 190
Upper Clyde Shipbuilders, 44–5, 128

Varley, Eric, Labour minister of state for the Ministry of Technology 1969–70, 33
victimization, 62, 109, 112, 137, 157–8, 256–7, *see also* blacklisting
Vietnam, 140, 195–7, 216–22, *see also* internationalism

Warsaw Pact, 197, *see also* Cold War
Watson, Sam, general secretary of the Durham miners' association, 1947–63, 218
Wegg, Jerry, Cardowan miner, 123
Wegg, Margaret, Cardowan pit canteen worker and 1984–5 strike activist, 96–7, 105–6, 123–4, 141–2, 148–50
welfare state, 119, 140, 160
Wester Auchengeich colliery, North Lanarkshire, *see* Auchengeich
Westwood, Joseph, Labour secretary of state for Scotland, 1945–7, 161
Wheeler, Albert, NCB Scottish Area director, 1980–5, 82–6, 228, 247–8
Whins, Clackmannanshire, 179
Wilkinson, Rhona, miners' granddaughter from Breich, 101, 125, 134
Williams, Emlyn, NUM South Wales Area president, 1973–86, 202
Wilson, Harold, Labour prime minister, 1964–70 and 1974–6, 32–3, 43–7, 196, 210, 216, 233

Wilson, Nicky, NUM Scotland president, 3, 82–4, 109, 191–2, 200–1, 228–9, 251
Windsor, Berkshire, 140
Wishaw, North Lanarkshire, 98
Wolfe, William, Scottish National Party leader, 1969–79, 37, 210
Women Against Pit Closures, 148–50
women's activism in miners' disputes, 130, 150–2
women's employment, 9, 17–18, 26–8, 42, 123–53, 159, 170

Woodmuir colliery, West Lothian, 101, 125
Workers Educational Association, 125, 191
World Federation of Trade Unions, 194–5
Wright, Maggie, film director, 148

Yorkshire, 29, 34, 91, 127, 129, 191–2, 249
Young, James D., historian, 199
Youth Training Scheme, 131

New Historical PERSPECTIVES

Recent and forthcoming titles

The Family Firm: Monarchy, Mass Media and the British Public, 1932–53 (2019)
Edward Owens

Cinemas and Cinema-Going in the United Kingdom: Decades of Decline, 1945–65 (2020)
Sam Manning

Civilian Specialists at War: Britain's Transport Experts and the First World War (2020)
Christopher Phillips

Individuals and Institutions in Medieval Scholasticism (2020)
edited by Antonia Fitzpatrick and John Sabapathy

Unite, Proletarian Brothers! Radicalism and Revolution in the Spanish Second Republic (2020)
Matthew Kerry

Masculinity and Danger on the Eighteenth-Century Grand Tour (2020)
Sarah Goldsmith

Coal Country: the Meaning and Memory of Deindustrialization in Postwar Scotland (2021)
Ewan Gibbs

Church and People in Interregnum Britain (2021)
edited by Fiona McCall

The Politics of Suffrage: Structures, Institutions and Practices (2021)
edited by Alexandra Hughes-Johnson and Lyndsey Jenkins